Dairy Goats Feeding and Nutrition

Dairy Goats Feeding and Nutrition

Edited by

Antonello Cannas and Giuseppe Pulina

Department of Animal Science
University of Sassari
Italy

Technical editor and reader of the English translation

Ana Helena Dias Francesconi

Department of Animal Science
University of Sassari
Italy

www.cabi.org

CABI is a trading name of CAB International

CABI Head Office
Nosworthy Way
Wallingford
Oxfordshire OX10 8DE
UK

Tel: +44 (0)1491 832111
Fax: +44 (0)1491 833508
Email: cabi@cabi.org
Web site: www.cabi.org

CABI
745 Atlantic Avenue
8th Floor
Boston, MA 02111
USA

Tel: +1 617 682-9015
Email: cabi-nao@cabi.org

Original edition published in Italian under the title: *L'alimentazione delle capre da latte* © Avenue Media 2005, Bologna, Italy.

A catalogue record for this book is available from the British Library, London, UK

Library of Congress Cataloging-in-Publication Data
Dairy goats feeding and nutrition / edited by Antonello Cannas and Giuseppe Pulina; technical editor and reader of the English translation, Ana Helena Dias Francesconi.
p. cm.
"Original edition published in Italian under the title:
L'alimentazione delle capre da latte, Avenue Media 2005, Bologna, Italy."
ISBN 978-1-84593-348-7 (alk. paper)
1. Goats—Feeding and feeds. 2. Goats—Nutrition. I. Cannas, Antonello.
II. Pulina, Giuseppe. III. Francesconi, Ana Helena Dias. IV. Title.

SF384.3.D35 2008
636.3'9142—dc22

2007024282

ISBN-13: 978-1-84593-348-7

First published 2008
Transferred to print on demand 2016

Typeset by AMA DataSet Ltd, UK.
Printed and bound by CPI Group (UK) Ltd, Croydon, CR0 4YY

Contents

Biographical Details of the Authors

Mauro Antongiovanni (mauro.antongiovanni@unifi.it) is Full Professor of Animal Nutrition and Feeding at the Department of Animal Science of the University of Florence, Italy, where he teaches a course on Animal Nutrition and Feeding. He has degrees in Chemistry and Agricultural Science from the University of Florence. He has conducted research on the associative effects among feeds in the diet of ruminants, using the comparative slaughter (Thomson and Camell) and gas production techniques, and on the modification of the fatty acid profile of milk and its products in dairy animals. He has also evaluated the *in vivo* digestibility and estimated the nutritional value, using the INRA method, of the most common ruminant feeds in Italy.

Alberto Stanislao Atzori (asatzori@uniss.it) has a degree in Agricultural Science and Technology and a PhD in Animal Husbandry from the University of Sassari, Italy. He has conducted research mainly on feeding of small ruminants and nutrient requirements of sheep and goats. He also had training on breeding and productive characteristics of the Iberian deer at the Instituto de Investigación en Recursos Cinegéticos (IREC, Experimental Institute on Cynegetic Resources), Section of Albacete (UCLM, University of Castilla-La Mancha), Spain, and on breeding systems of Sarda dairy sheep at the Istituto Zootecnico e Caseario della Sardegna (IZCS, Sardinian Institute for Animal Husbandry and Dairy Farming), Italy.

Marcella Avondo (mavondo@unict.it) is Full Professor of Animal Husbandry at the Faculty of Agricultural Science of Catania, Italy. Her main areas of research are methods to improve silage quality from forages and agro-industrial by-products, and the feeding behaviour of sheep and goats in extensive and semi-extensive systems, with particular interest in the factors that influence pasture intake and diet selection. At present, she is studying the effects of diet selection on sheep's and goat's milk quality.

Gianni Battacone (battacon@uniss.it) is Researcher at the Department of Animal Science of the University of Sassari, Italy. He has a degree in Agricultural

Science from the University of Sassari and a PhD in Animal Husbandry from the University of Perugia, Italy. His research focuses on the relationship between animal feeding and the safety and quality of animal products such as milk and meat.

Luciana Bava (luciana.bava@unimi.it) has a PhD in Animal Feeding and Technology and is Researcher in Animal Genetics and Breeding at the Department of Animal Science of the University of Milan, Italy. The main focus of her research has been the relationships between feeding systems, husbandry and milk quality of dairy goats and cows. She has also been involved in the evaluation of milk emission and milking hygiene.

Luisa Biondi (lubiondi@unict.it) is Associate Professor at the University of Catania, Italy, and has a PhD in Science and Technology of Animal Production in Disadvantaged Areas from the University of Bari, Italy. Her main research field is the relationship between animal production systems and product quality in sheep. Particular emphasis has been given to the use of local feed resources and their effects on the nutritional and nutraceutical characteristics of milk and meat.

Filippo Boe (f.boe@tiscali.it) has a degree in Agricultural Science and a PhD in Animal Husbandry from the University of Sassari, Italy. He had a fellowship from the Department of Animal Science, University of Sassari, to work on the nutrition of ruminants in organic farms. His research focuses on feeding and nutrition of small ruminants and on mathematical models for animal husbandry.

Adriana Bonanno (abonanno@unipa.it) is Associate Professor of Animal Science at the Faculty of Agriculture of the University of Palermo, Italy. Over time her research has dealt with various themes regarding dairy cows, sheep and goats of local breeds, and rabbit production. Her main fields of interest in small ruminants have been artificial insemination and oestrus synchronization, energy and nutritional value of conventional sources and by-products in lamb feeding, and the quality characteristics of dairy and meat products evaluated as a function of management system and feeding regimen. At present, her experimental activities are focused on studying the effects of conventional or organic grazing systems on selectivity behaviour, feeding intake, welfare status and quality of dairy products of sheep and goats.

Arianna Buccioni (arianna.buccioni@unifi.it) has a degree in Chemistry and a PhD in Animal Science from the Department of Animal Science of the University of Florence, Italy, where she currently has a research contract. She has conducted research on animal feeding and nutrition, with emphasis on the lipid metabolism of the rumen and the lipid components of milk and cheese from sheep, goats and cattle.

Andrea Cabiddu (acabiddu@tiscali.it) has a degree in Agricultural Science from the University of Piacenza, Italy, where he carried out studies on rumen metabolism, especially on fibre and protein degradation. He is Researcher at the Department of Animal Production of AGRIS Sardegna, the Research Agency of the Sardinian Region. He is presently working on the relationship between sheep and goats nutrition and milk composition, with particular attention to the fatty acid composition and other nutraceutical components of the milk.

Antonello Cannas (cannas@uniss.it) is Full Professor of Animal Science at the Department of Animal Science of the University of Sassari, Italy. He has a degree in Agricultural Science from the same university, and an MSc and PhD in Animal Science from Cornell University, Ithaca, New York, USA. He works on several aspects related to feeding and nutrition of small ruminants, including feed evaluation techniques and energy and protein requirements, with main focus on: the development of prediction equations and models to estimate requirements and animal performances; rumen turnover and its prediction; urea as a nutritional indicator; and comparative nutrition of small and large ruminants. He led the development of the Cornell Net Carbohydrate and Protein System for Sheep and of the Small Ruminant Nutrition System for sheep and goats. At present he teaches courses on sheep and goat production and on feed processing and evaluation at the Faculty of Agriculture of the University of Sassari.

Aldo Cappio-Borlino (genetica@uniss.it) is Associate Professor of Animal Breeding and Genetics at the Department of Animal Science of the University of Sassari, Italy. His main scientific interest lies in the mathematical modelling of the regular and continuous component of some dynamic processes of great importance in the field of animal science, such as the temporal evolution of milk production and rumen degradation of feeds. He proposed an original model for the mathematical description of the lactation curve of dairy sheep that has been cited in several international scientific papers and used by several researchers. Recently he has suggested a methodology for analysing time series of milk yield and quality, in order to disentangle deterministic and stochastic components.

Mauro Decandia (mdecandia@tiscali.it) has a degree in Agricultural Science from the University of Sassari, Italy, and is a Researcher at the Department of Animal Production of AGRIS Sardegna, the Research Agency of the Sardinian Region. He specialized in the feeding of farm animals at the Institut National Agronomique Paris-Grignon (INAPG). His expertise is in feeding behaviour, intake of nutrients, and milk yield and composition of grazing dairy sheep and browsing goats. He has studied the effects of different levels and sources of supplements on sheep and goat performances. He has worked on the chemical composition and nutritive evaluation of the main forage species grazed or browsed by small ruminants, measured with conventional methods and near infrared spectroscopy.

Antonino Di Grigoli (adigrigoli@unipa.it) has a degree in Agricultural Science from the University of Palermo, Italy, and a PhD in Animal Production. At present, he is Researcher in Animal Science at the Faculty of Agriculture of the University of Palermo and teaches a course on nutrition and animal feeding at the same faculty. He has carried out studies targeted at the exploitation of Sicilian cattle, sheep and goat breeds, and researches on the effects of technical and managerial factors on the milk and cheese production from grazing ruminants. His current experimental activities on sheep and goats are focused on studying the factors affecting feeding intake and behaviour at pasture.

Corrado Dimauro (dimauro@uniss.it) is Researcher at the Department of Animal Science of the University of Sassari, Italy. He has a degree in Physics from the University of Catania, Italy, and has a PhD in Animal Science from the

University of Sassari, Italy. His main scientific interest lies in the mathematical modelling of lactation curves using both empirical and mechanistic modelling approaches. At the present time he also works on the statistical analysis of microarray data.

Adriana Carmen Lucia Di Trana (adriana.ditrana@unibas.it) is Associate Professor of Animal Nutrition and Feeding at the Department of Animal Science of the University of Basilicata, Potenza, Italy. Her research activity concerns the feasibility of rearing Cashmere goats in typical systems of southern Italy, the quantitative and qualitative assessment of the fibre production and skin trait in different goat physiological conditions and feeding, as well as the links among growth, moulting of cashmere fibre and hormone profile. She has been responsible for the development and execution of research projects on Cashmere goats, supported by the Italian Ministry of University and Scientific and Technological Research (MURST). She has carried out research on the influence of feeding system, season and goat breed on fatty acid profile, conjugated linoleic acids and *n*–3 fatty acid content in milk and cheese. Her current research also includes the assessment of productive efficiency in goats on pasture with different types of supplies in shed, and the relationships among nutrition, hot season and oxidative status of lactating goats.

Stefania Fancellu (fancellu_stefania@tiscali.it) has a degree in Agricultural Science and a PhD in Animal Husbandry from the University of Sassari, Italy. Her research focuses on dietary strategies to improve milk quality concerning conjugated linoleic acids and the fatty acid profile of milk and meat in sheep and goats.

Vincenzo Fedele (vincenzo.fedele@entecra.it) is Researcher at CRA – Unità di Ricerca per la Zootecnia Estensiva (Experimental Unit of Extensive Animal Production) of Bella, Potenza, Italy. He has a degree in Agricultural Sciences from the University of Naples, Italy. His main areas of research are grazing behaviour and estimation of the herbage intake, digestibility, energy and protein requirements of small ruminants. In recent years, he has worked on product quality, with a main focus on the relationships among feeding systems and aromatic and nutritional characteristics of goat's milk and cheese.

Riccardo Fortin (riccardo.fortin@unimi.it) has a degree in Biological Sciences from the University of Pavia, Italy. He is a PhD student in Veterinary Hygiene and Animal Pathology at the Faculty of Veterinary Medicine of the University of Milan, Italy, where he is doing research on milk proteome. Previously he conducted research in the fields of biocrystallography at the University of Pavia and enzymology at the University of Groningen, The Netherlands.

Ana Helena Dias Francesconi (france@uniss.it) has a degree in Agricultural Science from the University of Sassari, Italy. She also has a degree in Agricultural Science and an MSc from the Universidade Federal do Rio Grande do Sul (UFRGS), Porto Alegre, Brazil. She has a PhD in Pomology from Cornell University, Ithaca, New York, USA. She conducted research on plant production, plant physiology and postharvest for several years. She is a technician at the Department of Animal Science of the University of Sassari, Italy. Since 1996, she has translated and edited papers on several topics (animal science, plant production, entomology) for publication in international scientific journals. Currently, her main research interests are animal science and bibliometry.

Matteo Gianesella (matteo.gianesella@unipd.it) is a PhD student in Veterinary Sciences specializing in Clinical Veterinary Sciences at the Faculty of Veterinary Medicine of the University of Padova, Italy. He collaborates with Professor Massimo Morgante to study the metabolic–nutritional diseases of ruminants. He is author of some publications in national and international journals and proceedings of meetings in his field of expertise.

Gian Franco Greppi (gianfranco.greppi@unimi.it) has a degree in Agriculture Sciences from the University of Milan, Italy. He is Associate Professor at the Department of Veterinarian Clinical Sciences of the University of Milan, where he teaches a course on statistics and bioinformatics. He is Scientific Director of the Proteomics Laboratory of the Istituto Sperimentale Italiano 'Lazzaro Spallanzani' of Milan, Italy. He is President of Lea Biotech, a spin-off of the University of Milan. He has participated in or has been responsible for many national and international research projects.

Lucio Lutri (llutri@inwind.it) has a degree in Agricultural Science from the University of Palermo. Italy. He is Researcher at the Istituto Sperimentale Zootecnico per la Sicilia, in Palermo. He has been working since 1984 as Technical Assistant in Agriculture for the Regional Agriculture and Forestry Department of Sicily, focusing on ruminant nutrition in intensive and extensive livestock systems. At the moment he is working mainly on feeding and management of dairy sheep and goats, with a main focus on the prediction of nutrient requirements of lactating and growing animals.

Nicolò Pietro Paolo Macciotta (macciott@uniss.it) is Associate Professor of Animal Breeding and Genetics at the Department of Animal Science of the University of Sassari, Italy. His main scientific interest lies in the applications of mathematical modelling to several fields of animal science, with particular reference to animal breeding. He has carried out works on lactation curve modelling, problems in quantitative trait loci detection in farm animals, optimization of selection schemes for ruminants, and applications of bioeconomic models to estimate economic values of milk production traits in sheep.

Marcello Mele (mmele@agr.unipi.it) is Researcher at the Faculty of Agriculture of the University of Pisa, Italy. He has a degree in Agricultural Science from the University of Pisa and a PhD in Animal Science from the University of Perugia, Italy. His main research field is the lipid metabolism of ruminants in relation to improvement of milk and meat quality, in particular of their nutritional and nutraceutical properties.

Giovanni Molle (gmolle@tiscalinet.it) has a degree in Agricultural Science from the University of Sassari, Italy. He is a researcher at the Department of Animal Production of AGRIS Sardegna, the Research Agency of the Sardinian Region. Using techniques learned from the Institute of Grassland and Environmental Research of Great Britain and the Agricultural Research Institute of Israel, he has carried out studies on the intake, feeding behaviour, energy costs and body composition of grazing sheep and goats. He has also studied the relationship between feeding and reproduction in ruminants and, in particular, in dairy sheep.

Massimo Morgante (massimo.morgante@unipd.it) is Full Professor of Veterinary Medical Pathology at the Faculty of Veterinary Medicine of the University of Padova, Italy. He has a degree in Veterinary Medicine from the University of Perugia, Italy, and is a diplomate of the European College of Bovine Health Management (ECBHM). He has studied the metabolic and nutritional diseases of ruminants for many years. He is author of several publications in national and international journals and proceedings of meetings in his area of expertise.

Anna Nudda (anudda@uniss.it) is Researcher at the Department of Animal Science of the University of Sassari, Italy. She has a degree in Agricultural Science from the University of Sassari, and a PhD in Animal Husbandry from the University of Perugia, Italy. She has worked on several aspects of quality of sheep's and goat's milk, and lactation physiology. Currently her research is focused on dietary strategies to improve the fatty acid profile of milk and meat, from a nutritional point of view. She teaches courses on animal production and on morphological and physiological evaluation of farm animals at the Faculty of Agriculture of the University of Sassari.

Renato Italo Pagano (renato.pagano@tiscali.it) is a PhD student at the Faculty of Agricultural Science of Catania, Italy. His studies have focused on the behaviour of grazing goats, with the aim of developing a model to predict pasture intake and diet selection. He has worked on the effect of diet quality on nutritional and nutraceutical characteristics of sheep's and goat's milk.

Giuliano Pinna (biglianu@tiscali.it) has a degree in Agricultural Science from the University of Sassari. From 1999 to 2004, he worked at the Istituto Zootecnico e Caseario per la Sardegna (IZCS, Sardinian Institute for Animal Husbandry and Dairy Farming), where he was involved in different projects aimed at the study of the technological, chemical, physical and sensorial characteristics of dairy sheep and goat products. He presently works in a private cheese factory (Pinna SpA) in Thiesi, Sardinia, Italy.

Giuseppe Pulina (gpulina@uniss.it) has a PhD in Animal Husbandry from the University of Rome, and is Full Professor at the Department of Animal Science of the University of Sassari, Italy. At present he is General Director of AGRIS Sardegna, the Research Agency of the Sardinian Region. He has carried out research on sheep and goats, focusing on the quality of animal products, on mathematical models for animal husbandry techniques, and on genetic selection and reproduction. He has been a lecturer at the Faculty of Agriculture of the University of Sassari since 1988. He is referee of several national and international journals in his field.

Luca Rapetti (luca.rapetti@unimi.it) has a PhD in Animal Feeding and Technology from the University of Pisa, Italy, and is Associate Professor at the Department of Animal Science of the University of Milan, Italy. He has conducted research on feeding and energy metabolism of ruminants, especially goats. His main area of expertise is animal feeding and nutrition.

Paola Roncada (paola.roncada@unimi.it) has a degree in Chemistry and Pharmaceutical Technology from the University of Milan, Italy. She has a PhD in Biochemistry from the Faculty of Sciences of the University of Sassari, Italy. She is head of the Proteomics Laboratory of the Istituto Sperimentale Italiano

'Lazzaro Spallanzani' of Milan. She has worked on the proteomic characterization of milk proteins of different animal species, with focus on the proteins of goat's milk and its products, and on the allergy and intolerance to milk from various species in the newborn.

Pierlorenzo Secchiari (psecchia@agr.unipi.it) is Full Professor of Animal Genetics and Breeding at the Faculty of Agriculture of the University of Pisa, Italy. His main areas of research in animal science are: animal reproduction, animal husbandry, breeding of quantitative characters, genetic variability of cattle and sheep breeds, and quality of products of animal origin. He is a member of the Committee for the Protection of Autochthonous Genetic Resources of Tuscany and, since 2004, has been President of the Italian Scientific Association of Animal Science (ASPA).

Lucia Sepe (lucia.sepe@entecra.it) is Researcher at CRA – Unità di Ricerca per la Zootecnia Extensiva (Experimental Unit of Extensive Animal Production) of Potenza, Italy. She has a degree in Agricultural Science from the University of Reggio Calabria, Italy. Until 1998, as Director of the Centro di Collegamento Ricerca-Divulgazione (Centre of Connection Research-Divulgation) of Bella, Italy, she was responsible for the organization of national and international courses and meetings aimed at transfering research results to extension technicians and farmers. She has conducted research on Cashmere goats, including a training period at the Macaulay Land Use Research Institute in Aberdeen, Scotland. Since 1999 she has been on the editorial staff of *Caseus*, a journal specializing on cheese. She was also editor of the *Caseus International* journal. Her main research fields are feeding of goats, especially floristic factors influencing the quality of milk and cheese.

Andrea Serra (aserra@agr.unipi.it) has a degree in Agricultural Science from the University of Pisa, Italy, and a PhD in Animal Science from the University of Perugia, Italy. His research has focused on the quality of animal products, i.e. milk and meat, especially of cattle meat. At present he has a research contract from the Section of Animal Science of the Department of Agronomy and Management of Agroecosystems of the Faculty of Agriculture of the University of Pisa.

Calogero Stelletta (calogero.stelletta@unipd.it) has a degree in Veterinary Medicine and a PhD in Veterinary Public Health and Food Safety from the Faculty of Veterinary Medicine of the University of Perugia, Italy. At the present time, he works as Researcher in Veterinary Obstetrics and Gynaecology at the Faculty of Veterinary Medicine of the University of Padova, Italy. His main interests are the mechanisms of neuroendocrine control of the metabolic–nutritional and reproductive diseases of ruminants. He has published several papers in national and international journals and proceedings of meetings in his field of expertise.

Roberto Steri (r.steri@tiscali.it) has a degree in Agricultural Science from the University of Sassari, Italy, and is a PhD student in Animal Science at the same university. He had a fellowship from the Department of Animal Science of the University of Sassari to study the protection of animal farm species at risk of extinction and of autochthonous races and breeds. He has conducted research on the somatic variability of the Sarda sheep breed and mathematical models applied to sheep and goats.

Izabelle Auxiliadora Molina de Almeida Teixeira (izabelle@fcav.unesp.br) has a degree in Animal Science from the Universidade Federal de Lavras, Brazil, and an MSc and PhD in Animal Production from the Universidade Estadual Paulista (UNESP), Brazil. She is Professor at the Department of Animal Science of UNESP, where she teaches goat production to undergraduate students and feed evaluation and nutritional requirements for ruminants to graduate students. She works on several aspects of small ruminant nutrition, with a main focus on protein, energy and mineral requirements for goats.

Maria Dimitios Yiakoulaki (yiak@for.auth.gr) has a degree in Forest and Natural Environment, a MSc in Animal Science and a PhD in Range Management from the Aristotle University of Thessaloniki, Greece. Her main areas of research are: goat nutrition, feeding behaviour and intake of goats grazing on Mediterranean shrublands, nutritive values of forages, and grazing management systems. Currently she teaches courses on range management, range animal nutrition and range ecology at the Faculty of Forestry and Natural Environment at the Aristotle University of Thessaloniki, Greece.

1 Nutrition and Quality of Goat's Milk

G. PULINA, A. NUDDA, G. BATTACONE, S. FANCELLU AND A.H.D. FRANCESCONI

Dipartimento di Scienze Zootecniche, Università di Sassari, Via E. De Nicola 9, 07100 Sassari, Italy

Introduction

Goat's milk is used for drinking and making cheese and yoghurt. Among non-genetic factors, feeding is the main factor influencing milk composition and transformation properties. Feeds can also transfer toxic or undesirable substances to milk. Many studies on the effects of nutrition on goat's milk quality have focused on milk fat content and milk fatty acid (FA) profile (Schmidely and Sauvant, 2001; Chilliard *et al.*, 2003; Chilliard and Ferlay, 2004; Morand-Fher *et al.*, 2007; Sanz Sampelayo *et al.*, 2007). This chapter deals with feeding effects on the contents of fat, protein, somatic cells, minerals and undesirable substances in goat's milk. Other aspects related to the relationship between nutrition and lipid composition, aromatic compounds and nutritional properties of goat's milk are discussed by Mele *et al.* (Chapter 3, this volume) and Fedele (Chapter 5, this volume).

Milk Composition Affects Cheese Yield

The yield of cheese depends mainly on milk fat and milk total protein (milk TP = N × 6.38), and varies depending on the type of cheese produced (Fekadu *et al.*, 2005). A major factor which influences milk fat and protein concentration is milk yield. In goats, as in other dairy ruminants, genetic and phenotypic correlations between milk yield and fat and protein concentrations are negative (Emery, 1988), highlighting the so-called 'dilution effect' (Table 1.1). Such genetic correlations are stronger than the phenotypic ones in a local breed (Murciano-Granadina: Analla *et al.*, 1996), while this is not observed in highly selected breeds (Saanen, Alpine: Boichard *et al.*, 1989; Barbieri *et al.*, 1995).

The reduction of fat and protein content of milk as milk yield increases is well known (Emery, 1988). As milk yield increases, lactose is synthesized and secreted at the same rate as milk, while fat and protein synthesis generally

Table 1.1. Genetic and phenotypic correlations between milk yield and fat and total protein (TP) concentration in goat's milk.

Breed	Genetic fat	Genetic TP	Phenotypic fat	Phenotypic TP	Reference
Alpine	0.07	–0.30	–0.13	–0.36	Boichard *et al.* (1989)
Saanen	–0.16	–0.08	–0.17	–0.35	Boichard *et al.* (1989)
Alpine	–0.08	–0.42	–0.16	–0.44	Barbieri *et al.* (1995)
Murciano-Granadina	–0.89	–0.65	–0.48	–0.47	Analla *et al.* (1996)

increase at a slower rate. This phenomenon can be described by the allometric model:

$$y = ax^b$$

where y is fat or protein yield (g/day); x is milk yield (kg/day); and a and b are equation coefficients. Application of this model to data of two breeds highly selected for milk yield (Saanen and Alpine) and of a local breed not highly selected for this trait (Sarda) showed that fat and TP synthesis varied proportionally to output of milk (or synthesis of lactose), with an exponent of 0.95 in the local breed for both parameters, and 0.83 and 0.87 in the highly selected breeds for milk fat and milk TP, respectively (Pulina *et al.*, 2003) (Fig. 1.1). Therefore, the higher the milk yield, the more cheese produced per goat, even though each additional unit of production results in a lower increase in cheese yield per litre (Table 1.2).

Due to the fact that b is almost one in the local goat breed, the 'dilution effect' is much weaker than in highly selected breeds and other ruminants (e.g. cattle and sheep) (Pulina *et al.*, 2003). Because the relationship between milk fat concentration and yield has a higher variability than that between milk protein concentration and yield, modification of milk composition by nutritional means should be easier to achieve for fat than for protein, as observed in sheep (Pulina *et al.*, 2006).

The yield of cheese depends mainly on milk fat and protein concentrations and somatic cell content (SCC), and varies depending on the type of cheese produced. However, in all types of cheese, milk fat and TP (or casein) are good predictors of cheese yield (Table 1.3).

During the cheese-making process, removed whey is heated at 80°C for the production of ricotta, whose residue is 'scotta'. The composition of these by-products depends on milk composition and technological factors (Table 1.4).

Increased SCC in goat's milk reduces cheese yield, expressed on a dry matter (DM) basis, by increasing the moisture in curd and cheese (Leitner *et al.*, 2004; Albenzio *et al.*, 2006; Raynal-Ljutovac *et al.*, 2007). No effects of SCC on goat's cheese yield were observed for milk with SCC ranging from 250,000 to

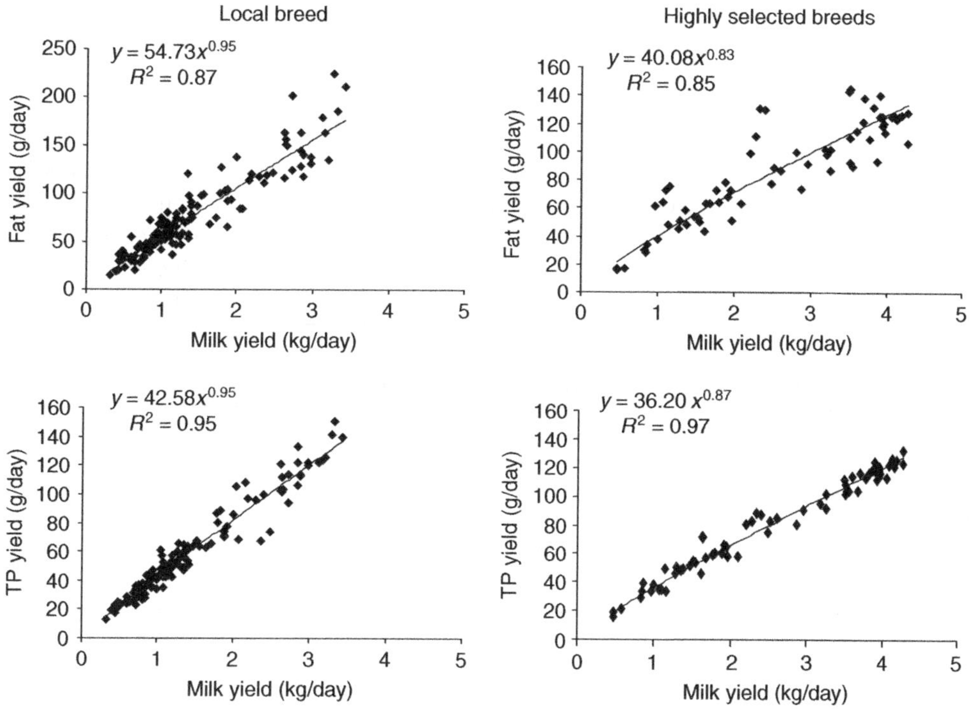

Fig. 1.1. Relationships between milk yield and milk fat yield and milk total protein (TP) yield of Sarda (n = 135; local breed) and Saanen and Alpine goats (n = 69; highly selected breeds). Data refer to the whole lactation. (Modified from Pulina *et al.*, 2003.)

Table 1.2. Coefficients of the allometric model $y = ax^b$ to estimate fat and total protein (TP) yield (y, in g) from milk yield (x, in kg) in a local breed (Sarda) and highly selected dairy breeds (Saanen and Alpine).

	Fat			TP		
Breed	*a*	*b*	R^2	*a*	*b*	R^2
Local	54.73	0.952	0.87	42.58	0.947	0.95
Highly selected	40.08	0.829	0.85	36.20	0.865	0.97
Cattle[a]	57.82	0.875	0.64	40.50	0.946	0.90
Sheep[a]	73.35	0.876	0.76	58.17	0.904	0.89

[a]Coefficients in cattle and sheep are reported for comparison (Pulina *et al.*, 2003).

>1,000,000 cells/ml (Galina *et al.*, 1996). These observations were confirmed in other studies cited by Raynal-Ljutovac *et al.* (2007), suggesting that factors linked to technology or cheese ripening could minimize the negative effects of high SCC on goat's milk. However, fresh cheese made from high-SCC milk presented higher proteolytic activity and a higher population of coagulase-positive

Table 1.3. Equations estimating cheese yield (CY, g cheese/100 g milk) as a function of fat (Fat), total protein (TP = N × 6.38), casein (CN) or total solids (TS) concentration in goat's milk.

Equation	R^2	Cheese	Reference
CY = 5.72 Fat + 0.29 TP + 0.76	0.81	Soft	Zeng *et al.* (2007)
CY = 2.16 TS – 4.85 Fat – 0.09	0.53	Semi-hard	Zeng *et al.* (2007)
CY = 8.61 TP – 12.76	0.79	Fresh	Guo *et al.* (2004)
CY = 3.85 Fat + 3.31	0.75	Fresh	Guo *et al.* (2004)
CY = 2.31 Fat + 5.77 CN – 5.97	0.80	Fresh	Guo *et al.* (2004)

Table 1.4. Mean composition (%) of whey and 'scotta' of goat's milk.

By-product[a]	Dry matter	Fat	TP	Ash
Whey (cheese plant 1)	6.58	0.57	1.03	0.52
Whey (cheese plant 2)	7.57	1.12	1.04	0.48
Scotta (cheese plant 1)	5.70	0.10	0.69	0.50
Scotta (cheese plant 2)	5.91	0.15	0.72	0.49

TP, total protein
[a]Samples were collected from two milk-processing plants, located in Sardinia (Italy), every 2 weeks from early spring (March–April) to early summer (July).

staphylococci than that made from low-SCC milk (Raynal-Ljutovac *et al.*, 2007). Ultra heat treatment of goat's milk with very high SCC may lead to lower heat stability of milk, due to an increase in whey protein content (Guthy, 1979, cited by Raynal-Ljutovac *et al.*, 2007), or to gelification problems, due to an increased content of plasmin enzyme (Battacone *et al.*, 2003).

Milk Fat and Protein Content

Among milk components, fat is the most sensitive to nutritional changes of the animals, even if fat variation in goat's milk (within one percentage point) is much lower than that observed in cow's milk. Milk protein content can be modified only slightly in goats, since it is highly influenced by polymorphism at the αS_1-casein locus (see Greppi *et al.*, Chapter 4, this volume). The concentrations of lactose, minerals and other solid components in milk are barely (mostly indirectly) influenced by diet.

Factors influencing milk fat content

Milk fat concentration and yield are affected by several nutritional factors. The most important are: (i) concentration, intake and source of dietary non-fibre carbohydrates; (ii) particle size of feeds and fibre; (iii) use of probiotics, such as yeast, in the diet; (iv) amount, physical characteristics and FA composition of dietary fat supplements; and (v) presence of precursors of *trans*-10, *cis*-12 conjugated linoleic acid, which can cause milk fat depression. Interactions among these factors and other less important ones make prediction of milk fat concentration complex.

Fibre and non-structural carbohydrates

In dairy cows (Emery, 1988; Sutton, 1989) and dairy sheep (Pulina *et al.*, 2006), a positive correlation between milk fat concentration and diet neutral-detergent fibre has often been reported. However, in sheep, energy balance is probably the factor that has the largest effect on milk fat concentration (Cannas *et al.*, 2002). When summarizing data from ten published experiments on lactating goats fed a variety of diets, mostly indoors, we found no relationship between milk fat concentration and dietary neutral-detergent fibre concentration ($R^2 = 0.09$). On the contrary, Santini *et al.* (1992) found a significant increase of milk fat concentration, from 2.48 to 3.32%, as acid-detergent fibre (ADF) increased from 14 to 26% (on a DM basis) in high-producing goats in early lactation. In this trial high fibre supply did not decrease milk production significantly, but 14% ADF in the diet was not enough for adequate fat synthesis, leading to a production of milk with less than 3% fat. The negative effects of low ADF on milk fat content did not depend on the milk production level of the animals. A reduction of milk fat content as dietary ADF decreases has also been observed in goats with medium (about 2 kg/day per head) (Mele *et al.*, 2005) and low (about 0.5 kg/day per head) (Kawas *et al.*, 1991) milk production levels.

Schmidely and Sauvant (2001) estimated milk fat content as a function of the percentage of concentrates in total mixed rations (from 20 to 70%, DM basis) (Eqn 1.1) or in traditional diets (from 15 to 100%) (Eqn 1.2):

$$\text{Fat (g/kg)} = 44.6 - 0.128 \times \text{Concentrate (\% DM)} \quad (R^2 = 0.98) \qquad (1.1)$$

$$\text{Fat (g/kg)} = 35.7 - 0.050 \times \text{Concentrate (\% DM)} \quad (R^2 = 0.97) \qquad (1.2)$$

Milk fat content decreased slightly (especially in traditional diets) as concentrate in the diet increased.

Morand-Fehr and Sauvant (1980) found that increased concentrate supply during late gestation influenced milk fat content and yield positively in early lactation goats. In that study, diets composed of lucerne hay, fed *ad libitum*, and 600 g of concentrate (grains, middlings, soybean and groundnut oil meals, minerals, and vitamins), during the last 6 weeks of gestation, increased milk fat percentage (+11%) and milk fat yield (+36%) compared with goats fed hay and 150 g of concentrate. Dietary concentrate increase, from 20 to 65%, in late lactation and dry goats increased milk fat content and yield in the subsequent early lactation phase (Goetsch *et al.*, 2001). Diets with very low forage/concentrate ratios, such as 30:70 (Rapetti *et al.*, 1997) or 12:88 (Bailoni and Andrighetto, 1995), had no negative effects on milk fat content in goats. A recent study on grazing goats evidenced that increasing amounts of concentrate in the diet (0, 0.33 and 0.66 kg per kilogram of milk produced above 1.5 kg/day) influenced milk yield positively, without effects on milk fat content and yield (Min *et al.*, 2005).

Based on the findings described above, goats seem to be less sensitive to fibre deficiency in their diet than cows, and, when they are fed diets that would cause the so-called 'low-fat syndrome' in cows, their milk fat content remains unchanged. Furthermore, diets with various forage/concentrate ratios, but with similar energy intake, cause small changes in milk fat content (see review of Sanz Sampelayo *et al.*, 2007).

Quality, composition in structural carbohydrates, density and size of fibrous particles are important aspects of dietary fibre. In primiparous Alpine does fed high concentrate rations, an increase of bermuda grass hay mean particle length from 2.38 to 3.87 mm resulted in higher total chewing and rumination times, milk fat content and fat-corrected milk production (Lu, 1987). Increases in milk fat content in goats due to increases in ingested fibre particle size were also observed by Sanz Sampelayo *et al.* (1998b). Furthermore, a non-forage diet characterized by small particle size and lack of long forage fibre did not influence milk fat content and yield in goats (Bava *et al.*, 2001). Goats fed different sources of carbohydrate from concentrates did not differ in milk and fat yield, which instead varied with the energy balance of the animals (Hadjipanayiotou, 2004; Sanz Sampelayo *et al.*, 2007).

More frequently than in other species, goats produce milk with a higher percentage of protein than fat (inversion of fat and protein percentage). This phenomenon often occurs in goats fed diets with very high concentrate content (Kawas *et al.*, 1991; Morand-Fehr *et al.*, 2000), but may also occur in those fed more traditional diets (Bailoni and Andrighetto, 1995; Rapetti *et al.*, 1997).

Effect of yeast on milk fat

The inclusion of yeast (*Saccharomyces cerevisiae*) in diets, at two levels of protein digested in the intestine (7.8 versus 10.8% of PDI), fed to goats during the first 6 weeks after parturition markedly increased milk fat content and yield, probably because yeast increased mobilization of body fat reserves (Giger-Reverdin *et al.*, 1996). Similarly, goats supplied with 6 g of yeast per day had a higher milk fat content than control goats (no yeast) and those fed 3 g of yeast daily (Abd El-Ghani, 2004). On the contrary, Stella *et al.* (2007) observed that diet supplementation with *S. cerevisiae* significantly reduced milk fat content from 4.46 to 4.32% ($P < 0.01$), even if treated goats had a slightly higher milk fat yield than control, due to a higher milk production.

Fat supplements

Milk fat concentration may be influenced by: (i) lipid content in the diet; (ii) level of fat supplementation; and (iii) quality of ingested fat. Many fat sources can be used as supplements to increase dietary net energy concentration or modify milk fat content or its FA profile in goats. In addition, fat increases the energetic efficiency of lactating ruminants, by increasing net energy intake and favouring nutrient partitioning towards milk production (Palmquist, 1994). The effects of lipid supplementation on goat's milk fat secretion have been studied by Chilliard *et al.* (2003) and Sanz Sampelayo *et al.* (2007), and are discussed in detail by Mele *et al.* (Chapter 3, this volume).

Factors influencing milk total protein content

Milk TP concentration is influenced by many nutritional factors. However, as previously noted, the range of variation of milk protein concentration is much narrower than that of milk fat, sharply reducing the possibilities of modifying its concentration by nutritional means.

Dietary energy and forage physical form

The use of carbohydrate supplements as an energy source in diets can improve N utilization by reducing ammonia levels in the rumen and increasing microbial protein synthesis. The presence of adequate carbohydrate fractions in diets with highly rumen-degradable protein increases the development of rumen microbes, thus improving the utilization of N for bacterial protein. Moreover, a balanced availability of carbohydrates in the rumen facilitates the production of propionate, thus reducing the requirement of amino acids for gluconeogenesis, which then become more available for intestinal uptake and mammary gland protein synthesis. This explains why goat's milk TP concentration is positively influenced by energy concentration of the diet, particularly when it comes from soluble carbohydrates (Sanz Sampelayo *et al.*, 1998b; Schmidely *et al.*, 1999). In high-concentrate diets, the substitution of barley grain with maize grain, both differing in rumen protein degradability (50 versus 32%, respectively), did not influence milk yield and protein content, but the diet containing barley caused intestinal disturbances (Hadjipanayiotou, 2004).

When lactating Granadina goats were fed diets with two physical forms of lucerne hay (long versus chopped to 3–4 cm particle size and pelleted), milk protein content and yield were not influenced. However, milk protein of goats fed the diet with pelleted lucerne hay was higher in casein and lower in whey protein and serum albumin than that of goats fed long lucerne hay (Sanz Sampelayo *et al.*, 1998b). This might be attributed to a lower rate of ruminal protein degradation, which could have led to a greater efficiency of N utilization by the mammary gland for milk production and protein synthesis in the goats fed pelleted lucerne hay. However, since the genotype for αS_1-casein in milk was not determined in that trial, this result could be due to the genetic interference of milk protein polymorphism, instead of the hypothesized nutritional effect.

Dietary protein

In several studies, the crude protein (CP) concentration in the diet did not influence milk protein content (Andrighetto *et al.*, 1989; Hadjipanayiotou and Photiou, 1995). However, when dietary CP concentration was raised from low (below 14%, on DM basis) to adequate level (17%, on DM basis), daily milk protein yield increased, without changing the milk TP concentration (Schmidely *et al.*, 2002).

Dietary protein sources can influence milk protein content and yield (Table 1.5). This could be explained by the rumen degradability of dietary CP and by the availability of essential amino acids of the escape protein fraction. The replacement of a dietary CP source highly degradable by the rumen (e.g. soybean meal) by a source having very low rumen degradability (e.g. hydrolysed feather meal)

Table 1.5. Influence of different protein sources on milk protein content and yield.

Breed	Lactation stage	Protein source	Dietary CP (%)	Milk yield (kg/day)	Protein content (%)	Protein yield (g/day)	Reference
Alpine	14 DIM	SBM	16.0	3.20	3.0	97	Lu *et al.* (1990)
		Feather meal	16.0	3.16	2.8	87	
Damascus	136 DIM	SBM	15.3	1.95	3.96	77.2	Hadjipanayiotou and Koumas (1991)
		Fish meal	14.6	1.94	3.93	76.2	
	165 DIM	SBM	15.3	1.66	4.07	67.6	
		Fish meal	14.6	1.60	4.04	64.6	
Damascus	66 DIM	SBM	15.9	2.38	4.34	103.3	Hadjipanayiotou (1992)
		SBM-P	16.0	2.63	4.12	108.4	
		Fish meal	16.1	2.36	4.36	102.9	
	108 DIM	SBM	15.9	2.11	4.10	86.5	
		SBM-P	16.0	2.08	4.16	86.5	
		Fish meal	16.1	2.05	4.36	89.4	
Damascus	66–118 DIM	SBM	15.1	2.34	4.23	99.0	Hadjipanayiotou (1995)
		Heat-treated SBM	15.5	2.24	4.34	97.2	
Damascus	91 DIM	SBM	13.0	1.31	4.2	55.0	Hadjipanayiotou and Photiou (1995)
		SBM	15.0	1.35	4.1	55.4	
		SBM	18.0	1.52	4.2	63.8	
		SBM-P	13.0	1.28	4.3	55.0	
		SBM-P	15.0	1.37	4.1	56.2	
		SBM-P	18.0	1.52	4.0	60.8	

Damascus	15 days post-kidding	SBM	17.0	2.61	4.3	112.2	Hadjipanayiotou *et al.* (1996)
		SBM + meat meal	17.2	2.33	4.2	97.9	
Alpine primiparous	21–42 DIM	Control	9.5	1.55	2.73	42.3	Fernandez *et al.* (1997)
		Protein	14.0	1.79	2.70	48.3	
		Protein (plus urea)	14.0	1.63	2.81	45.8	
Alpine pluriparous	21–42 DIM	Control	12.5	2.49	3.11	77.4	Fernandez *et al.* (1997)
		Protein	17.0	2.46	2.85	70.1	
		Protein + urea 25	17.0	2.61	2.98	77.8	
		Protein + urea 50	16.0	2.23	2.62	58.4	
Granadina	80 DIM	Bean	18.0	1.02	3.19	32.5	Sanz Sampelayo *et al.* (1998a)
		Sunflower	19.1	1.05	2.85	29.9	
		Maize gluten	17.6	0.95	3.55	33.7	
		Cottonseed	17.6	0.96	3.27	31.4	
Murciano-Granadina	160 DIM	SBM	15.7	1.84	3.78	69.6	Fernández *et al.* (2003)
		Sunflower	16.3	2.14	3.38	72.3	
		Fish meal	15.0	2.23	3.31	73.8	

CP, crude protein; DIM, days in milk; SBM, soybean meal; SBM-P, protected soybean meal (formaldehyde-treated).

decreased milk TP content and yield, milk true protein and casein (Lu *et al.*, 1990). When goats were fed diets with similar CP content, but different rumen degradability due to the different protein sources used (beans, sunflower cake, maize gluten, cottonseed), those fed maize gluten showed the highest content of protein, casein, β-casein and cheese yield (Sanz Sampelayo *et al.*, 1999). However, differences in degradability of the CP of the diet cannot alone explain modifications in milk CP concentration, since this milk component did not change when soybean meal in the diet was formaldehyde-treated (Hadjipanayiotou and Photiou, 1995), extruded (Schmidely *et al.*, 2005) or heat-treated (Hadjipanayiotou, 1995) in order to reduce rumen degradation. This suggests that, for a good understanding of the process of milk protein production from the CP ingested by goats, it is necessary to take into account not only the kinetics of the ruminal degradation of the protein supplied but also the amino acid profile of the rumen-undegradable protein fractions (Sanz Sampelayo *et al.*, 1999). These results suggest that the relationship between the CP in the diet and milk protein production is due to a multivariate process and is not the same for all milk protein fractions. These aspects have been studied by Sanz Sampelayo *et al.* (1999), who reported that the rumen-undegradable protein and its essential amino acid content are weakly associated with milk protein and with its αs-casein and β-casein fractions, while milk protein and its casein fractions seem to be mainly associated with quickly degradable protein in the rumen. All of these results, which are not always univocal, confirm that the relationship between protein concentration in milk and that in the diet of goats must be better studied, in order to determine if dietary protein could be a valid tool to improve the protein content in goat's milk.

Dietary fat

In general, the use of rumen-protected or inert forms of fat supplements in the diets of dairy goats, aiming to increase dietary energy intake in high-producing goats, especially in early lactation, did not markedly influence milk protein content (Table 1.6). The strongest reduction in milk protein percentage (–4.1%) occurred when calcium soap of polyunsaturated FAs was used (Sanz Sampelayo *et al.*, 2002) compared with the diet without added fat. These results are in contrast with several observations in sheep (Pulina *et al.*, 2006) and cows (Wu and Huber, 1994; Onetti and Grummer, 2004), where the addition of calcium soap of FAs reduced milk protein concentration.

The level of reduction in milk protein yield observed with supplemental fat ranged from –3.2% to about –15% when calcium soap of palm oil (Teh *et al.*, 1994) or polyunsaturated fat oil (Sanz Sampelayo *et al.*, 2002), respectively, was used in high doses compared with the non-supplemented group. The reduction of milk protein yield due to supplemental fat could be a consequence of the reduced milk yield observed in some experiments. Again, these observations contrast with observations in cows, where milk protein yield is usually increased or not affected by supplemental fat (Wu and Huber, 1994).

Unprotected fat supplements from whole seeds, free oil or animal sources did not influence milk protein content and yield negatively (Table 1.7) (see also the review of Chilliard *et al.*, 2003). The mechanism involved in such responses is not known. One possible explanation might be the influence of the fat

Table 1.6. Effects of rumen-protected or inert fat supplements on milk protein content and yield in goats. The data are expressed as differences in percentage between supplemented and basal diet.

Fat source	Dietary fat (%)	Lactation stage	Duration (weeks)	Milk (%)	Protein content (%)	Protein yield (%)	Reference
Protected oil[a,b]	7.0	104 DIM	7	0.0	–0.02	–0.02	Lanzani *et al.* (1985)
CaS-PUFAs[a]	6.0	5 weeks	12	3.0	0.0	3.0	Baldi *et al.* (1992)
Megalac[a]	3.0	3–12 weeks	10	2.1	–0.4	1.8	Teh *et al.* (1994)
	6.0			–7.5	4.6	–3.2	
	9.0			–14.3	–0.4	–14.7	
CaS-palm oil[a]	4.0	12–41 weeks	29	14.6	2.5	17.5	Rousselot *et al.* (1995)
CaS-PUFAs[c]	9.0	Mid	4	–6.6		–5.2	Pérez *et al.* (2000)
RP-tuna oil[a]	6.9	12 weeks		–5.9	4.1	0.3	Kitessa *et al.* (2001)
PUFA-RP[c]	9.0	Mid	3	2.9	–4.1	–1.7	Sanz Sampelayo *et al.* (2002)
	12.0			–12.0	–1.0	–13.7	

CaS, calcium soap; PUFA, polyunsaturated fatty acid; RP, rumen-protected; DIM, days in milk.
[a]Dietary fat (% of concentrate).
[b]25% soybean oil was homogenized with 75% maize meal treated with formaldehyde.
[c]Dietary fat (% of ration).

Table 1.7. Effect of unprotected fat supplements on milk protein content and yield in goats. Data are expressed as differences in percentage between supplemented and basal diet.

Fat source	Dietary fat (%)	Lactation stage	Milk (%)	Protein content (%)	Protein yield (%)	Reference
Animal fat[a]	5	1–17 weeks	−10.4	3.6	−7.2	Lu (1993)
Animal fat[a,b]	1.5	3–13 weeks	8.1	4.6	13.1	Brown-Crowder *et al.* (2001)
	3		10.2	1.8	12.1	
	4.5		12.6	3.6	16.7	
	6		−5.6	3.6	−2.2	
Tuna oil[c]	7.1		−28.0	10.9	−20.3	Kitessa *et al.* (2001)
Linseed oil[a]	5–6		9.1	8.8	18.7	Chilliard *et al.* (2003)
Linseed[a]	5–6		1.7	10.3	12.2	
Sunflower oil[a]	5–6		10.1	3.7	14.2	
Sunflower seeds[a]	5–6		8.7	7.0	16.3	
Lupin seeds[a]	5–6		10.5	6.6	17.8	
Soybeans[a]	5–6		17.8	5.5	24.3	
Extruded linseed[c]	5	Late	−12.5	−0.7	−14.2	Nudda *et al.* (2006)
	10		−4.2	4.9	−1.4	
Extruded soybeans[a]	10	12–25 weeks	3.3	0.6	4.0	Schmidely *et al.* (2005)
	20		14.2	−0.3	13.9	

[a]Dietary fat (% of ration).
[b]Partially hydrogenated fat.
[c]Dietary fat (% of concentrate).

supplement on rumen metabolism, i.e. the defaunating action of FAs (both as free FAs or calcium salt of FAs) on rumen protozoa observed in rumen samples of goats (Matsumoto *et al.*, 1991) may increase microbial protein availability in the intestine. Fujihara *et al.* (2003) observed that the mean urinary excretion of TP derivatives decreased significantly in defaunated goats compared with that in faunated goats. Previous trials in sheep showed that the supplementation of linseed oil in increasing doses (13, 26 or 40 ml/day) increased the duodenal flow of total N and bacterial N and the efficiency of bacterial protein synthesis (Ikwuegbu and Sutton, 1982).

A strong negative effect of fat sources on milk protein yield occurred only when tuna oil was used in an unprotected form (Kitessa *et al.*, 2001), probably because of the inhibitory effects of the main FA of fish oil (i.e. 20:5*n*–3) on microbial protein synthesis (Wachira *et al.*, 2000).

In a trial in which milk protein fractions were determined (Sanz Sampelayo *et al.*, 2002), the content of α-casein was higher and that of seroalbumin was lower for the basal diet than for rumen-protected fat added to concentrate at

9–12%. This is supported by previous observations of Lanzani *et al.* (1985), in which protected soybean oil increased non-casein N and non-protein N (NPN) in treated goats compared with control.

Effect of yeast on milk protein

Yeast cultures have positively influenced animal performance in dairy cows, because of their positive effects on the number of rumen bacteria and on the digestion of feedstuffs (Arambel and Kent, 1990; Harrison *et al.*, 1998).

In goats, the inclusion of yeast (*S. cerevisiae*) in diets with two levels of PDI (7.8 versus 10.8% of PDI) during the first 6 weeks after parturition tended to increase yield of milk and protein compared with control, but only in the diet with low PDI. Milk protein content was not affected by yeast use, whereas milk urea content was higher for the diet with 10.8% PDI, with stronger effects when this diet was supplemented with yeast (Giger-Reverdin *et al.*, 1996). Supplementation with a mixture of yeast culture with malate had no beneficial effects on the performance of dairy goats (Salama *et al.*, 2002).

Abd El-Ghani (2004) found in goats in the first and second month of lactation fed concentrate and lucerne hay that the daily administration of 6 g of yeast (*S. cerevisiae*) significantly increased milk yield, milk protein and total solid contents compared with non-supplemented goats. Use of live *S. cerevisiae* as a supplement to Saanen goats in the first 3 weeks of lactation increased milk yield (+0.3 kg/day) but did not change protein content and yield compared with control animals (Stella *et al.*, 2007).

Increase of milk production, without changes in milk protein content, observed in goats fed yeast supplements is probably due to the greater flow of microbial protein and amino acids to the duodenum compared with control, as observed previously in dairy cows (Erasmus *et al.*, 1992).

Effect of tanniniferous feeds on milk protein content

The influence of tanniniferous feeds on milk protein content in sheep and goats has recently been reviewed (Vasta *et al.*, 2007). Browse species are rich in tannins and several studies have evaluated the impact of such tanniniferous species on goat's milk yield and composition. For example, Decandia *et al.* (2000) studied milk yield and composition of Sarda goats in late lactation browsing a maquis dominated by lentisk (*Pistacia lentiscus* L.) of which the leaves were extremely rich in condensed tannins (22% DM), and fed small amounts of concentrates with or without polyethylene glycol (PEG). Goats fed 50 g of PEG per day had higher dietary protein digestibility (53 versus 40%) and milk yield (755 versus 645 g/day) than goats fed no PEG. Despite this, no differences in milk protein percentage were observed between the two groups. In Maber goats grazing Mediterranean woodland (rich in *Quercus* spp., *Pistacia palestina*, *P. lentiscus*, *Sarcopoterium spinosa*, *Rhamnus palatina* and *Calicitome villosa*) and supplemented with 200 g of concentrate daily, the addition of 10 g of PEG in the diet markedly increased milk yield (1.46 versus 1.00 kg/day) without influencing milk protein concentration throughout lactation (Gilboa *et al.*, 2000).

In conclusion, high concentrations of condensed tannins in alternative feed resources markedly reduce rumen microbial activity and digestion of bacterial and feed amino acids in the intestine. However, their deactivation by PEG increases milk yield, sometimes markedly, and maintains milk fat and protein concentration, thus counterbalancing the dilution effect caused by the higher milk yield.

Factors influencing milk urea and other nitrogen compounds

The effects of nutrition on milk protein profile have been little studied in goats. In general, an increase of dietary CP did not affect milk TP content, but influenced the milk N fraction, with increases in NPN at the expense of casein N (Polidori *et al.*, 1991), in urinary N excretion (Schmidely *et al.*, 2002) and in blood urea (Andrighetto *et al.*, 1989). Different protein sources used in isonitrogenous diets modified the milk protein fraction (Pailan and Kaur, 1996; Sanz Sampelayo *et al.*, 1998a).

The over-consumption of soluble protein sources (i.e. young ryegrass and diets rich in NPN) by ruminants can cause hyperammonaemia (also known as ammonia or urea toxicosis), due to increased blood ammonia, generally associated with increased plasma and milk urea content. In primiparous goats fed a diet with 14% CP, the use of urea as a feed contributing 33% of total dietary N did not influence milk yield and milk protein content, but increased non-casein N and serum protein N in milk and blood urea (Fernandez *et al.*, 1997). The same authors observed that in pluriparous goats fed 17% dietary CP, urea supplementation (50% of the total dietary N) significantly decreased milk protein content and casein compared with the control diet (no urea). Supplemented goats also showed subclinical hyperammonaemia, as evidenced by high ammonia and urea N in the blood.

Plasma concentration of urea and urinary excretion of N were higher in goats fed diets with rapidly degraded starch and N than in goats fed slowly degraded diets, indicating an inefficient use of ruminal ammonia (Schmidely *et al.*, 1999).

Since only a few studies have dealt with milk urea content in dairy goats, the possibility of using this molecule as an indicator of their protein and energy nutritional status is limited.

In lactating goats, when the dietary amount of soluble carbohydrates was low, urinary level of allantoin was low but milk urea content was above 30 mg/100 ml, whereas when the dietary amount of rumen-degradable N was low, milk urea content was below 28 mg/100 ml (Brun-Bellut *et al.*, 1984).

In goats fed diets with three levels of dietary CP (8.5, 14 and 20%, on DM basis), a close positive relationship between blood urea N (BUN) (8.3, 22 and 33 mg/dl, respectively) and dietary CP concentration was observed (Sahlu *et al.*, 1993). This probably happened because the efficiency of utilization of dietary N for milk protein synthesis decreased and N losses increased, as CP content in the diets increased, as observed previously in sheep (Cannas *et al.*, 1998). Indeed, in lactating goats increased rumen-degradable protein in the diet (6.7 versus 9.7% of DM), at constant levels of non-degradable protein, did not influence milk protein, but increased milk urea N (MUN) (Brun-Bellut *et al.*, 1990).

BUN concentration is proportional to MUN in dairy cows (Baker *et al.*, 1995) and in dairy sheep (Cannas *et al.*, 1998). In goats fed diets with 13.6% CP, BUN and MUN were highly correlated ($r = 0.88$) (Khaled *et al.*, 1999). In two groups of goats fed a silage-based diet (15% CP on DM basis) and a non-forage diet (21% CP on DM basis), a similar regression coefficient between BUN and MUN was found for each of the two diets (Bava *et al.*, 2001). The use of MUN concentration to estimate dietary CP content in goats fed on pasture is reported by Decandia *et al.* (Chapter 8, this volume), even if a previous estimation of DM intake is required. A direct estimation of dietary CP as a function of MUN is done in goats grazing herbaceous pasture by Bonanno *et al.* (Chapter 9, this volume).

Despite the importance of milk urea as a nutritional indicator in dairy cows and sheep, very few studies have evaluated its use in dairy goats. By pooling 76 data of eight different studies, Cannas (unpublished data) calculated the regressions between dietary CP concentration or daily CP intake and milk or blood urea. These two variables were weakly correlated with dietary CP concentration (Fig. 1.2) or not correlated at all with daily CP intake, in contrast to what is found in sheep (Fig. 1.3). Such differences are probably due to the fact that urea recycling in goats is more efficient than in sheep.

In addition, since high-energy diets reduce the necessity of goats to use amino acids as an energy source, less ammonia is produced from amino acid catabolism.

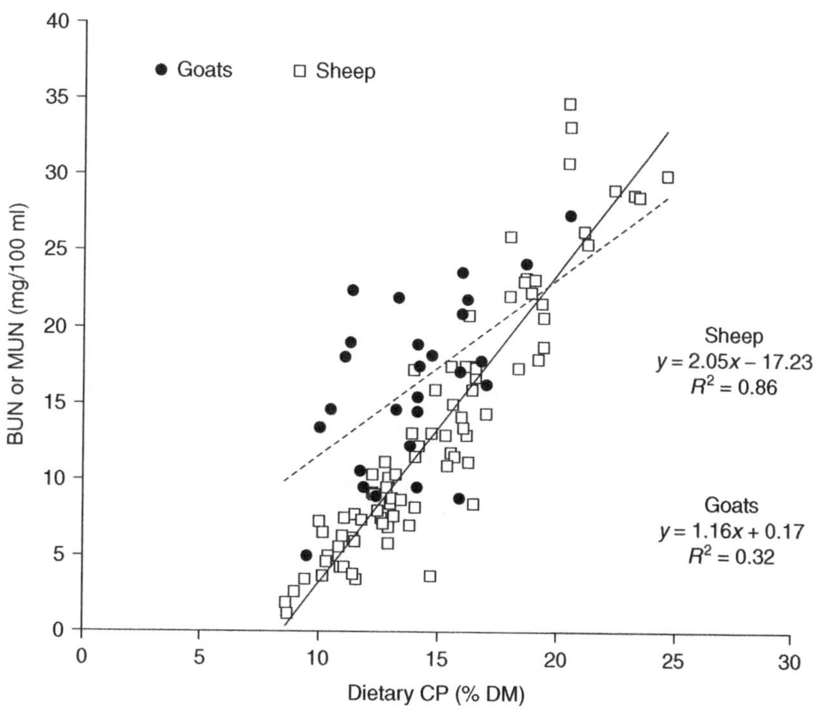

Fig. 1.2. Relationship between dietary crude protein (CP; percentage on dry matter, DM, basis) and urea concentration in blood (BUN) and milk (MUN) of goats and sheep. (Adapted from Cannas, unpublished data.)

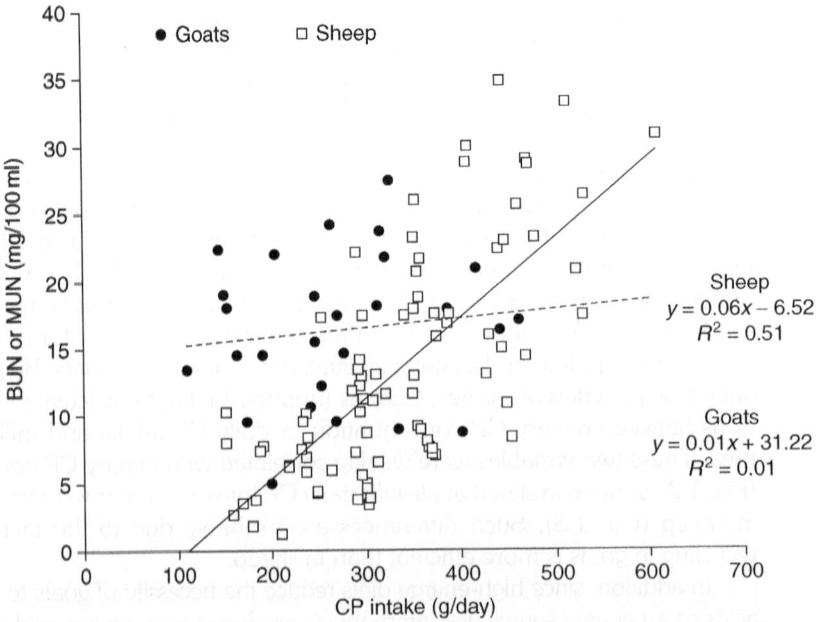

Fig. 1.3. Relationship between daily crude protein (CP) intake (g/day) and urea concentration in blood (BUN) and milk (MUN) of goats and sheep. (Adapted from Cannas, unpublished data.)

Degradability of dietary CP seems to be less important in goats than in dairy cattle, probably because the higher rumen passage rate of goats reduces the proportion of CP degraded in the rumen compared with cattle, thereby increasing the amount of escape protein and avoiding urea production.

Milk Mineral Content

Usually, macro minerals (Ca, P, Mg, K, Na, Cl, S) in milk are not modified by feeding. Mineral deficiency in goats can have long-term negative effects on milk production level, rather than influencing mineral concentration in milk. In goats, Ca and P requirements are high, because their prolificacy is higher than that of sheep. Requirements of Ca and P for growing kids are higher than for growing cattle. Nevertheless, Ca (1.3 g/l) and P (0.9 g/l) concentrations in goat's milk (Table 1.8) (Meschy, 2000) are similar to those of cow's milk (Greer, 1989). The knowledge of the variations in Ca and P concentration during different lactation stages can help to determine the requirements for these and, consequently, the appropriate amount of mineral dietary supplement.

In lactating goats, increasing dietary CP concentration (from 13 to 20%) and dietary P concentration, owing to the substitution of barley (4 g P/kg DM) with soybean meal (7 g P/kg DM), increased Ca and P content in milk, probably because about 60–70% of milk Ca is included into casein as calcium phosphocaseinate (Schmidely *et al.*, 2002).

Table 1.8. Effect of lactation stage on calcium and phosphorus concentration in milk of Saanen and Alpine goats (Meschy, 2000).

Lactation stage	Ca (g/l)	P (g/l)
Early	1.40	1.05
Mid	1.26	0.95
Late	1.15	0.89

The supplementation of I orally in the form of iodinated FA esters (480 mg I/ml iodinated oil), approximately 11 days before kidding, significantly increased I concentration in colostrum and milk of goats. Iodine level returned to the control level 75 days after the administration (Herzig *et al.*, 2003).

Somatic Cell Content

The somatic cell content (SCC) in milk is indirectly affected by nutrition. For example, poor nutrition may predispose goats to metabolic and infectious health disorders, which increase susceptibility of the mammary gland to inflammation. In sheep (Pulina *et al.*, 2006), correct integration of vitamin A or β-carotene, and vitamin E and Se in the ration may help to maintain immune responses of mammary gland cells, thereby reducing the incidence of infections that could otherwise increase milk SCC. When evaluating the effects of feeding systems on SCC in goats, Fedele *et al.* (1996) found that diets unbalanced in energy, CP and minerals increased SCC in milk.

The use of officinal plants to enhance the immune system has been studied in goats. In a study aiming to reduce mammary infections and SCC in goat's milk, the effects of *Echinacea purpurea* (100 mg of extract, corresponding to 2 mg of active ingredient) on the secretion of lactoferrin in milk, which is an anti-bacterial, anti-viral and immunostimulating compound, were evaluated. Treatment with *E. purpurea* extract increased lactoferrin and progressively decreased SCC and microbial cell count in milk (Reklewska *et al.*, 2004). In organic farms, Saanen goats supplemented with leaves of rosemary (*Rosmarinus officinalis* L.), from 10 days before kidding to 7 weeks of lactation, did not show changes in SCC content but showed significant changes in neutrophils and macrophages distribution in milk and blood (Savoini *et al.*, 2003).

The effects of dietary supplements of zinc methionine on milk SCC and udder health were evaluated in goats (Salama *et al.*, 2003). In spite of the lack of significant differences in SCC between supplemented and control goats, resistance to udder stress was enhanced in the treated group, probably because a higher level of available Zn might have improved keratin synthesis in the teat canal tissue and decreased the incidence of new infections. Milk SCC was not influenced by the administration of betaine, a by-product of choline metabolism essential for the immune system, for 5 months in lactating goats (Fernández *et al.*, 2004).

Undesirable Substances

The whole dairy goat sector is responsible for the safety of its food products and feeds. Goat feeds, including herbage, may be subjected to contamination from diverse sources, such as environmental pollution and activities of insects and microbes. Undesirable substances that can occur in feedstuffs include organic compounds such as plant toxins, mycotoxins, dioxins and polychlorinated biphenyls (PCBs), and inorganic compounds, such as heavy metals. The effects of undesirable substances in feedstuffs range from reduced intake to reproductive dysfunction and increased incidence of diseases in animals. Moreover, in many cases they can be transferred to edible animal products such as milk.

Mycotoxins

Mycotoxins are low-molecular-weight secondary metabolites, produced by filamentous fungi, which affect human and animal health. Mould growth and mycotoxin contamination of feedstuffs can occur at all stages of the production cycle (cropping, harvest, transport and storage). Mycotoxins can be found in feed ingredients, such as maize, sorghum grain, barley, wheat, rice meal, cottonseed meal, groundnuts and other legumes. Most of them are relatively stable, not destroyed and, sometimes concentrated, during feed processing. The possibility of transferring mycotoxins or their metabolites to milk represents a potential risk to dairy consumers.

Aflatoxins, which are included among the substances carcinogenic to humans by the International Agency for Research on Cancer of the World Health Organization (WHO) (IARC, 2002), represent the family of mycotoxins of major concern in goat husbandry. The ingested aflatoxin B1 (AFB1) is biotransformed by hepatic microsomal cytochrome P450 into aflatoxin M1 (AFM1), which is then excreted in the milk of lactating animals. The occurrence of AFM1 in goat's milk has been reported in several studies (Roussi *et al.*, 2002; Oliveira and Ferraz, 2007; Virdis *et al.*, 2008). The concentration of AFM1 in goat's milk reached a steady-state condition within days from the start of AFB1 intake (Smith *et al.*, 1994). The carry-over value of AFB1 from feed into AFM1 was about 0.40% for goats receiving 124.4 μg of AFB1 daily for 2 weeks (Rao and Chopra, 2001). In ruminants, the addition of sequestering agents, which bind to AFB1, reduced its absorption by the gastrointestinal tract and attenuated the transfer of AFM1 to milk (Diaz *et al.*, 2004). In lactating goats, the addition of aluminosilicate or activated charcoal to feeds dramatically decreased AFM1 concentration in milk (Table 1.9; Fig. 1.4). The AFM1 concentration in unprocessed milk was linearly related to AFM1 concentrations in curd and whey (Battacone *et al.*, 2005). In a study on the AFM1 distribution in different protein fractions in goat's milk and during the production of goat's cheese, neither ultrafiltration nor acidic or enzymatic treatments influenced the interaction between the toxin and casein or whey proteins (Barbiroli *et al.*, 2007).

Several studies have reported that small ruminants appear to be tolerant to mycotoxins such as fumonisins (Gurung *et al.*, 1998), ochratoxin A (Blank *et al.*, 2003), zearalenone and trichothecenes (Kiessling *et al.*, 1984).

Table 1.9. Effect of different adsorbents on transfer of aflatoxin B1, as aflatoxin M1 (AFM1), in goat's milk. The data are expressed as mean ± standard error.

Aflatoxin	Contamination (ppb)	Adsorbent	Addition (%)	AFM1 in milk (ppb)	Length of administration (days)	Reference
Mix[a]	100	HSCAS	0	0.481 ± 0.17	6	Smith *et al.* (1994)
	100	HSCAS	1	0.244 ± 0.01		
	100	HSCAS	2	0.099 ± 0.02		
	200	HSCAS	0	1.619 ± 0.23		
	200	HSCAS	4	0.187 ± 0.02		
Aflatoxin B1, pure	100		0	0.498 ± 0.085	14	Rao and Chopra (2001)
	100	SB	1	0.164 ± 0.041		
	100	AC	1	0.091 ± 0.028		

HSCAS, hydrated sodium calcium aluminosilicate; SB, sodium bentonite; AC, activated charcoal.
[a]79, 16, 4 and 1% of aflatoxins B1, G1, B2 and G2, respectively.

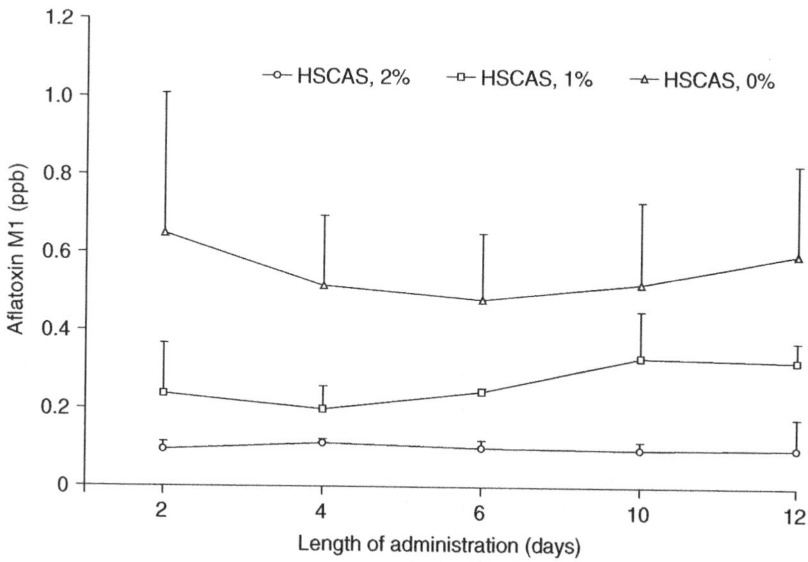

Fig. 1.4. Effect on aflatoxin M1 concentration in milk of goats fed diets containing aflatoxins at 100 ppb in which 0%, 1% or 2% of hydrated sodium calcium aluminosilicate (HSCAS) was added. (Data from Smith *et al.*, 1994.)

Dioxins and polychlorinated biphenyls

'Dioxins' is a term referring to the categories of polychlorinated dibenzo-*p*-dioxins (PCDDs) and polychlorinated dibenzofurans (PCDFs). These substances occur mainly as by-products of incomplete combustion and of certain chemical processes. The toxicity of individual dioxin and dibenzofuran compounds (or congeners) varies considerably. The PCDD and PCDF congeners that are likely to be of toxicological significance are those with Cl atoms at the 2, 3, 7 and 8 positions, of which the most toxic is 2,3,7,8-tetrachlorodibenzo-*p*-dioxin. PCBs are chlorinated hydrocarbons synthesized by direct chlorination of biphenyl. Because there are PCB congeners that show similar toxicological properties to dioxins, these compounds are often termed 'dioxin-like PCBs'.

PCDDs, PCDFs and PCBs are hazardous organic compounds of great importance to animal agriculture, because these compounds are persistent and tend to bioconcentrate in the lipids of tissues and products. All dioxins enter the food chain through animals that are exposed to contaminated surroundings and feeds, and tend to concentrate mainly in body and milk fat. Among foods of animal origin, dairy products have the highest level of contamination (Bocio *et al.*, 2003). The application of sewage sludge to agricultural land is a potential way of transporting the lipophilic halogenated hydrocarbon contaminants in animal tissues and products. When soil and pasture are contaminated by polychlorinated compounds, grazing animals (e.g. goats) are the most exposed to them, especially when pasture is the major component of the diet, since soil ingestion is an additional contaminant source. Animal uptake of soil and plant contaminants is related to: (i) rates of soil ingestion; (ii) feeding and management systems; and (iii) bioavailability of contaminant molecules and their kinetics in animals (Fries, 1996). Dairy goats treated with PCB 126 and PCB 153 at doses of 49 and 98 ng/kg body weight (BW) per day, respectively, showed milk concentrations of PCB 126 and PCB 153 of 0.24 ng/g and 507 ng/g (wet weight), respectively (Lyche *et al.*, 2004). The different kinetic properties observed for these two congeners may have been due to hepatic sequestration of PCB 126 and accumulation of PCB 153 in fat tissue and milk in goats. In sheep, Vrecl *et al.* (2005) showed that PCBs were transferred from blood to either milk or faeces by two different pathways. In particular, an enrichment of lipophilic precursors was observed in milk, because of its higher lipid content compared with faeces. These results suggest that in lactating animals the higher chlorinated, coplanar and metabolically stable PCBs are preferentially excreted in milk. Recently, Costera *et al.* (2006) determined the transfer of 17 PCDDs and PCDFs and 18 PCBs from feed to milk in Alpine lactating goats (mean BW: 50 ± 5 kg) fed 800 g of contaminated hay daily (2 ng/kg DM for WHO-PCDD/PCDF-TEQ versus 0.38 ng/kg DM for WHO-dioxin-like PCBs-TEQ, where TEQ is a dioxin toxic equivalent, calculated by looking at all toxic dioxins and furans and measuring them in terms of the most toxic form of dioxin, (2,3,7,8-tetrachlorodibenzo-*p*-dioxin)) for 10 weeks. The concentrations of the various compounds in milk were similar, with an increasing phase until a plateau was reached. For PCDDs/ PCDFs the plateau was reached after at least 35 days of contaminated hay intake. For PCBs, the steady state was reached very quickly, the latest being 22 days after the beginning of the experiment. The concentration of PCDDs and PCDFs ranged from 0.16 to 3.83 ng/g milk fat, while that of PCBs ranged from 2 to 3524 ng/kg milk fat. The carry-over rates of PCDDs, PCDFs and PCBs appeared to be a function of their physical and chemical properties, such as the number of Cl atoms and the metabolism of these compounds.

Polycyclic aromatic hydrocarbons (PAHs) are a large group of organic pollutants with two or more fused aromatic rings. They are mainly by-products of pyrolytic processes of organic materials during human activities, such as combustion of natural gas, processing of coal and crude oil, and vehicle traffic. These organic pollutants are ubiquitous and can occur at a low level of contamination in water, air or soil.

Owing to their physical and chemical properties, PAHs move through hydrophobic compartments of the food chain. In lactating animals, after ingestion of contaminated feeds (e.g. pasture and forage), pollutants can be transferred to milk. The selectivity of the mammary epithelial barrier, which favours the passage of lipophilic, low-molecular-weight compounds, suggests that mammary epithelium could play a key role in the selective transfer of PAHs from food to milk (Cavret *et al.*, 2005). Grovaa *et al.* (2006) reported a rapid transfer of phenanthrene, pyrene and benzo(*a*)pyrene (PAH compounds), administered in a single dose, from feed to blood of lactating goats, and similar plasma kinetics for the three compounds. Both phenanthrene and pyrene showed a low milk transfer (1.6 and 1.9%, respectively), while benzo(*a*)pyrene showed a very low milk transfer (0.2%). Chronic oral exposure of lactating goats to different PAHs also confirmed that their transfer through the organism and excretion in milk is affected by the selectivity of the intestinal or mammary epithelial barrier for PAHs in lactating goats or by their capacity to metabolize them (Grovaa *et al.*, 2006).

Heavy metals

Anthropogenic processes such as the combustion of coal and mineral oil, smelting, mining, alloy processing and the paint industry are the major sources of heavy metals to animals in the vicinity of industrial areas. Animals allowed to graze on or fed contaminated pastures or fodders are continually exposed to these pollutants, leading to various health hazards. Also, agricultural activity may result in elevated levels of metals in soils. In particular, use of sewage sludge and long-term addition of inorganic fertilizers and pesticides can increase total and available metal contents in the upper soil horizons. Heavy metal accumulation in agricultural crops is particularly important in goat farming, because they are a source of toxic elements to the human food chain. Uptake and accumulation by crop plants represent the main entry pathway for potentially health-threatening toxic metals into human and animal food. In many cases, the relationship between metal content in the soil and in the crop is positive and linear (Grytsyuk *et al.*, 2006).

For Cd, which is one of the most dangerous metals for animal and human health, the passage through the dairy food chain is not limited by factors such as the absorption capacity of the soil, soil pH and salinity, and metal phytotoxicity, differently to what happens to other heavy metals (Li *et al.*, 2005). High metal concentration in kidneys and liver, as well as histopathological changes in kidneys, were found in goats that were given lead acetate 50 mg/kg BW, cadmium chloride 10 mg/kg BW, or lead acetate 50 mg + cadmium chloride 10 mg/kg

BW, orally and daily for 42 days (Haneef *et al.*, 1998). Milhaud *et al.* (2000) studied the transfer of Cd from feed to goat's milk in lactating goats that received cadmium chloride 2 mg/kg BW or 4 mg/kg BW for 12 weeks. The pattern of Cd concentration in milk approached a steady-state condition after 10 weeks in both treated groups. The kinetics of Cd showed a non-linear relationship with Cd intake. In particular, the excretion of Cd in milk of goats fed 4 mg Cd/kg BW was threefold higher than that in goats fed 2 mg Cd/kg BW. This kinetic behaviour is explained by the high affinity of this heavy metal for some serum components, such as metallothioneins, which play an important role in metabolism and kinetics of the toxic metal (Lu *et al.*, 2005). Transfer of trace heavy metals to milk is a result of complex bio-mechanisms that require the activity of carrier proteins. The pattern of Cd concentration in serum was similar to that in milk, even if the values were slightly higher. This suggests that the mammary gland did not act as a barrier to protect the suckling offspring, or other milk consumers, from exposure to the metal. Concentration of Cd in fresh curd was strongly associated with the concentration of this heavy metal in the unprocessed milk and was about 3.5–5.5 times higher than that found in goat's milk (Milhaud *et al.*, 2000), similar to what is observed in cheese-making with sheep's milk (Mehennaoui *et al.*, 1999).

Practical Recommendations

Several actions can be adopted by goat farmers to improve the quality of goat's milk.

In order to increase the production of milk fat and protein:

- Rations should have a balanced energy/protein ratio, in terms of quality and rumen degradability, to optimize rumen bacterial fermentations and allow maximum production of milk and milk components.
- ADF content (on DM basis) in the diet should be higher than 18%, in order to maintain milk production level with adequate fat content (>3%).
- Forage/concentrate ratio should be higher than 45–55 to maintain milk fat content above 3%.
- Part of the fibre should have adequate particle length (>3 mm) to stimulate chewing activity.
- Yeast might be used in the diet, due to its positive influence on milk fat content.
- In goats farmed on extensive areas, the addition of PEG to the diet has a positive influence on milk protein content and yield.
- Excess protein should be avoided to reduce urinary N excretion.
- Fat supplements can be used in goat rations without substantial negative effects on milk protein content.

In order to minimize the transfer of undesirable substances to goat's milk:

- The feedstuffs should be controlled to eliminate those contaminated with natural (e.g. mycotoxins) or artificial (e.g. pesticides) contaminants.

- Grazing on pastures or fodders that are grown or harvested in areas continually exposed to pollutants should be managed carefully.
- When diets are supposed to be contaminated by mycotoxins and analysis is not feasible, sequestering agents should be mixed with concentrates to reduce the negative effects of the mycotoxin on animal health and the potential transfer of toxic metabolites to milk.

After all, good goat's milk is the ultimate result of the right blend of high-quality feedstuffs.

References

Abd El-Ghani, A.A. (2004) Influence of diet supplementation with yeast culture (*Saccharomyces cerevisiae*) on performance of Zaraibi goats. *Small Ruminant Research* 52, 223–229.

Albenzio, M., Caroprese, M., Marino, R., Muscio, A., Santillo, A. and Sevi, A. (2006) Characteristics of Garganica goat milk and Cacioricotta cheese. *Small Ruminant Research* 64, 35–44.

Analla, M., Jiménez-Gamero, I., Muñoz-Serrano, A., Serradilla, J.M. and Falagán, A. (1996) Estimation of genetic parameters for milk yield and fat and protein contents of milk from Murciano-Granadina goats. *Journal of Dairy Science* 79, 1895–1898.

Andrighetto, I., Ramanzin, M., Tealdo, E. and Dalvit, P. (1989) Impiego di diete complete a diverso tenore proteico nell'alimentazione di capre da latte. *Zootecnica e Nutrizione Animale* 15, 661–671.

Arambel, M.J. and Kent, B.A. (1990) Effect of yeast culture on nutrient digestibility and milk yield response in early to mid-lactation dairy cows. *Journal of Dairy Science* 73, 1560–1563.

Bailoni, L. and Andrighetto, I. (1995) Effetto del rapporto foraggi:concentrati sui parametri digestivo-metabolici e sulle prestazioni produttive di capre primipare in lattazione. *Zootecnica e Nutrizione Animale* 21, 333–344.

Baker, L.D., Ferguson, J.D. and Chalupa, W. (1995) Responses in urea and true protein of milk to different protein feeding schemes for dairy cows. *Journal of Dairy Science* 78, 2424–2434.

Baldi, A., Cheli, F., Corino, C., Dell'Orto, V. and Polidori, F. (1992) Effects of feeding calcium salt of long chain fatty acids on milk yield, milk composition and plasma parameters of lactating goats. *Small Ruminant Research* 6, 303–310.

Barbieri, M.E., Manfredi, E., Elsen, J.M., Ricordeau, G., Bouillon, J., Grosclaude, F., Mahè, M.F. and Bibé, B. (1995) Influence du locus de la caséine $\alpha s1$ sur les performances laitières et les paramètres génétiques des chèvres de race alpine. *Genetics, Selection, Evolution* 27, 437–450.

Barbiroli, A., Bonomi, F., Benedetti, S., Mannino, S., Monti, L., Cattaneo, T. and Iametti, S. (2007) Binding of aflatoxin M1 to different protein fractions in ovine and caprine milk. *Journal of Dairy Science* 90, 532–540.

Battacone, G., Nudda, A., Pulina, G. and Enne G. (2003). The effects of parity and SCC on plasmin and plasminogen content in goat milk. In: *Proceedings of the 54th International Congress of European Association of Animal Production*, Rome. Wageningen Academic, Wageningen, The Netherlands, p. 183.

Battacone, G., Nudda, A., Palomba, M., Pascale, M., Nicolussi, P. and Pulina, G. (2005) Transfer of aflatoxin B1 from feed to milk and from milk to curd and whey in dairy sheep fed artificially contaminated concentrates. *Journal of Dairy Science* 88, 3063–3069.

Bava, L., Rapetti, L., Crovetto, G.M., Tamburini, A., Sandrucci, A., Galassi, G. and Succi, G. (2001) Effects of a nonforage diet on milk production, energy, and nitrogen metabolism in dairy goats throughout lactation. *Journal of Dairy Science* 84, 2450–2459.

Blank, R., Rolfs, J.P., Sudekum, K.H., Frohlich, A.A., Marquardt, R.R. and Wolffram, S. (2003) Effects of chronic ingestion of ochratoxin a on blood levels and excretion of the mycotoxin in sheep. *Journal of Agricultural and Food Chemistry* 51, 6899–6905.

Bocio, A., Llobet, J.M., Domingo, J.L., Corbella, J., Teixido, A. and Casas, C. (2003) Polybrominated diphenyl ethers (PBDEs) in foodstuffs: human exposure through the diet. *Journal of Agricultural and Food Chemistry* 51, 3191–3195.

Boichard, D., Bouloc, N., Ricordeau, G., Piacere, A. and Barrillet, F. (1989) Genetic parameters for first lactation dairy traits in the Alpine and Saanen goat breeds. *Genetics, Selection, Evolution* 21, 205–215.

Brown-Crowder, I.E., Hart, S.P., Cameron, M., Sahlu, T. and Goetsch, A.L. (2001) Effects of dietary tallow level on performance of Alpine does in early lactation. *Small Ruminant Research* 39, 233–241.

Brun-Bellut, J., Laurent, F. and Vignon, B. (1984) Taux d'ureé du lait, allantoine urinaire, temoins de la nutrition azoteé chez la chévre en lactation. *Canadian Journal of Animal Science* 64, 281–282.

Brun-Bellut, J., Blanchart, G. and Vignon, B. (1990) Effect of rumendegradable protein concentration in diet on digestion, nitrogen utilization and milk yield by dairy goats. *Small Ruminant Research* 3, 575–581.

Cannas, A., Pes, A., Mancuso, R., Vodret, B. and Nudda, A. (1998) Effect of dietary energy and protein concentration on the concentration of milk urea nitrogen in dairy ewes. *Journal of Dairy Science* 81, 499–508.

Cannas, A., Nudda, A. and Pulina, G. (2002) Nutritional strategies to improve lactation persistency in dairy ewes. In: Thomas, D.L. and Porter, S. (eds) *Proceedings of the 8th Great Lakes Dairy Sheep Symposium*, Cornell University, Ithaca, New York, pp. 17–59.

Cavret, S., Feidt, C., Le Roux, Y. and Laurent, F. (2005) Study of mammary epithelial role in polycyclic aromatic hydrocarbons transfer to milk. *Journal of Dairy Science* 88, 67–70.

Chilliard, Y. and Ferlay, A. (2004) Dietary lipids and forages interactions on cow and goat milk fatty acid composition and sensory properties. *Reproduction, Nutrition, Development* 44, 467–492.

Chilliard, Y., Ferlay, A., Rouel, J. and Lamberet, G. (2003) A review of nutritional and physiological factors affecting goat milk lipid synthesis and lipolysis. *Journal of Dairy Science* 86, 1751–1770.

Costera, A., Feidt, C., Marchand, P., Le Bizec, B. and Rychen, G. (2006) PCDD/F and PCB transfer to milk in goats exposed to a long-term intake of contaminated hay. *Chemosphere* 64, 650–657.

Decandia, M., Sitzia, M., Cabiddu, A., Kababya, D. and Molle, G. (2000) The use of polyethylene glycol to reduce the anti-nutritional effects of tannins in goats fed woody species. *Small Ruminant Research* 38, 157–164.

Diaz, D.E., Hagler, W.M. Jr, Blackwelder, J.T., Eve, J.A., Hopkins, B.A., Anderson, K.L., Jones, F.T. and Whitlow, L.W. (2004) Aflatoxin binders II: reduction of aflatoxin M1 in milk by sequestering agents of cows consuming aflatoxin in feed. *Mycopathologia* 157, 233–241.

Emery, R.S. (1988) Milk fat depression and the influence of diet on milk composition. *The Veterinary Clinics of North America. Food Animal Practice* 4, 289–305.

Erasmus, L.J., Botha, P.M. and Kistner, A. (1992) Effect of yeast culture supplement on production, rumen fermentation, and duodenal nitrogen flow in dairy cows. *Journal of Dairy Science* 75, 3056–3065.

Fedele, V., Claps, S. and Rubino, R. (1996) Effect of feeding systems on somatic cell count in goats. In: Rubino, R. (ed.) *Proceedings of the International Symposium 'Somatic Cells and Milk of Small Ruminants'*. European Association of Animal Production, Publication No. 77. Wageningen Pers, Wageningen, The Netherlands, pp. 167–172.

Fekadu, B., Soryal, K., Zeng, S., Van Hekken, D., Bah, B. and Villaquiran, M. (2005) Changes in goat milk composition during lactation and their effect on yield and quality of hard and semi-hard cheeses. *Small Ruminant Research* 59, 55–63.

Fernández, C., Sánchez-Seiquer, P. and Sánchez, A. (2003) Use of a total mixed ration with three sources of protein as an alternative feeding for dairy goats on southeast of Spain. *Pakistan Journal of Nutrition* 2, 18–24.

Fernández, C., Sánchez-Seiquer, P., Sánchez, A., Contreras, A. and de la Fuente, J.M. (2004) Influence of betaine on milk yield and composition in primiparous lactating dairy goats. *Small Ruminant Research* 52, 37–43.

Fernandez, J.M., Sahlu, T., Lu, C.D., Ivey, D. and Potchoiba, M.J. (1997) Production and metabolic aspects of nonprotein nitrogen incorporation in lactation rations of dairy goats. *Small Ruminant Research* 26, 105–117.

Fries, G.F. (1996) Ingestion of sludge applied organic chemicals by animals. *Science of the Total Environment* 185, 93–108.

Fujihara, T., Todoroki, M. and Nakamura, K. (2003) The effect of rumen protozoa on the urinary excretion of purine derivatives in goats. *Journal of Agricultural Science* 140, 101–105.

Galina, M.A., Morales, R., Lopez, B. and Carmona, M.A. (1996) Effect of somatic cell count on lactation and soft cheese yield by dairy goats. *Small Ruminant Research* 21, 251–257.

Giger-Reverdin, S., Bezault, N., Sauvant, D. and Bertin, G. (1996) Effects of probiotic yeast in lactating ruminants: interaction with dietary nitrogen level. *Animal Feed Science and Technology* 63, 149–162.

Gilboa, N., Perevolotsky, A., Landau, S., Nitsan, Z. and Silanikove, N. (2000) Increasing productivity in goats grazing Mediterranean woodland and scrubland by supplementation of polyethylene glycol. *Small Ruminant Research* 38, 183–190.

Goetsch, A.L., Detweiler, G. and Sahlu, T. (2001) Dairy goat performance with different dietary concentrate levels in late lactation. *Small Ruminant Research* 41, 117–125.

Greer, F.R. (1989) Calcium, phosphorus, and magnesium: how much is too much for infant formulas? *Journal of Nutrition* 119, 1846–1851.

Grovaa, N., Rychena, G., Monteaub, F., Le Bizecb, B. and Feidta, C. (2006) Effect of oral exposure to polycyclic aromatic hydrocarbons on goat's milk contamination. *Agronomy for Sustainable Development* 26, 195–199.

Grytsyuk, N., Arapis, G., Perepelyatnikova, L., Ivanova, T. and Vynogradska, V. (2006) Heavy metals effects on forage crops yields and estimation of elements accumulation in plants as affected by soil. *Science of the Total Environment* 354, 224–231.

Guo, M., Park, Y.W., Dixon, P.H., Gilmore, G.A. and Kindstedt, P.S. (2004) Relationship between the yield of cheese (Chevre) and chemical composition of goat milk. *Small Ruminant Research* 52, 103–107.

Gurung, N.K., Rankins, D.L. Jr, Shelby, R.A. and Goel, S. (1998) Effects of fumonisin B1 contaminated feeds on weanling Angora goats. *Journal of Animal Science* 76, 2863–2870.

Hadjipanayiotou, M. (1992) Effect of protein source and formaldehyde treatment on lactation performance of Chios ewes and Damascus goats. *Small Ruminant Research* 8, 185–197.

Hadjipanayiotou, M. (1995) Effect of feeding heat treated soybean meal on the performance of lactating Damascus goats. *Small Ruminant Research* 18, 105–111.

Hadjipanayiotou, M. (2004) Replacement of barley grain for corn in concentrate diets fed to dairy Damascus goats at different frequencies. *Small Ruminant Research* 51, 229–233.

Hadjipanayiotou, M. and Koumas, A. (1991) Effect of protein source on performance of lactating Damascus goats. *Small Ruminant Research* 5, 319–326.

Hadjipanayiotou, M. and Photiou, A. (1995) Effects of protein source and level on performance of lactating Damascus goats in negative energy balance. *Small Ruminant Research* 15, 257–263.

Hadjipanayiotou, M., Koumas, A., Hadjigavriel, G., Antoniou, I., Photiou, A. and Theodoridou, M. (1996) Feeding dairy ewes and goats and growing lambs and kids mixtures of protein supplements. *Small Ruminant Research* 21, 203–211.

Haneef, S.S., Swarup, D., Dwivedi, S.K. and Dash, P.K. (1998) Effects of concurrent exposure to lead and cadmium on renal function in goats. *Small Ruminant Research* 28, 257–261.

Harrison, G.A., Hemken, R.W., Dawson, K.A., Harmon, A.J. and Barber, K.B. (1998) Influence of addition of yeast culture supplement to diets of lactating cows on ruminal fermentation and microbial populations. *Journal of Dairy Science* 71, 2967–2975.

Herzig, I., Poul, J., Pisarikova, B. and Gopfert, E. (2003) Milk iodine concentration in cow treated orally or intramuscularly with a single dose of iodinated fatty acid esters. *Veterinary Medicine Czech* 48, 155–162.

IARC (2002) *Some Traditional Herbal Medicines, Some Mycotoxins, Naphthalene and Styrene. Summary of Data Reported and Evaluation*. Monographs on the Evaluation of the Carcinogenic Risk to Humans, Vol. 82. International Agency for Research on Cancer, Lyon, France.

Ikwuegbu, O.A. and Sutton, J.D. (1982) The effect of varying the amount of linseed oil supplementation on rumen metabolism in sheep. *British Journal of Nutrition* 48, 365–375.

Kawas, J.R., Lopes, J., Danelon, D.L. and Lu, C.D. (1991) Influence of forage-to-concentrate ratios on intake, digestibility, chewing and milk production of dairy goats. *Small Ruminant Research* 4, 11–18.

Khaled, N.F., Illek, J. and Gajdusek, S. (1999) Interactions between nutrition, blood metabolic profile and milk composition in dairy goats. *Acta Veterinaria Brunensis* 68, 253–258.

Kiessling, K.H., Pettersson, H., Sandholm, K. and Olsen, M. (1984) Metabolism of aflatoxin, ochratoxin, zearalenone, and three trichothecenes by intact rumen fluid, rumen protozoa, and rumen bacteria. *Applied and Environmental Microbiology* 47, 1070–1073.

Kitessa, S.M., Gulati, S.K., Ashes, J.R., Fleck, E., Scott, T.W. and Nichols, P.D. (2001) Utilisation of fish oil in ruminants. II. Transfer of fish oil fatty acids into goats' milk. *Animal Feed Science and Technology* 89, 201–208.

Lanzani, A., Bondioli, P., Mariani, C., Fedeli, E., Polidori, F., Dell'Orto, V., Corino, C., Guisi, A., Contarini, G. and Ferro, E. (1985) Prove di impiego di un Liquid Feed addizionato con grassi protetti nell'alimentazione di capre in lattazione. In: *Atti 6° Congresso Nazionale Associazione Scientifica di Produzioni Animali*. Fondazione Iniziative Zooprofilattiche e Zootecniche, Brescia, Italy, pp. 87–92.

Leitner, G., Merin, U. and Silanikove, N. (2004) Changes in milk composition as affected by subclinical mastitis in goat. *Journal of Dairy Science* 87, 1719–1726.

Li, Y., McCrory, D.F., Powell, J.M., Saam, H. and Jackson-Smith, D. (2005) A survey of selected heavy metal concentrations in Wisconsin dairy feeds. *Journal of Dairy Science* 88, 2911–2922.

Lu, C.D. (1987) Implication of forage particle length on chewing activities and milk production in dairy goats. *Journal of Dairy Science* 70, 1411–1416.

Lu, C.D. (1993) Implication of feeding isoenergetic diets containing animal fat on milk composition of Alpine does during early lactation. *Journal of Dairy Science* 76, 1137–1147.

Lu, C.D., Potchoiba, M.J., Sahlu, T. and Fernandez, J.M. (1990) Performance of dairy goats fed isonitrogenous diets containing soybean meal or hydrolyzed feather meal during early lactation. *Small Ruminant Research* 3, 425–434.

Lu, J., Jin, T., Nordberg, G. and Nordberg, M. (2005) Metallothionein gene expression in peripheral lymphocytes and renal dysfunction in a population environmentally exposed to cadmium. *Toxicology and Applied Pharmacology* 206, 150–156.

Lyche, J.L., Skaare, J.U., Larsen, H.J.S. and Ropstad, E. (2004) Levels of PCB 126 and PCB 153 in plasma and tissues in goats exposed during gestation and lactation. *Chemosphere* 55, 621–629.

Matsumoto, M., Kobayashi, T., Takenaka, A. and Itabashi, H. (1991) Defaunation effects of medium-chain fatty acids and their derivatives on goat rumen protozoa. *Journal of General and Applied Microbiology* 37, 439–445.

Mehennaoui, S., Delacroix-Buchet, A., Duche, A., Enriquez, B., Kolf-Clauw, M. and Milhaud, G. (1999) Comparative study of cadmium transfer in ewe and cow milks during rennet and lactic curds preparation. *Archives of Environmental Contamination and Toxicology* 37, 389–395.

Mele, M., Serra, A., Rafanelli, M.R., Conte, G. and Secchiari, P. (2005) Effect of forage: concentrate ratio and soybean oil supplementation on milk yield and quality from dairy goats. *Italian Journal of Animal Science* 4(Suppl. 2), 392–394.

Meschy, F. (2000) Recent progress in the assessment of mineral requirements of goats. *Livestock Production Science* 64, 9–14.

Milhaud, G.E., Delacroix-Buchet, A., Han, M., Mehennaoui, S., Duché, A., Enriquez, B. and Kolf-Claw, M. (2000) Devenir du cadmium du lait de chèvre dans la crème et les caillés présure et lactique. *Lait* 80, 277–288.

Min, B.R., Hart, S.P., Sahlu, T. and Satter, L.D. (2005) The effect of diets on milk production and composition, and on lactation curves in pastured dairy goats. *Journal of Dairy Science* 88, 2604–2615.

Morand-Fehr, P. and Sauvant, D. (1980) Composition and yield of goat milk as affected by nutritional manipulation. *Journal of Dairy Science* 63, 1671–1680.

Morand-Fehr, P., Tessier, J., Meschy, F. and Sauvant, D. (2000) Effect of roughage level and source in diets on the risk of reversing fat and protein percentages in goat milk. *Cahiers Options Méditerranéennes* 52, 115–118.

Morand-Fehr, P., Fedele, V., Decandia, M. and Le Frileux, Y. (2007) Influence of farming and feeding systems on composition and quality of goat and sheep milk. *Small Ruminant Research* 68, 20–34.

Nudda, A., Battacone, G., Usai, M.G., Fancellu, S. and Pulina, G. (2006) Supplementation with extruded linseed cake affects the concentrations of conjugated linoleic acid and vaccenic acid in goat milk. *Journal of Dairy Science* 89, 277–282.

Oliveira, C.A.F. and Ferraz, J.C.O. (2007) Occurrence of aflatoxin M1 in pasteurised, UHT milk and milk powder from goat origin. *Food Control* 8, 375–378.

Onetti, S.G. and Grummer, R.R. (2004) Response of lactating cows to three supplemental fat sources as affected by forage in the diet and stage of lactation: a meta-analysis of literature. *Animal Feed Science and Technology* 115, 65–82.

Pailan, G.H. and Kaur, H. (1996) Influence of dietary protein content and digestibility on milk yield and blood constituents in lactating goats. *Small Ruminant Research* 20, 47–51.

Palmquist, D.L. (1994) The role of dietary fats in efficiency of ruminants. *Journal of Nutrition* 124(Suppl. 8), 1377 S–1382 S.

Pérez, L., Sanz Sampelayo, M.R., Gil Extremera, F. and Boza, J. (2000) Effects of soap supplies on goat milk production and composition. *Cahiers Options Méditerranéennes* 52, 103–106.

Polidori, F., Baldi, A., Cheli, F. and Pulina, G. (1991) Alimentazione e qualità del latte caprino. In: Enne, G. and Greppi, G.F. (eds) *Atti III Simposio Internazionale 'Qualità del Latte Ovino-Caprino'*, Varese, Italy. Istituto Sperimentale Italiano L. Spallanzani, Milan, Italy, pp. 105–134.

Pulina, G., D'Andrea, F., Dimauro, C. and Cappio-Borlino, A. (2003) Kinetics of fat and protein secretion in dairy cattle, sheep, goats and buffaloes. *Italian Journal of Animal Science* 2(Suppl. 1), 296–298.

Pulina, G., Nudda, A., Battacone, G. and Cannas, A. (2006) Effects of nutrition on sheep milk content of fat, protein, somatic cell count, aromatic compounds and undesirable substances. *Animal Feed Science and Technology* 131, 255–291.

Rao, N.S.B. and Chopra, R.C. (2001) Influence of sodium bentonite and activated charcoal on aflatoxin M1 excretion in milk of goats. *Small Ruminant Research* 41, 203–213.

Rapetti, L., Tamburini, A., Provetto, G.M., Galassi, G. and Succi, G. (1997) Energy utilization of diets with different hay proportions in lactating goats. *Zootecnica e Nutrizione Animale* 23, 317–328.

Raynal-Ljutovac, K., Pirisi, A., de Crémoux, R. and Gonzalo, C. (2007) Somatic cells of goat and sheep milk: analytical, sanitary, productive and technological aspects. *Small Ruminant Research* 68, 126–144.

Reklewska, B., Bernatowicz, E., Ryniewicz, Z., Rua Pinto, R. and Zdziarski, K. (2004) Preliminary observations on the *Echinacea*-induced lactoferrin production in goat milk. *Animal Science Papers and Reports* 22, 17–25.

Rousselot, M.C., Broqua, C.B., de Araujo, C. and Borgida, L.P. (1995) Effets des fibres et des matières grasses protégées sur la composition du lait de chèvre. *Annales de Zootechnie* 44, 376 (Abstr.).

Roussi, V., Govaris, A., Varagouli, A. and Botsoglou, N.A. (2002) Occurrence of aflatoxin M1 in raw and market milk commercialized in Greece. *Food Additives and Contaminants* 19, 863–868.

Sahlu, T., Hart, S.P. and Fernandez, J.M. (1993) Nitrogen metabolism and blood metabolites in three goat breeds fed increasing amounts of protein. *Small Ruminant Research* 10, 281–292.

Salama, A.A.K., Caja, G., Garína, D., Albanella, E., Such, X. and Casals, R. (2002) Effects of adding a mixture of malate and yeast culture (*Saccharomyces cerevisiae*) on milk production of Murciano-Granadina dairy goats. *Animal Research* 51, 295–303.

Salama, A.A., Caja, G., Albanell, E., Such, X., Casals, R. and Plaixats, J. (2003) Effects of dietary supplements of zinc-methionine on milk production, udder health and zinc metabolism in dairy goats. *Journal of Dairy Research* 70, 9–17.

Santini, F.J., Lu, C.D., Potchoiba, M.J. and Fernandez, J.M. (1992) Dietary fiber and milk yield, mastication, digestion, and rate of passage in goats fed alfalfa hay. *Journal of Dairy Science* 75, 209–219.

Sanz Sampelayo, M.R., Amigo, L., Ares, J.L., Sanz, B. and Boza, J. (1998a) The use of diets with different protein sources in lactating goats: composition of milk and its suitability for cheese production. *Small Ruminant Research* 31, 37–43.

Sanz Sampelayo, M.R., Perez, L., Boza, J. and Amigo, L. (1998b) Forage of different physical forms in the diets of lactating Granadina goats: nutrient digestibility and milk production and composition. *Journal of Dairy Science* 81, 492–498.

Sanz Sampelayo, M.R., Perez, M.L., Gil Extremera, F., Boza, J.J. and Boza, J. (1999) Use of different dietary protein sources for lactating goats: milk production and composition as functions of protein degradability and amino acid composition. *Journal of Dairy Science* 82, 555–565.

Sanz Sampelayo, M.R., Pérez, L., Martín Alonso, J.J., Gil Extremera, F. and Boza, J. (2002) Effects of concentrates with different contents of protected fat rich in PUFAs on the performance of lactating Granadina goats. Part II. Milk production and composition. *Small Ruminant Research* 43, 141–148.

Sanz Sampelayo, M.R., Chillard, Y., Schmidely, P. and Boza, J. (2007) Influence of type of diet on the fat constituents of goat and sheep milk. *Small Ruminant Research* 68, 42–63.

Savoini, G., Cattaneo, D., Paratte, R., Varisco, G., Bronzo, V., Moroni, P. and Pisoni, G. (2003) Dietary rosemary extract in dairy goat organically managed: effects on immune response, mammary infection and milk quality. *Italian Journal of Animal Science* 2(Suppl. 1), 548–550.

Schmidely, P. and Sauvant, D. (2001) Taux butyreux et composition de la matière grasse du lait chez les petits ruminants: effets de l'apport de matières grasses ou d'aliment concentré. *INRA Productions Animales* 14, 337–354.

Schmidely, P., Lloret-Pujol, M., Bas, P., Rouzeau, A. and Sauvant, D. (1999) Influence of feed intake and source of dietary carbohydrate on milk yield and composition, nitrogen balance, and plasma constituents of lactating goats. *Journal of Dairy Science* 82, 747–755.

Schmidely, P., Meschy, F., Tessier, J. and Sauvant, D. (2002) Lactation response and nitrogen, calcium, and phosphorus utilization of dairy goats differing by the genotype for αs1-casein in milk, and fed diets varying in crude protein concentration. *Journal of Dairy Science* 85, 2299–2307.

Schmidely, P., Morand-Fehr, P. and Sauvant, D. (2005) Influence of extruded soybeans with or without bicarbonate on milk performance and fatty acid composition of goat milk. *Journal of Dairy Science* 88, 757–765.

Smith, E.E., Phillips, T.D., Ellis, J.A., Harvey, R.B., Kubena, L.F., Thompson, J. and Newton, G. (1994) Dietary hydrated sodium calcium aluminosilicate reduction of aflatoxin M1 residue in dairy goat milk and effects on milk production and components. *Journal of Animal Science* 72, 677–682.

Stella, A.V., Paratte, R., Valnegri, L., Cigalino, G., Soncini, G., Chevaux, E., Dell'Orto, V. and Savoini, G. (2007) Effect of administration of live *Saccharomyces cerevisiae* on milk production, milk composition, blood metabolites, and faecal flora in early lactating dairy goats. *Small Ruminant Research* 67, 7–13.

Sutton, J.D. (1989) Altering milk composition by feeding. *Journal of Dairy Science* 72, 2801–2814.

Teh, T.H., Trung, L.T., Jia, Z.H. and Gipson, T.A. (1994) Varying amount of rumen-inert fat for high producing goats in early lactation. *Journal of Dairy Science* 77, 253–258.

Vasta, V., Nudda, A., Cannas, A., Lanza, M. and Priolo, A. (2007) Alternative feed resources and small ruminants meat and milk quality. A review. *Animal Feed Science and Technology*, in press.

Virdis, S., Corgiolu, G., Scarano, C., Pilo, A.L. and De Santis, E.P.L. (2008) Occurrence of aflatoxin M1 in tank bulk goat milk and ripened goat cheese. *Food Control* 19, 44–49 (in press).

Vrecl, M., Ursic, M., Pogacnik, A., Zupancic-Kralj, L. and Jan, J. (2005) Excretion pattern of co-planar and non-planar tetra- and hexa-chlorobiphenyls in ovine milk and faeces. *Toxicology and Applied Pharmacology* 204, 170–174.

Wachira, A.M., Sinclair, A., Wilkinson, R.G., Hallett, K., Enser, M. and Wood, J.D. (2000) Rumen biohydrogenation of *n*–3 polyunsaturated fatty acids and their effects on microbial efficiency and nutrient digestibility in sheep. *Journal of Agricultural Science* 135, 419–428.

Wu, Z. and Huber, J.T. (1994) Relationship between dietary fat supplementation and milk protein concentration in lactating cows: a review. *Livestock Production Science* 39, 141–155.

Zeng, S.S., Soryal, K., Fekadu, B., Bah, B. and Popham, T. (2007) Predictive formulae for goat cheese yield based on milk composition. *Small Ruminant Research* 69, 180–186.

2 Mathematical Modelling of Goat Lactation Curves

N.P.P. MACCIOTTA, C. DIMAURO, R. STERI AND A. CAPPIO-BORLINO

Dipartimento di Scienze Zootecniche, Università di Sassari, Via E. De Nicola 9, 07100 Sassari, Italy

Introduction

Milk production throughout lactation, in goats as in other domestic ruminants, is the result of the processes of synthesis and secretion of organic and inorganic compounds and of active and passive blood filtration by specialized epithelial cells of the mammary gland (Mepham, 1987). Starting from the end of gestation there is a phase of rapid cellular activation, followed by cellular regression (cellular remodelling), at varying rates, that ends with the cessation of lactation or dry-off (Hurley, 1989). All of these physiological mechanisms result in a typical pattern of milk yield over time (Fig. 2.1), characterized by an initial phase of increasing production which reaches a maximum (lactation peak) and then declines more or less rapidly until dry-off.

The lactation pattern depicted in Fig. 2.1 can be considered the result of two components. The first, regular and continuous, is the expression of the physiological mechanisms underlying milk production, and is named the lactation curve. The second, represented by deviations from the regular pattern, can be ascribed to several factors of variation such as the nutrition and health status of individual animals or environmental perturbations (Fig. 2.2).

Basic features of the lactation curve shape are: (i) days in milk (DIM) at which the lactation peak occurs; (ii) daily yield at peak; (iii) rate of decrease of yield after the peak, the inverse of which is known as persistency of lactation and measures the ability of the animal to maintain a constant yield after the lactation peak; and (iv) total lactation yield, which can be estimated from the area under the lactation curve. Knowledge of the main characteristics of the lactation curve is of great help to dairy producers and professionals in making management and breeding decisions. The prediction of total milk yield from a few tests in early lactation enables one to: (i) calculate the gross income that can be obtained from a goat; (ii) choose which animals have to be culled; (iii) identify sick animals on the basis of a decrease in yield before the appearance of clinical signs (e.g. subclinical

 Dairy Goats Feeding and Nutrition (eds A. Cannas and G. Pulina)

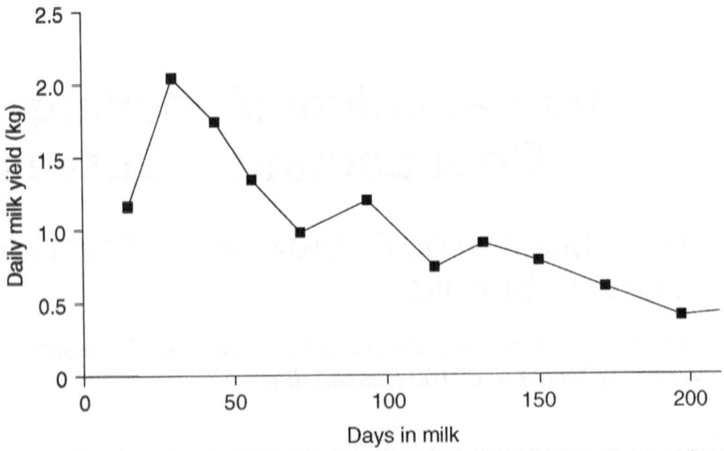

Fig. 2.1. Lactation pattern of a Sarda breed goat.

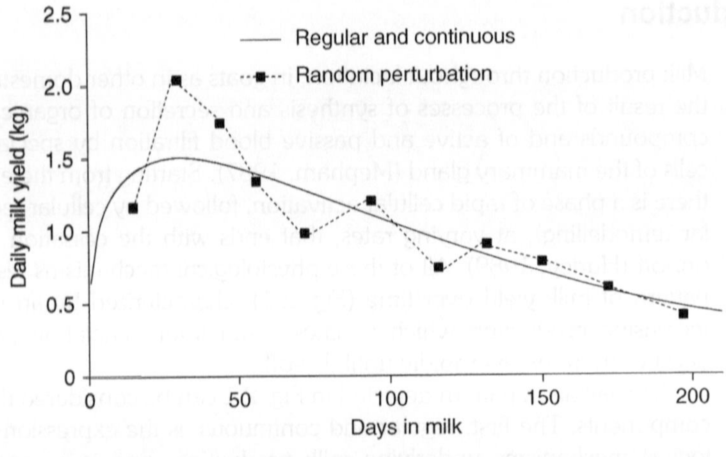

Fig. 2.2. The two main components of the lactation pattern of a Sarda breed goat: regular and continuous component and random perturbation.

mastitis); and (iv) identify high-producing animals that have higher dietary requirements than average-producing animals (Gipson and Grossman, 1989).

Mathematical Modelling of Lactation Curves

The description of the temporal evolution of milk production in domestic ruminants is one of the most important applications of mathematical modelling in animal science. Lactation curve models are implemented in feeding management software for livestock species (Boe *et al.*, 2005). They represent an essential component of random regression test day models, which are an upgraded version of genetic models used to predict breeding values and to estimate variance components for milk production, in selection schemes for the genetic improvement of milk production traits in dairy animals (Schaeffer, 2004).

Data to be modelled usually consist of daily test day records of milk yield (or composition, e.g. fat or protein content or somatic cell count, SCC) measured at different times from parturition on the same animal, thus representing a case of repeated measurements design.

The most common approach in lactation curve modelling is to fit suitable functions of time, $y = f(t)$, to test day records. Such an empirical approach is essentially aimed at describing the regular and continuous component of the lactation pattern. Some of the functions used to model goat lactation curves are reported in Table 2.1.

Table 2.1. Mathematical models used to fit goat lactation curves.

Model	Equation	Reference
Cappio-Borlino *et al.* (1995)	$y(t) = at^{be(-ct)}$	Todaro *et al.* (1999), Fernández *et al.* (2002), Guimarães *et al.* (2006)
Cobby and Le Du (1978)	$y(t) = a - bt - ae^{-ct}$	Fernández *et al.* (2002), Guimarães *et al.* (2006)
Dhanoa (1981)	$y(t) = at^{bc}e^{-ct}$	Fernández *et al.* (2002), Guimarães *et al.* (2006)
Grossman and Koops (1988)	$y(t) = \sum_{i=1}^{2}\{a_ib_i[1-\tan^2(b_i(t-c_i))]\}$	Gipson and Grossman (1989), Macedo *et al.* (2001)
Morant and Gnanasakthy (1989)	$y(t) = ae^{[b(1+t'/2)t']} + c(t')^2 - (1.01/t)$	Williams (1993), Macedo *et al.* (2001), Gonçalves *et al.* (2001), Guimarães *et al.* (2006)
Nelder (1966)	$y(t) = (t/a) + bt + ct^2$	Guimarães *et al.* (2006)
Wilmink (1987)	$y(t) = a + be^{-kt} + ct$	Macciotta *et al.* (2004a)
Wood (1967)	$y(t) = at^{b}e^{-ct}$	Gipson and Grossman (1990), Rota *et al.* (1993), Giaccone *et al.* (1995), Ruvuna *et al.* (1995), Montaldo *et al.* (1997), Andonov *et al.* (1999), Akpa *et al.* (2001), Fernández *et al.* (2002), Macciotta *et al.* (2003), McManus *et al.* (2003), Silva *et al.* (2005), Zambom *et al.* (2005), Guimarães *et al.* (2006)

y = daily milk yield measured at time; t = days in milk; $t' = (t - 150)/100$; a, b, c, k = function parameters.

It is worth noticing that most of the mathematical functions used to describe the goat lactation curve have three parameters. This is a direct consequence of the limited average number of records per lactation available for goats, not more than six for an average lactation length of 221 days for goat breeds raised in Italy during 2005 (based on data published by AIA, 2005).

Models with a large number of parameters, such as the five-parameter multiple regression of Ali and Schaeffer (1987), have been used to fit average curves in test day repeatability models (Schaeffer and Sullivan, 1994) or individual random curves in random regression models (Breda *et al.*, 2006).

The alternative and more complex approach is based on the use of mechanistic models, aimed at translating in mathematical terms a hypothesis about the physiological and biochemical processes that regulate the phenomenon of interest (Neal and Thornley, 1983). However, the application of mechanistic models in goat lactation curve modelling has not been successful, as happened in other dairy species such as cattle and sheep. This was due to their high theoretical complexity, the large number of input variables involved and the high computation requirements.

A peculiarity of goat lactation curve modelling, which can also be found in dairy sheep, is with regard to the first phase of the lactation pattern. In intensive high-producing farming systems (typical of France, north European countries and northern Italy), kids are artificially reared and milking starts immediately after kidding. By contrast, in extensively farmed flocks (typical of southern Mediterranean countries, Africa and Latin America), milk of the first month of lactation is usually suckled by the kid. In the second situation, early lactation yields can be obtained from partial milking (i.e. after the kid has suckled) (Wahome *et al.*, 1994; Ruvuna *et al.*, 1995) or estimated from the average daily gain of the kid (Giaccone *et al.*, 1995). In any case, the scarce availability of data points during the first month of lactation often prevents an accurate modelling of this phase and may lead to unexpected outcomes, such as the estimation of curves without the lactation peak or a great variability in the time at which the lactation peak occurs.

Besides the prediction of total lactation yield from few test days, the estimation of lactation curve parameters may allow one to calculate its main characteristics. Some of the mathematical functions reported in Table 2.1 have parameters that have a clear meaning in terms of shape of the lactation curve. An example is the Wood (1967) equation, which is probably the most popular function of the lactation curve. In the Wood model, a is a scale parameter that regulates the general level of the curve, b controls the type and the magnitude of the curvature of the function, and c regulates the decrease of yield after the lactation peak. Table 2.2 reports values of parameters a, b and c of the Wood function estimated for some goat breeds. Values of these parameters can be used to calculate some essential features of lactation curve shape such as the DIM at which the lactation curve is attained (t_m), the peak yield (y_m) and lactation persistency (p) (France and Thornley, 1984):

$$t_{\mathrm{m}} = -\frac{b}{c}; \quad y_{\mathrm{m}} = \frac{a}{(c/b)^b e^b}; \quad p = -(b+1)\ln c$$

Some characteristics of the lactation curve shape for different goat breeds, estimated with the Wood function, are reported in Table 2.3. The time at which the

Table 2.2. Estimates of parameters *a*, *b* and *c* of the Wood function obtained in several goat breeds.

Breed	*a*	*b*	*c*	Reference
Alpine, La Mancha, Nubian, Saanen, Toggenburg	2.316	0.230	–0.005	Gipson and Grossman (1990)
Derivata di Siria	1.388	0.163	–0.005	Giaccone *et al.* (1995)
East African, Galla	0.345	0.149	–0.082	Ruvuna *et al.* (1995)
Crosses European × local Mexican breeds	3.756	0.641	–0.0109	Montaldo *et al.* (1997)
Crosses Saanen × local Brazilian breeds	1.056	0.383	–0.0123	Macedo *et al.* (2001)
Murciano-Granadina	2.287	0.129	–0.029	Fernández *et al.* (2002)
Red Sokoto	0.586	0.316	–0.023	Akpa *et al.* (2001)
Sarda	1.007	0.182	–0.007	Macciotta (unpublished data)
Small East African	0.333	0.227	–0.0052	Wahome *et al.* (1994)
Verata	1.290	0.207	–0.0052	Rota *et al.* (1993)

lactation peak occurs is highly variable, ranging from about 2 weeks to more than 60 days. Such differences could be mainly due to breed characteristics, even if the role of sampling effects (considering the great difference in the number of lactations among the considered studies) should not be neglected. The lactation peak occurs earlier in local and low selected breeds (at about 15–30 DIM) than in highly selected breeds such as Saanen, Alpine and their crosses (about 2 months). These results are in agreement with those observed in tropical breeds, which have lactation peaks at about 3 weeks from parturition (Akpa *et al.*, 2001), and in their crosses with high-producing breeds, which show a delayed peak (Ruvuna *et al.*, 1995). First-kidding goats tend to have later peaks than higher parities in Saanen and Murciano-Granadina breeds (Table 2.3), in agreement with results reported by Gipson and Grossman (1990). Peak milk yields reflect the different productive characteristics of the breeds.

Values of lactation persistency (Table 2.3) express a non-dimensional measure of the time interval during which milk yield is maintained at a value similar to the peak (Cappio-Borlino *et al.*, 1989). It can be observed that such values tend to decrease from first to greater parities. These results are in agreement with those reported in other dairy species such as cattle, sheep and buffaloes: younger animals show lower peaks and greater persistency than older animals, because the maturation process which is still in progress during the first lactation counteracts the normal decline in milk yield (Stanton *et al.*, 1992).

Most studies on goat lactation curve modelling deal with average curves of homogeneous groups of animals (parity order, kidding season, number of kids at parturition). When average curves are modelled, almost all functions give good fitting performances, with R^2 values often higher than 0.80 (Montaldo *et al.*, 1997;

Table 2.3. Days in milk at which the lactation peak occurs (t_m), peak yield (y_m) and lactation persistency (p) calculated from parameters estimated by fitting the Wood function in different goat breeds.

Breed	Observations	Parity	t_m (days)	y_m (kg)	p	Reference
Alpine, La Mancha,	–	1	53	3.0	6.68	Gipson and
Nubian, Saanen,	–	≥2	46	4.4	6.51	Grossman (1990)
Toggenburg						
Derivata di Siria	6	1	45	1.65	6.63	Giaccone
	22	2	63	1.89	6.82	*et al.* (1995)
	38	≥3	47	2.05	6.40	
Crosses Saanen,	49	1	58	1.78	7.53	Montaldo
Alpine,						*et al.* (1997)[b]
Toggenburg × local	221	2	61	2.38	7.57	
Mexican goats	355	3	54	2.54	7.17	
Crosses Saanen ×	31	1–4	34	2.24	6.27	Macedo
local Brazilian breeds						*et al.* (2001)[b]
Florida	968	1–7	13	2.68	4.32	Pena Blanco
						et al. (1999)
Murciano-Granadina	190	1	4.5[a]	1.66	3.75	Fernández
	167	2	2.4	2.35	3.81	*et al.* (2002)[b]
	376	≥3	2	1.78		
Red Sokoto	22	1	20	0.76	5.45	Akpa *et al.*
	17	2	20	0.95	5.44	(2001)[b]
	13	3	19	1.64	5.39	
Saanen	150	1	64.6	3.22	7.04	Groenewald and
	211	2	54.4	4.21	6.84	Viljoen (2003)
	253	3	58.8	4.53	7.02	
Sarda	161	1	32	1.35	6.07	Macciotta
	222	2	31	1.57	6.04	(unpublished
	152	3	38	1.53	6.01	data)

[a]Weeks.
[b]Values of y_m, t_m and p in the works by Akpa *et al.* (2001) and Macedo *et al.* (2001), and p in the works by Montaldo *et al.* (1997) and Fernández *et al.* (2002), have been calculated by using Wood parameter values reported by the authors.

Todaro *et al.*, 2000; Macedo *et al.*, 2001; Fernández *et al.*, 2002). These performances are no longer maintained when individual patterns are fitted, despite the increasing interest in this topic, particularly for breeding purposes. Actually the genetic modification of the lactation curve shape, to obtain economical advantages, is one of the major challenges for geneticists and professionals interested in the genetic improvement of milk production traits (Rekaya *et al.*, 2000).

A study carried out on individual lactation curves of Frisa goats using the Wilmink function (Macciotta *et al.*, 2004a) showed a great range of R^2 values

(Table 2.4). This result is in agreement with findings in dairy cattle, where the great variability of individual lactation curve shapes has been ascribed to both biological differences among cows and the interaction between the structure of the data analysed and the mathematical properties of the models used (Olori *et al.*, 1999; Landete-Castillejos and Gallego, 2000; Macciotta *et al.*, 2005b). In the same study, some atypical shapes (Fig. 2.3), i.e. without the lactation peak, were also observed. Detection of atypical shapes, also reported in cattle (Shanks *et al.*, 1981) and sheep (Cappio-Borlino *et al.*, 1997), is based on the signs of the estimated function parameters. In the case of the Wood function, atypical shapes are characterized by negative values of the parameter *b* (*a* being positive and *c* negative), whereas for the Wilmink model atypical shapes have positive values of *b* (*a* being positive and *c* negative).

The frequency of atypical shapes (about 30%) observed by Macciotta *et al.* (2004a) for goats is similar to that reported for dairy cattle and sheep. Fernández *et al.* (2002) found an atypical average lactation curve for goats at third or

Table 2.4. Distribution of individual lactation curves of Frisa goats among different R^2 classes. (From Macciotta *et al.*, 2004a.)

R^2 class	Absolute frequency	Relative frequency
<0.20	65	0.14
0.20–0.40	35	0.08
0.40–0.60	52	0.12
0.60–0.80	106	0.24
>0.80	190	0.42

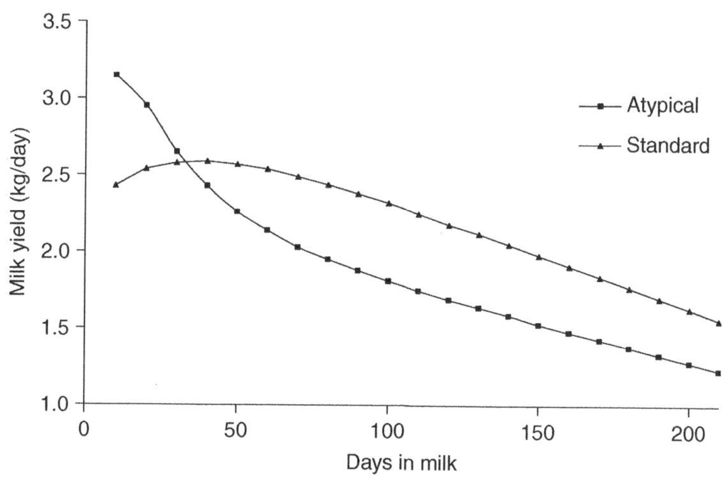

Fig. 2.3. Example of standard and atypical lactation curves for milk yield of Frisa breed goats estimated by the Wilmink function. (Adapted from Macciotta *et al.*, 2004a.)

greater kidding and attributed it to the occurrence of the lactation peak very close to parturition, which therefore could not be recognized by the mathematical function used. Reasons for the atypical lactation curve are: (i) biological differences between animals; (ii) mathematical properties of the model used; and (iii) the already mentioned structure of the data analysed (mainly the distance from parturition of the first record available and the distribution of records throughout lactation). The main consequence of an atypical shape is the change in sign and, therefore, in meaning of lactation curve parameters. This makes the interpretation of outcomes very difficult when individual values of lactation curve parameters are analysed with linear models to estimate the effect of fixed and random factors on lactation curve shape traits (Shanks *et al.*, 1981), or when mathematical functions are used to fit individual curves in random regression test day models (Jamrozik and Schaeffer, 1997).

Finally, although most studies on goat lactation curve modelling have dealt with milk yield, some work on modelling of the lactation pattern of milk components has also been done. For example, Rota *et al.* (1993) fitted the Wood function to curves for fat and protein content and for SCC in Verata goats, obtaining good fitting performances also for SCC ($R^2 = 0.97$).

Factors Affecting Lactation Curve Shape

Fitting regular functions of time to milk production data is essentially aimed at describing its regular and continuous component, but studies on the effects of environmental and non-genetic factors in general on daily milk production and lactation curve shape are also of great interest.

The effect of parity on milk yield has been reported for Mediterranean (Rota *et al.*, 1993; Fernández *et al.*, 2002; Macciotta *et al.*, 2005a), north European and Mexican (Bouloc, 1992; Montaldo *et al.*, 1997; Crepaldi *et al.*, 1999) goat breeds. Milk yield tends to increase from the first up to the fourth lactation, while fat and protein content are less affected by parity.

The number of kids at parturition has been found to be a relevant factor for variation of milk yield in goats. Does with two kids show a higher milk production than those with one kid, both on extensive (Zygoyannis, 1994; Peris *et al.*, 1997; Kominakis *et al.*, 2000; Fernández *et al.*, 2002; Macciotta *et al.*, 2005a) and in intensive (non-suckled) (Crepaldi *et al.*, 1999; Goonewardene *et al.*, 1999) farming systems. These results confirm the role of both the suckling reflex and the physiological mechanism during pregnancy that prepares the udder to produce more milk for does carrying multiple fetuses.

Kidding season affects milk yield. However, a meaningful comparison among different studies cannot be performed due to great differences in climate and farming systems across countries.

Fitting the same function separately on data grouped according to a classification factor allows for the evaluation of effects of non-genetic variability factors on lactation curve shape. As an example, average lactation curves of Sarda goats grouped according to parity and number of kids at parturition are reported in Figs 2.4 and 2.5, respectively. The clear separation of the lactation curve of

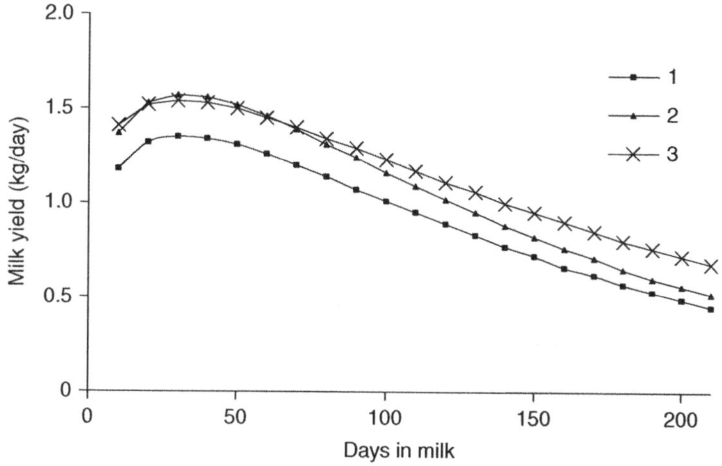

Fig. 2.4. Average lactation curves for milk yield of Sarda breed goats of different parities estimated with the Wood model.

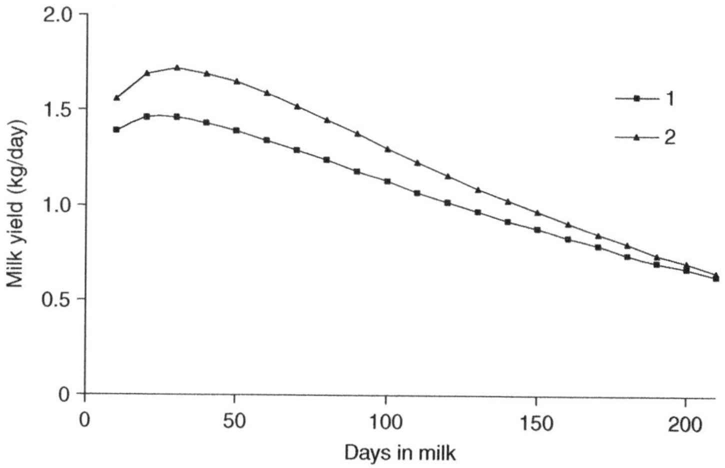

Fig. 2.5. Average lactation curves for milk yield of Sarda breed goats with different number of kids at parturition estimated by the Wood model.

first-kidding goats from higher parities (Fig. 2.4) highlights the difference in production level among different parities. The different curves reported in Fig. 2.5 highlight the higher production level of goats with two kids at parturition. Moreover, it can also be observed that the effect of type of kidding is essentially concentrated in the first 120 days of lactation.

Besides the visual inspection of the estimated curve, the functional approach allows evaluation of the effect of different sources of variation on the lactation curve by analysing values of function parameters estimated for each individual pattern (Gipson and Grossman, 1989; Wahome *et al.*, 1994; Giaccone *et al.*, 1995;

Ruvuna *et al.*, 1995; Montaldo *et al.*, 1997; Akpa *et al.*, 2001; McManus *et al.*, 2003). As an example, Table 2.5 reports least squares means for different parities of the parameters of the Wood function fitted to individual lactation curves of Derivata di Siria (Giaccone *et al.*, 1995) and Red Sokoto (Akpa *et al.*, 2001) goats. In both studies, a statistically significant effect of parity on parameter *a* was found, with increasing values going from primiparous to higher parities. This was expected since parameter *a* is responsible for the height of the curve, i.e. the level of production, which is generally lower in young animals. The effect of parity on parameter *b* reported by Giaccone *et al.* (1995) can be explained by a more pronounced curvature of the lactation curve in the first phase of lactation due to higher peak productions. Finally, the absolute value of parameter *c*, which has a minus sign, was lowest for first-kidding goats, evidencing the higher persistency of young animals.

However, a drawback of using a mathematical function of time to estimate the effect of non-genetic factors on the lactation curve is that it assumes a constant effect of each factor during the whole lactation, whereas there are effects, such as variation in feeds, climate conditions or health status, that may affect only part of the lactation period (Stanton *et al.*, 1992). As an alternative, short-term environmental effects can be conveniently modelled with the so-called test day models (Stanton *et al.*, 1992). These linear mixed models include a test date factor, which models effects observed at each date in which production is measured, and a DIM factor, whose estimates allow the construction of the lactation curve corrected by the effects of other factors included in the model. Test day models analyse test day milk yields according to a slit-plot in time statistical design (Diggle *et al.*, 2002), where the animal is the main plot and the different time intervals at which measurements are taken are the subplots (Macciotta *et al.*, 2004b).

Table 2.5. Least squares means of individual values of Wood parameters for different parities in goats.

	Parameter			
	a	*b*	*c*	Reference
Parity[a]				Akpa *et al.* (2001)
1	0.388	0.338	0.017	
2	0.471	0.355	0.018	
3	0.877	0.323	0.017	
	$P \leq 0.05$	NS	NS	
Parity[b]				Giaccone *et al.* (1995)
1	0.116	0.115	0.0029	
2	0.119	0.153	0.0031	
3	0.129	0.163	0.0045	
	$P \leq 0.01$	$P \leq 0.01$	$P \leq 0.01$	

NS, not significant.
[a]Red Sokoto.
[b]Derivata di Siria.

In goats, test day models have been used to estimate lactation curves in Murciano-Granadina, Girgentana and Sarda breeds (Todaro *et al.*, 1999; Fernández *et al.*, 2002; Macciotta *et al.*, 2005a). Figures 2.6, 2.7 and 2.8 report average lactation curves for milk yield, fat and protein content, respectively, for Sarda breed goats of three different parities, estimated with a test day model.

Lactation curves estimated by test day models are less regular and are characterized by the absence of the lactation peak in comparison with those reported in Figs 2.4 and 2.5. Actually, in the functional approach, the effect of lactation stage on milk yield is modelled with functions originally conceived to describe the typical shape of the lactation curve (Fig. 2.1), which tend to reconstruct the increasing phase until the lactation peak, even in the case of poor availability of data. By contrast, the use of DIM intervals in test day models to fit the effect

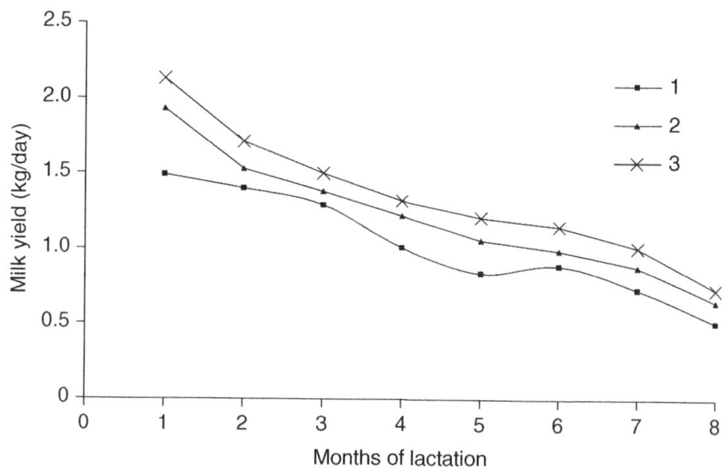

Fig. 2.6. Average lactation curves for milk yield of Sarda breed goats of different parities estimated with a test day model.

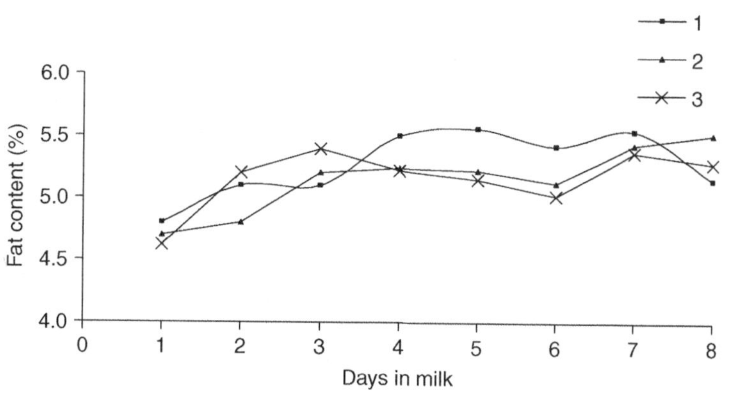

Fig. 2.7. Average lactation curves for milk fat content of Sarda breed goats of different parities estimated with a test day model.

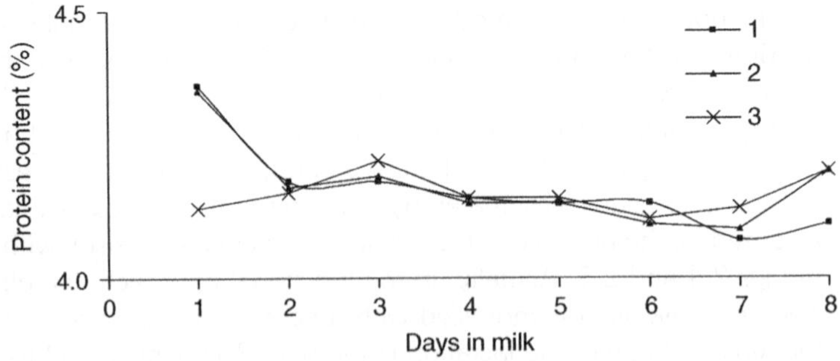

Fig. 2.8. Average lactation curves for milk protein content of Sarda breed goats of different parities estimated with a test day model.

of lactation stage allows for a greater flexibility and, therefore, waves in the middle of lactation may also occur (Druet *et al.*, 2003).

Lactation curves for fat and protein content show an opposite trend in comparison with milk yield. A more regular pattern for protein content in comparison with fat can be observed in goats, as observed previously in other ruminants.

Finally, an animal random factor is included in test day models, in order to account for individual variability. The ratio between the variance component pertaining to the animal factor and the total phenotypic variance (animal + residual) represents the average correlation among daily productions within each lactation, or repeatability. In Sarda goats, this ratio has been estimated to be 0.34, 0.17 and 0.45 for milk yield, fat and protein content, respectively (Macciotta *et al.*, 2005a).

Conclusions and Perspectives

Research on mathematical modelling of the lactation curve in goats mainly deals with empirical mathematical models applied to local and tropical goat breeds. Good results have been obtained in modelling average lactation curves, whereas a wide range of fitting performances have occurred when individual patterns have been fitted. The modelling of individual curves should be investigated further also taking into consideration a possible future use of random regression test day models to predict breeding values for selection purposes. Fewer studies have dealt with modelling lactation curves by test day models, even if the evaluation of non-genetic factors affecting specifically some part of the lactation has given interesting results.

Further development of models able to predict test day yields during lactation with reasonable accuracy is needed. Actually, an increase of the number of recorded animals is needed to enhance the impact of breeding programmes for dairy goats but this would increase the cost of selection programmes. A compromise between these two opposite needs can be achieved by a suitable mathematical model able to predict missing tests on the basis of a few tests actually recorded, with sufficient accuracy. Models based on the multivariate approach, such as the partial least square regression (Macciotta *et al.*, 2006) and neural networks (Fernández *et al.*, 2006), could be interesting options.

The great theoretical and technical complexity of the mechanistic approach has limited its use in animal science as in other fields of applied biology. This is particularly true for goats. On the other hand, greater effort in developing this approach could be repaid by more interesting results in terms of both the prediction of yields and the estimation of parameters directly related to biological processes underlying milk production. This could be of great help to researchers and professionals in the dairy goat industry.

References

AIA (2005) *Bulletin of Official Milk Records*. Associazione Italiana Allevatori, Rome.

Akpa, G.N., Asiribo, O., Oni, O.O. and Alawa, J.P. (2001) The influence of non-genetic factors on the shape of lactation curves in Red Sokoto goats. *Animal Science* 72, 233–239.

Ali, T.E. and Schaeffer, L.R. (1987) Accounting for covariances among test day milk yields in dairy cows. *Canadian Journal of Animal Science* 67, 637–644.

Andonov, S., Dzabirsky, V. and Srbinovska, S. (1999) Selection of dairy goats by using lactation curve model. In: *Proceedings of the 6th International Symposium on the Milking of Small Ruminants*. European Association of Animal Production, Publication No. 95. Wageningen Pers, Wageningen, The Netherlands, pp. 418–420.

Boe, F., Cannas, A., Rassu, S.P.G. and Pulina, G. (2005) CAPRISOFT, un software di programmazione per le aziende caprine. In: Pulina, G. (ed.) *L'Alimentazione della Capra da Latte*. Avenue Media, Bologna, Italy, pp. 347–364.

Bouloc, N. (1992) Courbes de lactation des chevres: quelques elements sur leur forme. *La Chevre* 193, 15–17.

Breda, F.C., Albuquerque, L.G., Yamaki, M., Reis Filho, J.C., Sarmento, J.L.R., Lopes, P.S. and Rodrigues, M.T. (2006) Estimação de parâmetros genéticos para produção de leite de cabras da raça Alpina. *Revista Brasileira de Zootecnia* 35, 396–404.

Cappio-Borlino, A., Pulina, G. and Rossi, G. (1989) La previsione della produzione lattea totale in pecore di razza sarda con l'uso di polinomi. *Produzione Animale* 35, 35–39.

Cappio-Borlino, A., Pulina, G. and Rossi, G. (1995) A non-linear modification of Wood's equation fitted to lactation curves of Sardinian dairy ewes. *Small Ruminant Research* 18, 75–79.

Cappio-Borlino, A., Macciotta, N.P.P. and Pulina, G. (1997) The shape of Sarda ewe lactation curve analyzed with a compartmental model. *Livestock Production Science* 51, 89–96.

Cobby, J.M. and Le Du, J.L.P. (1978) On fitting curves to lactation data. *Animal Production* 26, 127–133.

Crepaldi, P., Corti, M. and Cicogna, M. (1999) Factors affecting milk production and prolificacy of Alpine goats in Lombardy (Italy). *Small Ruminant Research* 32, 83–88.

Dhanoa, M.S. (1981) A note on an alternative form of the lactation model of Wood. *Animal Production* 32, 349–351.

Diggle, P.J., Heagerty, P., Liang, K.Y. and Zeger, S.L. (2002) *Analysis of Longitudinal Data*. Oxford University Press, Oxford, UK.

Druet, T., Jaffrezic, F., Boichard, D. and Ducorq, V. (2003) Modeling lactation curves and estimation of genetic parameters for first lactation test day records of French Holstein cows. *Journal of Dairy Science* 86, 2480–2490.

Fernández, C., Sánchez, A. and Garcés, C. (2002) Modeling the lactation curve for test-day milk yield in Murciano-Granadina goats. *Small Ruminant Research* 46, 29–41.

Fernández, C., Soria, E., Martin, J.D. and Serrano, A.J. (2006) Neural networks for animal science applications: two case studies. *Expert Systems with Applications* 31, 444–450.

France, J. and Thornley, J.H.M. (1984) *Mathematical Models in Agriculture*. Butterworths, London.

Giaccone, P., Portolano, B., Bonanno, A., Alicata, M.L. and Todaro, M. (1995) Aspetti quanti qualitativi della produzione lattea nella popolazione caprina derivata di Siria. *Zootecnica e Nutrizione Animale* 21, 94–107.

Gipson, T.A. and Grossman, M. (1989) Diphasic analysis of lactation curves in dairy goats. *Journal of Dairy Science* 72, 1035–1044.

Gipson, T.A. and Grossman, M. (1990) Lactation curves in dairy goats: a review. *Small Ruminant Research* 3, 383–396.

Gonçalves, J.C., Silva, M.A., Wechsleer, F.S. and Ramos, A.A. (2001) Fatores genéticos e de meio na produção de leite de caprinos leiteiros. *Revista Brasileira de Zootecnia* 30, 719–729.

Goonewardene, L.A., Okinea, E., Patrick, N. and Scheer, H.D. (1999) The relationship between multiple births and milk yields in non-suckled intensively managed dairy goats. *Small Ruminant Research* 32, 181–185.

Groenewald, P.C.N. and Viljoen, C.S. (2003) A Bayesian model for the analysis of lactation curves of dairy goats. *Journal of Agricultural, Biological and Environmental Statistics* 8, 75–83.

Grossman, M. and Koops, W. (1988) Multiphasic analysis of lactation curves in dairy cattle. *Journal of Dairy Science* 71, 1598–1608.

Guimarães, V.P., Rodrigues, M.T., Sarmento, J.L.R. and Rocha, D.T. (2006) Utilização de funções matemáticas no estudo da curva de lactação em caprinos. *Revista Brasileira de Zootecnia* 35, 535–545.

Hurley, W.L. (1989) Mammary gland function during involution. *Journal of Dairy Science* 72, 1637–1646.

Jamrozik, J. and Schaeffer, L.R. (1997) Estimates of genetic parameters for a test day model with random regressions for yield traits of first lactation Holsteins. *Journal of Dairy Science* 80, 762–770.

Kominakis, A., Rogdkis, E., Vasiloudis, C. and Liaskos, O. (2000) Genetic and environmental sources of variation of milk yield in Skopelos dairy goats. *Small Ruminant Research* 36, 1–5.

Landete-Castillejos, T. and Gallego, L. (2000) Technical note: the ability of mathematical models to describe the shape of lactation curves. *Journal of Animal Science* 78, 3010–3013.

Macciotta, N.P.P., D'Angelo, A., Cappio-Borlino, A., Gaviraghi, A., Gianoncelli, C. and Noè, L. (2003) The lactation curve of the Frontalasca breed goat. In: *Atti 38° Simposio Internazionale di Zootecnia 'Milk and Research'*, Lodi, Italy. MG Editori, Milan, Italy, pp. 191–197.

Macciotta, N.P.P., Gaviraghi, A., Cappio-Borlino, A. and Noè, L. (2004a) Adattamento del modello di Wilmink a curve di lattazione individuali di capre di razza Frontalasca. In: *Atti XVI Congresso Nazionale SIPAOC*, Siena, Italy. Kalb sas, Cagliari, Italy, pp. 294–295.

Macciotta, N.P.P., Pulina, G. and Cappio-Borlino, A. (2004b) Lactation and growth curves. In: Saxton, A.M. (ed.) *Genetic Analysis of Complex Traits using SAS*. SAS Institute Inc., Cary, North Carolina, pp. 97–147.

Macciotta, N.P.P., Fresi, P., Usai, G. and Cappio-Borlino, A. (2005a) Lactation curves of Sarda breed goats estimated with a test day model. *Journal of Dairy Research* 72, 470–475.

Macciotta, N.P.P., Vicario, D. and Cappio-Borlino, A. (2005b) Detection of different shapes of lactation curve for milk yield in dairy cattle by empirical mathematical models. *Journal of Dairy Science* 88, 1178–1191.

Macciotta, N.P.P., Dimauro, C., Bacciu, N., Fresi, P. and Cappio-Borlino, A. (2006) Use of partial least squares regression model to predict test day of milk, fat and protein yields in dairy goats. *Animal Science* 82, 463–468.

Macedo, V.P., Damasceno, J.C., Santos, G.T., Martins, E.N. and Macedo, F.A.F. (2001) Comportamento da curva de lactação de cabras mestiças Saanen em função da supplemetação de concentrado e do sistema de produção. *Revista Brasileira de Zootecnia* 30, 2093–2098.

McManus, C., Soares Filho, G., Mariante, A.S. and Louvandini, H. (2003) Fatores que influenciam os parâmetros das curvas de lactação em cabras no Distrito Federal. *Revista Brasileira de Zootecnia* 32(Suppl. 1), 1614–1623.

Mepham, T.B. (1987) *Physiology of Lactation*. Open University Press, Milton Keynes, UK.

Montaldo, H., Almanza, A. and Juarez, A. (1997) Genetic group, age and season effects on lactation curve shape in goats. *Small Ruminant Research* 24, 195–202.

Morant, S.V. and Gnanasakthy, A. (1989) A new approach to the mathematical formulation of lactation curves. *Animal Production* 49, 151–162.

Neal, H.D. and Thornley, J.H.M. (1983) The lactation curve in cattle: a mathematical model of the mammary gland. *Journal of Agricultural Science Cambridge* 101, 389–400.

Nelder, J.A. (1966) Inverse polynomials: a useful group of multi-factor response functions. *Biometrics* 22, 128–141.

Olori, V.E., Brotherstone, S., Hill, W.G. and McGuirk, B.J. (1999) Fit of standard models of the lactation curve to weekly records of milk production of cows in a single herd. *Livestock Production Science* 58, 55–63.

Pena Blanco, F., Vega Vilca, J., Sanchez Rodriguez, M., Martos Peinado, J., Garcia Martinez, A. and Domenech Garcia, V. (1999) Producción láctea y ajuste de la curva de lactación en caprinos de raza Florida. *Archivos de Zootecnia* 48, 415–424.

Peris, S., Caja, G., Such, X., Casals, R., Ferret, A. and Torre, C. (1997) Influence of kid rearing systems on milk composition and yield of Murciano-Granadina dairy goats. *Journal of Dairy Science* 80, 3249–3255.

Rekaya, R., Carabano, M.J. and Toro, M.A. (2000) Bayesian analysis of lactation curves of Holstein Friesian cattle using a non-linear model. *Journal of Dairy Science* 86, 2691–2701.

Rota, A.M., Gonzalo, C., Rodriguez, P.L., Rojas, A.I., Martin, L. and Tovar, J.J. (1993) Effects of stage of lactation and parity on somatic cell counts in milk of Verata goats and algebraic models of their lactation curves. *Small Ruminant Research* 12, 211–219.

Ruvuna, F., Kogi, J.K., Taylor, J.F. and Mkuu, S.M. (1995) Lactation curves among crosses of Galla and East African with Toggenburg and Anglo-Nubian goats. *Small Ruminant Research* 16, 1–6.

Schaeffer, L.R. (2004) Applications of random regression models in animal breeding. *Livestock Production Science* 86, 35–45.

Schaeffer, L.R. and Sullivan, P. (1994) Genetic evaluation of dairy goats using test day yields. In: *Proceedings of the 5th World Congress of Genetics Applied to Livestock Production*. University of Guelph, Guelph, Ontario, Vol. 18, pp. 182–185.

Shanks, R., Berger, P.J. and Freeman, A.E. (1981) Genetic aspects of lactation curves. *Journal of Dairy Science* 64, 1852–1860.

Silva, F.F., Muniz, J.A., Aquino, L.H. and Sáfadi, T. (2005) Abordagem Bayesiana da curva de lactação de cabras Saanen de primeira e segunda ordem de parto. *Pesquisa Agropecuária Brasileira* 40, 27–33.

Stanton, T.L., Jones, L.R., Everett, R.V. and Kachman, S.D. (1992) Estimating milk, fat, and protein lactation curves with a test day model. *Journal of Dairy Science* 75, 1691–1700.

Todaro, M., Madonia, G., Portolano, B., Margotta, S., Giaccone, P. and Console, A. (1999) Le curve di lattazione della capra Girgentana per latte, grasso, proteina e lattosio stimate con modelli test day. In: *Proceedings of the V Congress on 'Biodiversità: germoplasma locali e loro valorizzazione'*, Caserta, Italy. Imago Media, Piedimonte Matese, Italy, pp. 929–936.

Todaro, M., Madonia, G., Montalbano, L., Genna, G. and Giaccone, P. (2000) A non linear modification of the Wood model to estimate lactation curves of Girgentana goats. In: *Proceedings of the 7th International Conference on Goats*, Tours, France. LIR, Ivry sur Seine, France, pp. 245–246.

Wahome, R.G., Carles, A.B. and Schwartz, H.J. (1994) An analysis of the lactation curve of small east African goats. *Small Ruminant Research* 15, 1–7.

Williams, J.C. (1993) An empirical model for the lactation curve of white British dairy goats. *Animal Production* 57, 91–97.

Wilmink, J.B.M. (1987) Adjustment of test day milk, fat and protein yield for age, season and stage of lactation. *Livestock Production Science* 16, 335–348.

Wood, P.D.P. (1967) Algebraic model of the lactation curve in cattle. *Nature* 216, 164–165.

Zambom, M.A., Alcalde, C.R., Martins, E.R., Santos, G.T., Macedo, F.D.F., Horst, J.A. and Veiga, D.R. (2005) Lactation curve and milk quality of Saanen goats fed diets with different forage:concentrate ratios. *Revista Brasileira de Zootecnia* 34(Suppl. 1), 2515–2521.

Zygoyannis, D. (1994) A note on the effect of number and genotype of kids on milk yield and composition of indigenous Greek goats (*Capra prisca*). *Animal Production* 58, 423–426.

3 Lipids of Goat's Milk: Origin, Composition and Main Sources of Variation

M. MELE[1], A. BUCCIONI[2], A. SERRA[1], M. ANTONGIOVANNI[2] AND P. SECCHIARI[1]

[1]DAGA, Sezione di Scienze Zootecniche, Università di Pisa, Via del Borghetto 80, 56124 Pisa, Italy; [2]Dipartimento di Scienze Zootecniche, Università di Firenze, Via delle Cascine 5, 50100 Firenze, Italy

Introduction

The nutritional, technological and organoleptic characteristics of milk and its products are strongly influenced by milk lipids. Similar to other species, the milk of goats contains nutritional components with beneficial effects for human health, such as conjugated linoleic acid (CLA), vaccenic acid (VA), butyric acid, oleic acid, odd- and branched-chain fatty acids (OBCFAs), sphingomyelin and vitamin A. In addition, it is well known that lipids affect the yield of cheese and the colour and aroma of dairy products.

The quality and quantity of milk fat can be modified by changing the composition of the basal ration (forage/concentrate ratio, forage quality, pasture composition) and/or by supplementing the ration with different lipid sources. The modification of the quantity of lipids in goat's milk may solve some technological problems linked to its low fat content, particularly when it becomes lower than the protein content. In order to achieve that, it is possible to take advantage of the ability of small ruminants to respond positively to lipid supplementation of the diet. Unlike what has been observed in dairy cows, several studies on dairy goats have shown that lipid supplementation of diets increases the amount of milk fat, without decreasing the amount of milk proteins and worsening the milk's coagulation properties (Chilliard and Bocquier, 1993; Chilliard *et al.*, 2003).

Lipid Metabolism in the Rumen

The glycerol esters (glycerides) of dietary lipids undergo lipolysis and biohydrogenation by rumen microbes (Harfoot, 1978; Palmquist and Jenkins, 1980; Jenkins, 1993). Lipolysis results in the release of free fatty acids (FFAs) from esters, thus permitting biohydrogenation, which is the reduction of the number of double bonds present in the carbon chain of the fatty acid (FA). While feeds remain in the rumen, a very small amount of FAs are absorbed by the rumen wall or catabolized to form volatile fatty acids (VFAs) and CO_2, whereas remarkable amounts of FAs are synthesized *ex novo* from carbohydrate precursors by microbes. Therefore, the FAs which then reach the duodenum are in part of dietary origin and in part due to microbial activity.

Lipolysis

Shortly after ingestion, dietary glycerides are hydrolysed by microbial lipases which release the constituent FFAs (Fig. 3.1). Lipases are extracellular enzymes packaged in membranous particles composed of proteins, lipids and nucleic acids (Jenkins, 1993). Lipases hydrolyse triglycerides almost completely to FFAs and glycerol, except for small amounts of mono- and diglycerides. Glycerol is then rapidly fermented to propionic acid. Seventy-four bacterial strains capable of hydrolysing the ester bond of glycerides have been identified (Fay *et al.*, 1990). Some of them, such as *Anaerovibrio lipolytica* and *Anaerovibrio fibrisolvens*, have low hydrolysis activity, whereas many other strains characterized by esterase activity are not necessarily capable of hydrolysing lipid esters. In fact, only a few of them (e.g. 30 strains of *Butyrivibrio fibrisolvens*) can hydrolyse long-chain fatty acids (LCFAs). FAs may come not only from enzymatic hydrolysis of triglycerides, but also from the hydrolysis of galactolipids and phospholipids, and by the action of several galactosidases and phospholipases (phospholipase A, phospholipase C, phosphoesterase) produced by rumen microbes (Jenkins, 1993).

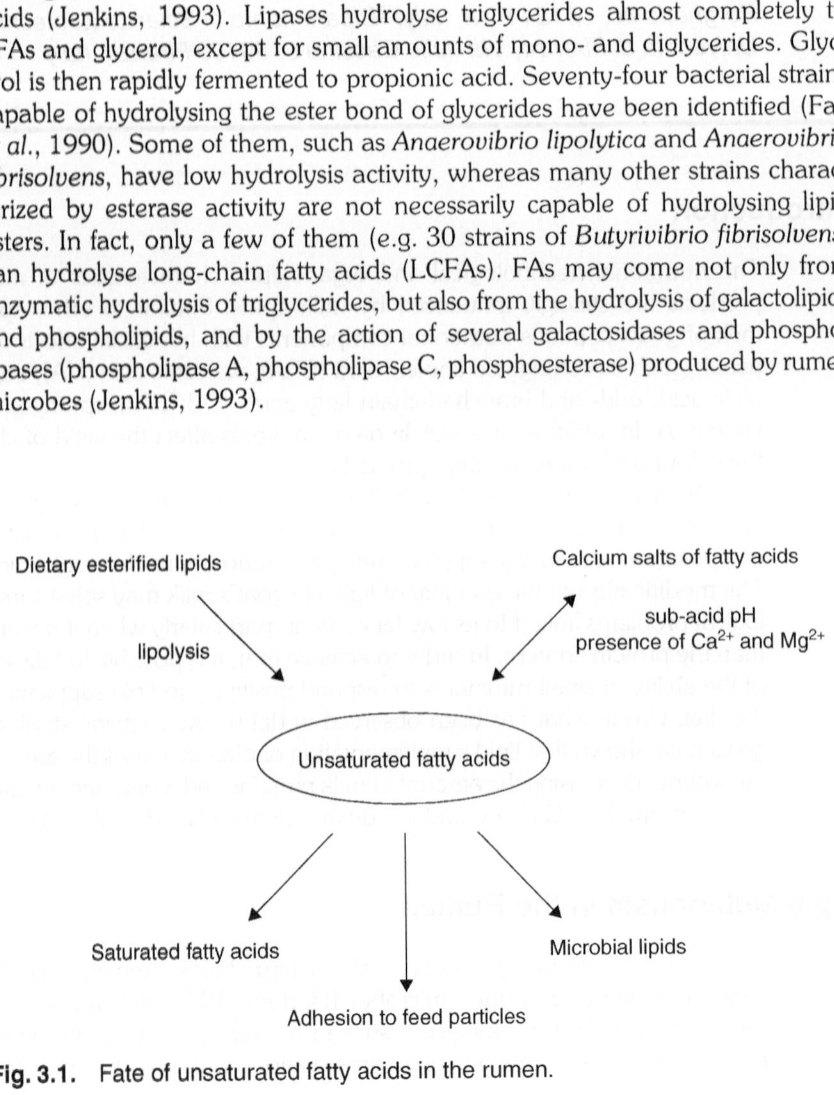

Fig. 3.1. Fate of unsaturated fatty acids in the rumen.

Biohydrogenation

The half-life of free unsaturated fatty acids (UFAs) in the rumen fluid is relatively short, due to their rapid hydrogenation to the saturated configuration by rumen microbes. The reduction of polyunsaturated fatty acids (PUFAs) ranges from 60 to 90% of total ingested PUFAs. It seems that this process protects microorganisms from the toxic effects of UFAs. This reduction process contributes little (1–2%) to the recycling of metabolic hydrogen (Czerkawski, 1984). When UFAs with a double bond in the *cis*-12 position (e.g. linoleic acid (LA), *cis*-9, *cis*-12 18:2 and α-linolenic acid (LNA), *cis*-9,*cis*-12,*cis*-15 18:3) are included in the diet, the first step of the biohydrogenation pathway consists of an enzymatic isomerization of such a bond to the *trans*-11 position (Fig. 3.2). In general, isomerase acts only in the presence of a free carboxylic function and, in the particular case of PUFAs, of the isolated *cis*-9,*cis*-12 diene. For this reason, lipolysis must occur before reduction and is considered the rate-determining step of the whole process, i.e. the stage determining the reaction kinetics. Therefore, the passage of small quantities of PUFAs through the rumen barrier may be due to incomplete lipolysis. After the *trans*-11 bond is formed, a microbial reductase

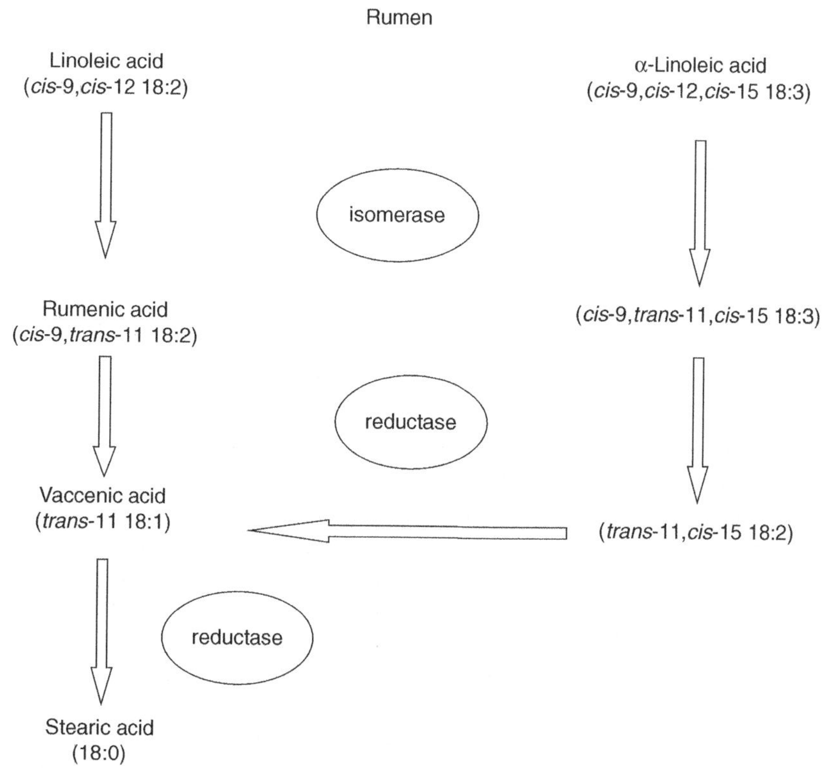

Fig. 3.2. Biohydrogenation of linoleic and linolenic acid. (Adapted from Harfoot and Hazlewood, 1988.)

hydrogenates the *cis*-9 bond. The amount of *trans*-11 18:1 hydrogenated to 18:0 is affected by rumen conditions and the concentration of LA, which irreversibly inhibits the process (Harfoot *et al.*, 1973). In fact, Moore *et al.* (1969) suggested that the presence of large amounts of unesterified LA blocks the second stage of biohydrogenation, while esterified LA does not. When the fibre content of the diet is lowered due to higher levels of concentrate, there is a reduction of the number of cellulolytic bacteria which are also responsible for lipolysis. Therefore, feed of such composition increases the fraction of dietary lipid that crosses the rumen barrier without being reduced; this is especially rich in oleic acid (*cis*-9 18:1) and LA, which are the most abundant FAs in cereal grains.

Other dietary factors that decrease lipolysis and rumen biohydrogenation are the utilization of excessively mature forages, rich in lignin, which decrease the feed degradability, and of feeds that are too finely ground. In the latter case, the adherence of bacteria on the surface of feed particles is poor and the flow of triglycerides through the rumen barrier is increased, thus shortening the time of exposure to the action of bacteria.

Finally, since the amount and type of fat added to the diet can also influence the modification of lipids in the rumen, this aspect will be dealt with separately.

Synthesis of microbial fatty acids

Rumen microbial cells contain about 10–15% of lipids (on a dry matter, DM, basis), which are characterized by a large portion of saturated fatty acids (SFAs), mainly palmitic acid (16:0) and stearic acid (18:0), originating from the uptake of dietary FAs and from *ex novo* synthesis in microbial cells. In addition, FAs of rumen bacteria contain various *trans* isomers of 18:1 and 18:2 derived from biohydrogenation of dietary LA and LNA (Vlaeminck *et al.*, 2006a). The contribution of each of these pathways depends on the fat content of the diet and the bacterial species present in the rumen liquor (Jenkins, 1993). High concentration of dietary lipids favours the transfer of lipids to the interior of the bacterial cell, forming small droplets in the cytosol. On the contrary, the *ex novo* synthesis of FAs leads prevailingly to the formation of 18:0 and 16:0 in the ratio 2:1 (Bauchart *et al.*, 1990). It must be considered that rumen microbes utilize FAs primarily to build up membrane phospholipids and, in small part, non-esterified FAs, while they do not store triglycerides (Viviani, 1970). Detailed studies on biosynthesis mechanisms have revealed that the absorption of ^{14}C-labelled acetate and glucose by microbes results in the synthesis of linear-chain FAs characterized by an even number of carbon atoms, whereas that of ^{14}C-labelled propionate and valerate leads to the synthesis of linear-chain FAs with an odd number of carbon atoms. By contrast, when isobutyrate, isovalerate and 2-methylbutyrate are used as precursors, branched-chain fatty acids (BCFAs) in the iso and anteiso forms are synthesized. In any case, the FA composition of the membrane of rumen bacteria is characterized by a large proportion of OBCFAs (Vlaeminck *et al.*, 2006b). In bacteria, the OBCFAs represent about 20% of total FAs, out of which 30% are contained in phospholipids. The monounsaturated fatty acids (MUFAs), which constitute about 15–20% of microbial FAs, are synthesized by anaerobic pathways in which the β-hydroxydecanoate is dehydrated in the β and γ positions, thus forming a double bond in position 3 and a geometric isomer of the *cis* type, instead of being dehydrated in α and β positions, which would lead to the formation of the *trans*-2 isomer. If the double bond is located in the *cis*-3 position, the successive reduction by C10-enoyl reductase cannot take place, thus permitting chain elongation up to 16:1 and 18:1. The latter FA may also be formed via desaturation by a desaturase enzyme present in the rumen liquor (Fig. 3.3) (Jenkins, 1993). In the rumen, only cyanobacteria can synthesize PUFAs, while other microorganisms can only absorb them from the rumen liquor and, in the case of protozoa, incorporate them under the form of membrane lipids. Recently, it has been suggested that protozoa preferentially incorporate CLA and VA (*trans*-11 18:1) formed by bacteria. Thus, the flow of UFAs, including CLA and VA, from the rumen could depend on the flow of protozoa rather than bacteria (Devillard *et al.*, 2006). Therefore, nutritional strategies aimed at modifying the composition of the FA pool escaping the rumen

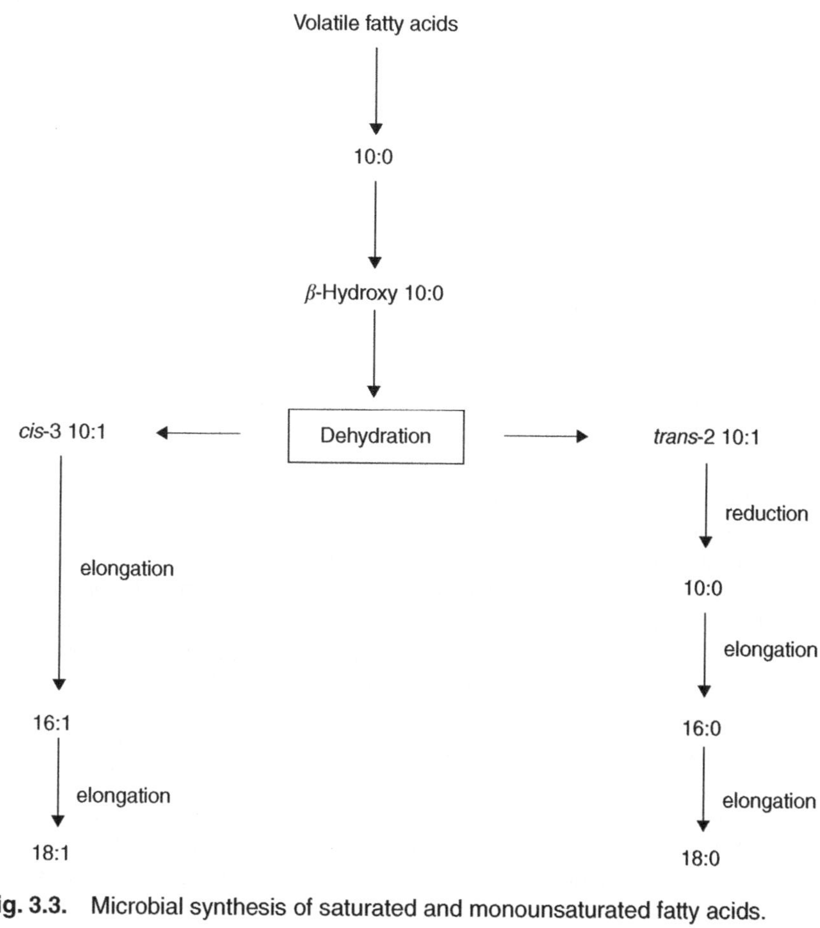

Fig. 3.3. Microbial synthesis of saturated and monounsaturated fatty acids.

should also take into consideration the interaction between rumen bacteria and rumen protozoa.

Lipid balance in the rumen

Each kilogram of DM consumed by a ruminant contains about 4% crude fat. Only 40% of the crude fat of forages is made up of FAs, whereas this percentage is higher than 70% in grains (Palmquist and Jenkins, 1980). Even if the loss of FAs in the rumen is negligible, some metabolic pathways responsible for it have been hypothesized. In fact, about 31.5% of oleic acid incubated in the epithelial cells of the rumen is absorbed by this tissue, while 8.2% is transferred to plasma (Jenkins, 1993). By contrast, palmitic acid is quickly metabolized to ketone-like substances and converted to 15:0 by α-oxidation and to 13:0 and 11:0 by β-oxidation (Jesse *et al.*, 1992). Moreover, more than 90% of FAs with fewer than 14 carbon atoms are absorbed through the rumen wall. It has been observed during the transit of feed boluses from mouth to duodenum that fat-supplemented diets usually lose more lipids than low-fat diets. The presence of lipids in the diet and the amount of digested organic matter could be the only factors affecting lipid synthesis by microorganisms. In conclusion, on average 87% of the ingested FAs reach the duodenum and the missing fraction is often compensated by the *ex novo* rumen microbial synthesis, of about 15 g/kg DM fermented in the rumen. The causes of such losses may be related to the lipid metabolism of rumen epithelial cells and to microbial degradation (Jenkins, 1994).

Effects of lipids on rumen fermentation

The supplementation of ruminant diets with lipids may have negative effects on rumen fermentation, resulting in a reduction in the digestibility of energy sources other than lipids. A ration containing approximately 10% fat decreases rumen degradation of structural carbohydrates by more than 50% (Ikwuegbu and Sutton, 1982; Jenkins and Palmquist, 1984). This is accompanied by a lower production of methane, hydrogen and VFA and by a decrease of the acetate/propionate ratio (Chalupa *et al.*, 1984; Czerkawski, 1984; Boggs *et al.*, 1987).

Usually, the effects of dietary lipid sources on rumen metabolism depend on differences in their chemical structure, such as the degree of unsaturation, since PUFAs inhibit fermentations more than SFAs (Palmquist and Jenkins, 1980; Chalupa *et al.*, 1984). Since the presence of the carboxyl function also seems to be important, several techniques for the protection of lipids have been developed, aiming to minimize the interactions between rumen bacteria and diet lipids. Calcium salts of FA and derivatives of carboxyl acids, such as amides, triglycerides and long-chain alcohols, are compounds which interfere less with fermentation and, at the same time, protect UFAs from biohydrogenation. By contrast, non-esterified PUFA is the lipid fraction which influences fermentation the most. The concentration of free UFAs in the rumen is regulated by: (i) amount and kind of fat in the diet; (ii) lipolysis rate; (iii)

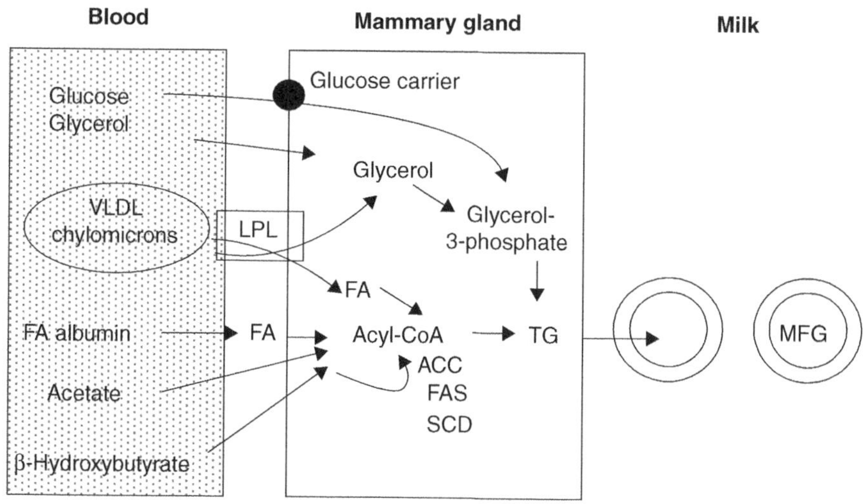

Fig. 3.4. Cell metabolism of fatty acids in the mammary gland. VLDL, very-low-density lipoprotein; FA, fatty acids; LPL, lipoprotein lipase; ACC, acetyl-CoA carboxylase; FAS, fatty acid synthetase; SCD, stearoyl-CoA desaturase; TG, triglycerides; MFG, milk fat globule. (Adapted from Chilliard *et al.*, 2000.)

biohydrogenation; and (iv) production of carboxyl salts (Fig. 3.1). In general, the major part of triglycerides is hydrolysed in a short time, even though this greatly depends on factors such as the maturation degree of forages, the nitrogen content of the ration and the particle size in the rumen.

Metabolism of the Mammary Gland and *ex novo* Synthesis of Fatty Acids

About 50% of milk FAs are synthesized in the mammary gland from blood acetate and *β*-hydroxybutyrate by the action of acetyl-CoA carboxylase and fatty acid synthetase (FAS) (Fig. 3.4). The remaining part comes both from the diet (about 40–45%) and from the mobilization of body lipid reserves of the animal, in variable proportions according to the lactation stage (Chilliard *et al.*, 2000). Such percentages may vary significantly with dietary manipulations. Since the glandular cells of the udder do not have the enzymatic pool necessary to elongate the carbon chain from 16 to 18 atoms, the neo-synthesized FAs are made up of short- and medium-chain FAs only (from 4:0 to 16:0). Half of milk palmitic acid (16:0) comes from endogenous synthesis and the other half from the diet. The presence of short- and medium-chain FAs in milk derives from an altered specificity of the FAS of lactating ruminants. In fact, in contrast to non-ruminant species, FAS of the ruminant mammary gland exhibits a transacylase with both loading and releasing activity for acyl chains from two to 12 carbon atoms in length (Barber *et al.*, 1997; Chilliard *et al.*, 2000). All milk FAs with chain length equal to or greater than 18 carbon atoms must be provided from the diet, which markedly influences the fraction of LCFAs in milk. The transfer of FAs from plasma to mammary tissue can be described by the kinetic model of Michaelis–Menten, with a maximum speed (V_m) and a Michaelis constant (K_m) of 22 and 32 mg/100 ml, respectively (Baldwin *et al.*, 1980). This process, together with rumen biohydrogenation and digestion in the gut, influences the efficiency of transfer of FAs from feeds to milk. Milk fat fluidity is affected by the desaturase activity of the mammary gland. In fact, the Δ^9-desaturase enzyme (which introduces a double bond in position 9 of the carbon chain) desaturates a large amount of stearic acid (18:0) into oleic acid (*cis*-9 18:1), reaching a ratio between the two FAs that optimizes the physicochemical traits of milk. The Δ^9-desaturase enzyme acts on other substrates such as myristic acid (14:0), palmitic acid (16:0) and VA as well. In the latter case, the conjugated *cis*-9,*trans* 11 isomer of linoleic acid (rumenic acid, RA) is formed.

The long-chain unsaturated fatty acids (LCUFAs) inhibit the lipogenic enzymes of the mammary gland and, in particular, interfere with the activity of acetyl-CoA carboxylase (which is the rate-limiting enzyme of FA synthesis). In fact, the administration of diets containing a large amount of fats rich in LCUFAs or the infusion of LCUFAs directly into the duodenum of cows and goats (Kitessa *et al.*, 2001) markedly decreased milk fat, due to the blocking of endogenous synthesis of the short- and medium-chain FAs. Moreover, the C20 and C22 PUFAs can reduce the mammary uptake of LCUFAs from plasma, thus influencing the lipoprotein lipase of the mammary gland (Storry *et al.*, 1974). Recent studies on 'milk fat depression' (MFD) in dairy cows have demonstrated a specific effect of the *trans*-10,*cis*-12 CLA isomer and of the *trans*-10 18:1 isomer on lipogenic enzymes of the mammary gland (Loor *et al.*, 2005). Although the direct administration of *trans*-10,*cis*-12 CLA to dairy cows and sheep inhibited milk fat synthesis (Baumgard *et al.*, 2000; Lock *et al.*, 2006), the responses of dairy goats were less clear. For example, Andrade and Schmidely (2006a) reported that mammary lipogenesis in dairy goats was not decreased by *trans*-10,*cis*-12 CLA infused post-ruminally. On the other hand, the administration of a lipid-encapsulated CLA supplement to lactating Alpine dairy goats resulted in reduced milk fat synthesis in a manner similar to that observed for lactating dairy cows and sheep. However, dose–response comparisons suggested that the degree of reduction in milk fat synthesis is less in goats than in sheep and dairy cows (Rovai *et al.*, 2007).

Lipids in Goat's Milk

The lipid fraction of goat's milk is organized in fat globules made up of glycerides (97–99%), mainly present within the globule core, and of phospholipids, glycolipids and sterols (1–3%), as an integral part of the globule membrane. This distribution remains almost unchanged even when milk fat content changes (Cerbulis *et al.*, 1982), and is very similar to that described for cow's milk. The fat globule core is made up of apolar molecules, such as triglycerides (96.8%), diglycerides (2.3%) and monoglycerides (0.9%), whereas the membranes contain both polar lipids, such as glycolipids (8.5%) and phospholipids (44.7%), and apolar lipids, such as triglycerides (26.5%), diglycerides and monoglycerides (4.7%) and cholesterol (15.6%). The polar lipids are the determinants of the main

Table 3.1. Phospholipids of the milk fat globule membrane of goat's milk.

Phospholipids (g/100 g phospholipids)	Cerbulis *et al.* (1982)	Patton *et al.* (1977)	Mele *et al.* (unpublished data)
Phosphatidylethanolamine	35.4	25.5	46.3
Phosphatidylcholine	28.2	27.6	20.7
Phosphatidylserine	3.2	9.6	14.8
Phosphatidylinositol	4.0	1.4	6.1
Sphyngomyelin	29.2	35.9	12.2

phospholipid fractions contained in the milk fat globule membrane of goat's milk (Table 3.1). Even though the distribution of phospholipids in the membrane of milk fat globules varies according to individual characteristics and stage of lactation, the main phospholipid classes remain phosphatidyl ethanolamine, phosphatidyl choline and sphyngomyelin.

Milk samples collected from five different goat flocks of Granadina breed were analysed (Fontecha *et al.*, 2000) by capillary gas chromatography and silver ion adsorption–liquid chromatography, to determine the composition of triglycerides of goat's milk by classifying them according to the number of carbon atoms (CN). It was observed that 55% of glycerides were composed of SFAs, 29% of MUFAs and 16% of PUFAs. Analysis of the same samples by silver ion adsorption–thin layer chromatography allowed them to be classified into four groups or elution bands, named A, B, C and D (Table 3.2). Fraction A, which represented 38.5% of total triglycerides, was composed mainly of SFAs (95.3%) and traces of MUFAs (0.9%). Fraction A contained 73%, 80% and 74% of total milk content of 8:0, 10:0 and 12:0, respectively. By contrast, the level of 4:0 was low in fraction A, being concentrated in fraction B (94%). The latter fraction accounted for 16.5% of triglycerides, of which 94.3% were SFAs and 1.5% MUFAs. Fraction C, containing 28.9% of total triglycerides, was mainly made up of palmitic acid, MUFAs and LA, in the form of isolated and conjugated diene. The last band, fraction D, was constituted by 16% triglycerides whose main FAs were MUFAs and PUFAs.

The main classes of triglycerides, subdivided by CN, were sorted out by gas chromatography. Triglycerides were mostly represented by CN_{40} and CN_{36} in fractions A and B, by CN_{38} and CN_{44} in fraction C, and by CN_{46} and CN_{54} in fraction D. As the degree of unsaturation increased, the number of carbon atoms of the apolar lipids increased. In goat's milk the main saturated glycerides present are (10:0; 14:0; 16:0), (10:0; 16:0; 16:0) and (10:0; 16:0; 18:1), whereas the main polyunsaturated triglyceride is (16:0; 18:1; 18:1), similar to bovine and human milk.

Conjugated linoleic acids

In goats, as well as other ruminant species, the lipid fraction of milk contains an appreciable amount of CLA. The most represented isomer in milk lipids of

Table 3.2. Percentage composition of fatty acids in the triglycerides of goat's milk, subdivided into four elution groups after silver ion adsorption–thin layer chromatography. (From Fontecha *et al.*, 2000.)

		Fraction			
Fatty acid	% of total	A	B	C	D
4:0	5.09	0.56	19.26	4.43	3.01
6:0	4.42	3.83	6.87	3.94	2.50
8:0	4.15	6.70	5.81	4.57	3.98
10:0	12.91	18.57	10.79	11.74	7.75
10:1	0.36			0.60	1.84
12:0	5.62	8.72	6.96	5.21	3.80
12:1	0.21		0.05	0.41	0.40
14:0	9.86	13.54	10.01	7.95	5.15
14:1 + 15:0	0.39	0.85	0.73	0.69	0.86
ai-15:0 + 15:0	0.83	2.18	1.67	1.06	1.08
15:1	0.09			0.37	0.53
i-16:0 + 16:0	25.38	30.54	24.26	19.64	12.30
16:1	1.41			2.79	3.98
i-17:0 + ai-17:0 + 17:0	1.26	0.88	1.05	0.66	0.98
17:1	0.34			0.60	0.73
18:0	7.17	9.57	7.17	6.58	4.47
18:1	15.46	0.90	1.47	24.24	27.35
18:2	2.83			1.19	13.80
18:3	0.35				1.44
CLA	0.57	0.06	0.12	0.76	0.76
20:0	0.11	0.18	0.42	0.11	0.28
20:1	0.05			0.03	0.18
Other acids	1.14	2.92	3.37	2.43	2.83

i, iso; ai, anteiso; CLA, conjugated linoleic acid.

ruminants is RA, which originates from two metabolic pathways: (i) the biohydrogenation of LA in the rumen (Fig. 3.2); and (ii) the desaturation of VA, formed in the rumen, within the mammary gland (Fig. 3.5). The first pathway involves the already described lipolysis and successive reductions of dietary UFAs in the rumen. For many years, *B. fibrisolvens* was the only bacterium thought capable of performing biohydrogenation (Kepler *et al.*, 1967). Successively, other microorganisms capable of reducing the double bond of FAs were identified: *Eubacterium lentum*, *Propionibacterium freudenreichi*, *Lactobacillus acidophilus*, *Lactobacillus reuteri*, *Megasphaera elsdeni* and *Bifidobacterium breve* (Fukuda *et al.*, 2005).

Studies on pure cellular cultures demonstrated that the whole biohydrogenation process is not performed by a single microorganism. Instead, it is coordinated by a pool of rumen bacteria, each controlling the various reaction steps, which can be divided into two groups: (i) group A hydrogenates LA and LNA to

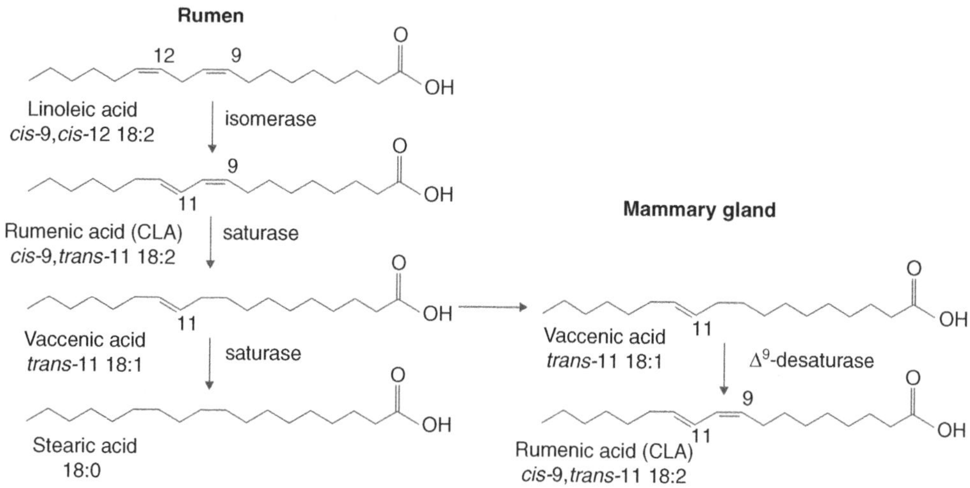

Fig. 3.5. Synthesis of rumenic acid, a conjugated linoleic acid (CLA), in the rumen and mammary gland.

VA; and (ii) group B concludes the sequence of biohydrogenation by reducing VA to 18:0 (Harfoot and Hazlewood, 1988). The synthesis of CLA in the rumen is illustrated in Fig. 3.2. The initial step is the isomerization of *cis*-9,*cis*-12 18:2 (LA) to *cis*-9,*trans*-11 18:2 (RA) (Harfoot, 1978). This passage is catalysed by the enzyme linoleic isomerase, which does not need the aid of cofactors and acts on double bonds located in the middle of the carbon chain and far from activating functional groups. The enzyme adheres to the bacterial membrane and is very selective, because it recognizes only dienes of the *cis*-9,*cis*-12 type present along the carbon chain of FAs with the free carboxyl function. The second step is the reduction of RA to *trans*-11 18:1 (VA), which is a fast reaction as demonstrated by *in vitro* studies incubating marked LA with rumen liquor. By contrast, the reduction of VA to 18:0 is much slower (rate-determining step), thus permitting the accumulation of VA in the rumen and its passage to plasma, via intestinal absorption. Recently, new strains of *B. fibrisolvens* that have high ability to isomerize LA to CLA or to hydrogenate LA to VA have been isolated, aiming to enhance the rumen production of VA and RA (Fukuda *et al.*, 2005, 2006). Once VA is absorbed into plasma and arrives at the mammary gland, it may be reconverted into RA by the action of Δ^9-desaturase (Fig. 3.5). As demonstrated in dairy cows, this second pathway of RA production provides more than 80% of the RA present in milk fat (Corl *et al.*, 1998). The pathway of LNA reduction is very similar to that of LA biohydrogenation. The most prevalent 18:3 isomer in feeds, i.e. α-LNA (*cis*-9,*cis*-12,*cis*-15 18:3), is hydrogenated to *trans*-11 18:1, the precursor of CLA synthesized by the mammary tissue (Fig. 3.2). γ-Linolenic acid (γ-LNA; *cis*-6,*cis*-9,*cis*-12 18:3), less common in feeds, is fermented in a similar way (Griinari and Bauman, 1999). Because of both processes, the most represented CLA isomer and *trans* 18:1 FA in ruminant milk fat are RA and VA, respectively, except under particular feeding conditions that markedly change the composition of the rumen bacterial population. In these cases, the biohydrogenation process shifts towards *trans*-10,*cis*-12 CLA and *trans*-10 18:1, instead of RA and VA. These first two FAs seem to be responsible for the MFD in dairy cows (Griinari *et al.*, 1998), but these aspects have not been confirmed yet in goats.

In addition to the transformation of VA to RA, Δ^9-desaturase is also responsible for the production of another CLA isomer (*trans*-7,*cis*-9) and a non-conjugated 18:2 (*cis*-9,*trans*-13) (Ulberth and Henninger, 1994; Yurawecz *et al.*, 1998). The mammary gland appears to be the tissue of greatest activity of Δ^9-desaturase in lactating ruminants (Kinsella, 1972). This enzyme, also named stearoyl-CoA desaturase (SCD), is expressed by the homonymous *Scd* gene located on chromosome 26 in cattle and goats and on chromosome 22 in sheep (Ward *et al.*, 1998; Bernard *et al.*, 2001; Taniguchi *et al.*, 2003). Among ruminants, the coding sequence of this gene is completely known for cattle and goats only. However, based on the information currently available, the structure of the *Scd* gene appears to be very well preserved in the ruminant species. The expression of the *Scd* gene and the activity of the related enzyme are sensitive to nutritional factors, such as the presence of PUFAs in feeds, and to endogenous factors, such as physiological stage (e.g. lactation phase) and hormonal balance (Ntambi, 1995). In a few cases, species-related differences in CLA content of milk have been reported for cows, goats and sheep grazing on the same pasture (Jahreis *et al.*, 1999; Nudda *et al.*, 2003). In these cases, variations were ascribed to differences among species in feeding habits and feed passage rate in the digestive system. On the other hand, recent studies have highlighted that factors other than diet may be associated with variations in milk FA composition, including VA and CLA, in both goats and sheep (Chilliard *et al.*, 2006; Mele *et al.*, 2007a). In particular, Chilliard *et al.* (2006) found that two groups of dairy goats with low or high αS_1-casein milk content had different proportions of at least 17 FAs in milk fat, particularly of saturated medium-chain FAs and stearic, oleic, linoleic and rumenic acids.

Branched fatty acids of goat's milk

Caproic acid (6:0), caprylic acid (8:0) and capric acid (10:0) are among the most characteristic FAs of goat's milk and are named after the species name. They are characterized by a sharp and persistent odour, typical of goats. In fact, a widely used parameter to detect the authenticity of goat's milk is the estimate of the ratio 12:0 to 10:0, as proposed by Ramos and Juarez (1986). The BCFAs are very important, because they are responsible for the typical aroma of goat's milk and cheese. There is an increasing interest in OBCFAs as potential diagnostic tools of rumen function (e.g. rumen fermentation pattern and bacterial nitrogen). Other reasons for the interest in OBCFAs are their anticarcinogenic effects on cancer cells, their influence on milk fat melting point and their potential as indicators of dairy product intake by humans (Vlaeminck *et al.*, 2006b). The iso and anteiso forms of 15:0, the iso 16:0 and the iso and anteiso forms of 17:0, which were the first ones to be identified, are the main BCFAs in milk of goats and cows (Massart-Leen *et al.*, 1981). Monomethyl BCFAs with chain length shorter than ten carbon atoms were identified, successively, only in goat's milk (Ha and Lindsay, 1993). Another 31 BCFAs, present at very low concentrations, were also identified: 25 of them are monomethyl branched, two are dimethyl and four are diethyl branched (Alonso *et al.*, 1999). Among the ethyl ones, 4-ethyloctanoate, together with 4-methyloctanoate, give characteristic goat-like or mutton-like flavours to dairy products. Even though 3-methylbutanoate, 4-methylpentanoate and 8-methylnonanoate have also been identified in goat's milk, they are not typical because they can be found in cow's milk as well.

Supplementation of the Diet with Lipid Sources

The introduction of lipid supplements into the diet of dairy goats enhances milk fat content and modifies milk FA profile, which is related to the lipid source. In addition, an adequate lipid supplementation can mend the inversion syndrome in which milk fat content goes below that of protein, with a consequent decrease of cheese yield (Morand-Fehr *et al.*, 1984a,b). Lipid sources can be added to the diet in different ways and their effects vary with the different type of lipid supplementation used. Moreover, for the same lipid source, the degree of protection from the rumen biohydrogenation processes formerly described can remarkably influence the effects on milk yield, and fat yield and composition.

Effects of lipid supplementation on milk yield and fat content

Based on a recent review of Chilliard *et al.* (2003) and on other data from the literature (Lu, 1993; Teh *et al.*, 1994; Hadjipanayiotou, 1999; Brown-Crowder *et al.*, 2001; Kitessa *et al.*, 2001; Chilliard *et al.*, 2002a; Rapetti *et al.*, 2002; Sanz-Sampelayo *et al.*, 2002; Bernard *et al.*, 2005; Mele *et al.*, 2005; Nudda *et al.*, 2006; Andrade and Schmidely, 2006b), a quadratic equation between the increase of the amount of supplemented fat and the increase of milk fat concentration was found, regardless of the type of supplemented fat and lactation stage (Fig. 3.6). However, the equation becomes linear if only animals at the initial phase of lactation are considered (Fig. 3.7), since lactation stage can affect the production response to dietary fat sources. At the beginning of lactation, lipid supplementation to the diet causes a general increase of milk yield and milk fat, regardless of the type of lipid source, whereas at mid-lactation it has a positive effect on milk fat only. By contrast, at the end of lactation, the effects of fat sources on production are much less evident (Chilliard *et al.*, 2001). In dairy sheep, the response of milk fat content and secretion to fat supplementation is generally similar to that of dairy goats, whereas in dairy cows milk fat content could often either decrease or not change (Chilliard *et al.*, 2003). A possible explanation for differences among ruminant species is that the high rumen solids and liquid passage rate of sheep and goats can reduce the interaction between

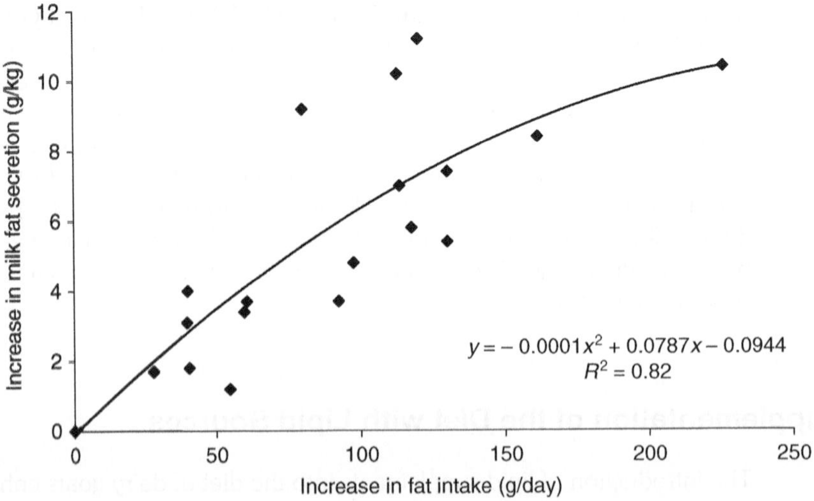

Fig. 3.6. Relationship between supplemented fat and milk fat secretion. Data represent means of dietary treatments and are referred to the whole lactation. (Data sources for dietary treatments: Lu, 1993; Teh *et al.*, 1994; Hadjipanayiotou, 1999; Brown-Crowder *et al.*, 2001; Kitessa *et al.*, 2001; Chilliard *et al.*, 2002a, 2003; Rapetti *et al.*, 2002; Sanz-Sampelayo *et al.*, 2002; Bernard *et al.*, 2005; Mele *et al.*, 2005; Schmidely *et al.*, 2005; Nudda *et al.*, 2006; Andrade and Schmidely, 2006b.)

fat supplements and rumen fermentation compared with what occurs in dairy cows (Pulina *et al.*, 2006).

Such variable responses are probably linked to the energy balance of the animal, which varies with lactation stage and influences lipogenesis enzymes as well. In fact, the activity of such enzymes is enhanced from the beginning to the end of lactation, when the dietary FAs are preferentially utilized for the storage of body fat rather than for the endomammary synthesis of milk fat (Chilliard *et al.*, 1991). The production responses to different types of fat supplements in the diet are quite uniform, except for unprotected fish oil, which has a negative influence on daily milk fat yield (Kitessa *et al.*, 2001).

Effects of lipid supplementation on milk fat composition

The composition of milk fat is markedly influenced by the type of supplemented fat. It is well known that the addition of adequate lipid sources to ruminant feeds can change the FA composition of milk, whose profile reflects that of the diet, modified by rumen biohydrogenation and microbial activity. A diet supplemented with palm oil, rich in palmitic acid (16:0), causes a remarkable increase in such FA and the corresponding MUFA (16:1) and a decrease in medium-chain FAs (from 10:0 to 14:0) and in oleic acid (*cis*-9 18:1). By contrast, diets rich in stearic acid (18:0) enhance the concentration of such FA

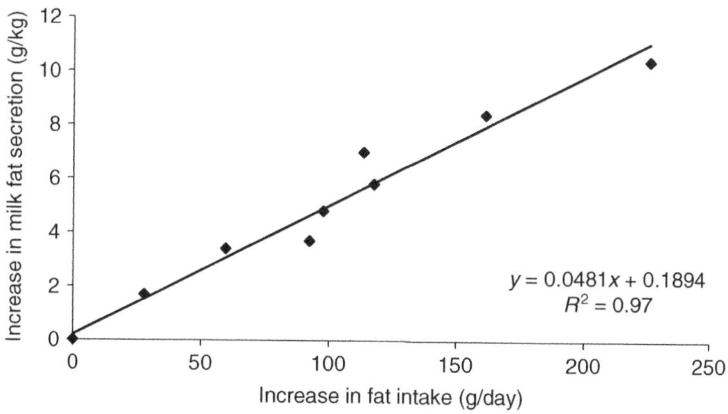

Fig. 3.7. Relationship between supplemented fat and milk fat secretion. Data represent means of dietary treatments and are referred to early lactation. (Data sources for dietary treatments: Lu, 1993; Teh *et al.*, 1994; Brown-Crowder *et al.*, 2001; Mele *et al.*, 2005; Nudda *et al.*, 2006; Andrade and Schmidely, 2006b.)

and of oleic acid, by the action of Δ^9-desaturase in the mammary gland and decrease the content of medium-chain FAs.

In general, when plant oils or oilseeds are fed to ruminants, the PUFAs contained in the supplement are quickly hydrolysed in the rumen from their glycerides (Chilliard *et al.*, 2002b; Antongiovanni *et al.*, 2003), unless they are rumen-protected, e.g. by encapsulation (McDonald and Scott, 1977). The lower the protection, the higher is the increase in the levels of 18:0 and 18:1, at the expense of short- and medium-chain FAs (Sanz-Sampelayo *et al.*, 2007). Some effects of protected and unprotected lipid sources in the diet (such as oil or raw or heat-treated oilseeds) on the FA composition of goat's milk are illustrated in Tables 3.3 and 3.4. It can be seen that supplementation with protected oils increases 18:1, 18:2 and 18:3 in milk fat, depending on the content of these FAs in the lipid source. In the absence of protection, 18:0 and 18:1 are particularly enhanced, due to rumen biohydrogenation of PUFAs (Gulati *et al.*, 1997; Mir *et al.*, 1999). With protected soybean oil, the increase in LA is also accompanied by an increase in 18:1 and 18:0, probably due to an incomplete protection of the oil. An efficient protection of plant oils supplemented to the diet enhances the absorption of PUFAs by the mammary gland from plasma.

As reported earlier in this chapter, PUFAs strongly limit the *ex novo* synthesis of short- and medium-chain FAs. Therefore, the modification of the milk FA profile is related not only to an increase in PUFAs and LCFAs, but also to a decrease in the saturated and short- and medium-chain FAs. The supplementation with protected cottonseeds increases the 18:2*n*–6 content and the 18:0 to 18:1 ratio in milk fat, although this supplement is poor in stearic acid and contains about 16% of 18:1. This fact may be due to the presence of cyclopropenoid acids in cottonseeds, which are strong inhibitors of the endomammary Δ^9-desaturase enzyme which catalyses the desaturation reaction of stearic acid to oleic acid.

Table 3.3. Effects of different lipid supplements supplied as oil on the fatty acid composition of milk (g/100 g lipids). Effects are expressed as the differences between fat-supplemented and non-supplemented control groups.

	Saturated fatty acids[a]		Soybean oil		Linseed oil	
Fatty acid	16:0	18:0	Protected[a]	Unprotected[b]	Protected[c]	Unprotected[d]
4:0–8:0	+1.5	+0.5	+1.4	–0.91	–0.3	–0.5
10:0–14:0	–4.4	–4.2	+1.9	–4.83	–7.4	–8.6
16:0	+6.4	–6.2	–8.7	–4.15	–7.7	–8.9
16:1	+2.4	–2.0	ND	–0.1	–0.3	ND
18:0	0.0	+7.0	+1.1	+1.5	+6.6	+4.8
18:1[e]	–2.0	+6.6	+2.5	+10.5	+7.6	+5.5
18:2	–0.5	–0.6	+4.4	+0.5	–0.1	0
18:3	+0.3	–0.6	ND	–0.1	+1.33	+1.27
cis-9, *trans*-11 CLA	ND	ND	ND	+3.2	+0.3	+0.82

CLA, conjugated linoleic acid; ND, not determined.
[a]Adapted from Chilliard *et al.* (2003).
[b]Adapted from Mele *et al.* (unpublished data).
[c]Adapted from Bernard *et al.* (2005).
[d]Adapted from Chilliard and Ferlay (2004).
[e]Sum of *cis* and *trans* isomers.

Usually, PUFAs, CLA and VA of milk fat are increased more by feeding unprotected plant oils than by feeding seeds of oil plants, whereas extruded or heat-treated seeds cause an intermediate effect (Martin *et al.*, 2004; Nudda *et al.*, 2006). When whole seeds are used, their oil becomes available in the rumen more gradually than when the unprotected oil is directly supplied. This leads to a more efficient reduction of PUFAs. Therefore, in this case, a higher amount of 18:0, which is the terminal product of the rumen biohydrogenation of PUFAs, is accumulated in the rumen. Even though such results were initially reported for dairy cows only, they have recently been confirmed for lactating goats supplemented with oil plant seeds, e.g. linseeds or sunflower seeds, or with their unprotected oils (Chilliard *et al.*, 2003). Milk fat content was increased (from 3 to 6 g/kg) by all treatments, while milk content of LA, LNA, VA and RA was markedly increased when goats were fed free oils rather than whole seeds. When lupin was added to the diet, the transfer of PUFAs to milk was less efficient compared with other whole seeds such as soybean or linseed. This could be due to the presence of a compound in lupin which results in a high degree of biohydrogenation (Chilliard *et al.*, 2003).

Forage source and lipid supplementation

The effects of vegetable oil supplemented to the diet may also depend on the forage base of the diet. For example, adding sunflower oil with high oleic acid content to a grass hay-based diet leads to an FA profile of milk different from that

Table 3.4. Effects of different lipid supplements supplied as oilseeds on the fatty acid composition of milk (g/100 g lipids). Effects are expressed as the differences between fat-supplemented and non-supplemented control groups.

Fatty acid	Rapeseed[a]		Cottonseeds[a]		Extruded linseed cake[b]	Rolled linseed[c]	Extruded soybean[d]
	Protected	Unprotected	Protected	Unprotected			
4:0–8:0	ND	ND	ND	–0.1	0	+0.3	ND
10:0–14:0	–0.6	–2.0	–2.1	–0.6	–2.0	–4.8	–5.8
16:0	–8.2	–4.1	–2.0	–4.3	–8.4	–10.2	–10.4
16:1	ND	ND	ND	ND	–0.2	–0.2	–0.1
18:0	–3.9	+5.8	+6.8	+6.6	+3.1	+5.0	+7.2
18:1[e]	+7.9	+6.9	–11.3	+4.0	+5.9	+10.6	+5.3
18:2	+5.9	+1.1	+14.3	+0.2	0	–0.4	+1.7
18:3	+3.1	+0.5	0.0	+0.2	+0.5	+0.3	0
Cis-9,*trans*-11 CLA	ND	ND	ND	ND	+0.5	+0.9	ND

CLA, conjugated linoleic acid; ND, not determined.
[a]Adapted from Chilliard *et al.* (2003).
[b]Adapted from Nudda *et al.* (2006).
[c]Adapted from Andrade and Schmidely (2006b).
[d]Adapted from Schmidely *et al.* (2005).
[e]Sum of *cis* and *trans* isomers.

obtained with the same supplement added to a maize silage-based diet. In the first case, oleic acid and RA in milk fat increase the most, whereas in the latter case *trans*-10 18:1, stearic acid and butyric acids show the highest increase. Similarly, linseed oil in association with maize silage increases the synthesis of 4:0 and *trans*-10 18:1, while when associated with lucerne hay it increases RA and VA in milk fat (Chilliard and Ferlay, 2004). In general, the use of linseed oil increases the *trans* forms of MUFAs and PUFAs with 18 carbon atoms. However, due to the interaction between this supplement and the forage base, the proportions between the different isomers change depending on the rumen metabolic pathways favoured by the different diets. Unlike the situation in dairy cows, few data are available on the influence of feeding on the various CLA isomers in goat's milk, whereas more information is available on goat's milk content of *trans* isomers of 18:1 (Alonso *et al.*, 1999; LeDoux *et al.*, 2002). In general RA is the main CLA isomer in milk, because of its mammary synthesis by Δ^9-desaturase. In addition, this enzyme also synthesizes *trans*-7,*cis*-9 CLA, which is quantitatively the second isomer present in milk. In general, high levels of RA in milk fat are accompanied by high levels of VA and of other *trans* isomers of 18:1 and conjugated or non-conjugated 18:2 and 18:3 (Chilliard and Ferlay, 2004).

The interaction between the forage/concentrate ratio of the diet and the kind of supplemented fat may also affect milk FA composition in dairy goats. Recently, Andrade and Schmidely (2006b) reported that feeding a high-concentrate diet combined with rolled canola seeds had a synergistic effect, which increased the proportion of *trans* 18:1, LNA and RA in milk fat at the expense of medium-chain FAs. In addition, this combination increased milk production and milk fat yield, without changing milk protein content. Similarly, when dairy goats were fed high- or low-concentrate diets, with or without supplementation of unprotected soybean oil (4% on DM basis), Mele *et al.* (2005) found no detrimental effect of soybean oil on milk fat yield and content. Moreover, soybean oil supplementation increased VA and RA milk content, regardless of the level of concentrate in the diet. These findings confirm that the influence of nutrition on milk fat content and secretion largely differs among ruminant species. In dairy cows, milk fat content is decreased by high-concentrate diets, especially if supplemented with unprotected PUFA-rich vegetable oils. By contrast, when dairy goats are fed vegetable oils, even added to low-fibre diets, milk fat content and yield almost always increase, and never decrease (Chilliard *et al.*, 2005; Mele *et al.*, 2005; Andrade and Schmidely, 2006b). According to the biohydrogenation theory of Bauman and Griinari (2001), MFD is caused by an increase of *trans*-10 18:1 and *trans*-10,*cis*-12 CLA, which sharply reduces mammary lipid secretion. In dairy goats, the lack of correlation between milk fat content and several CLA isomers, including *trans*-10,*cis*-12 CLA, could be related either to the fact that *trans*-10,*cis*-12 CLA did not increase above trace levels in milk, even when *trans*-10 18:1 increased, or to the failure of *trans*-10,*cis*-12 CLA to inhibit milk fat synthesis even when infused in the duodenum (Andrade and Schmidely, 2006a).

In order to increase *n*–3 FAs, such as eicosapentaenoic (EPA, 20:5) or docosahexaenoic (DHA, 22:5), in goat's milk, fish oil is commonly supplemented in the diet. The introduction of such lipid sources in protected form in the diet of lactating goats, at 3% of DM intake, allows one to: (i) enrich the milk in *n*–3 FAs; and (ii) avoid undesirable metabolic effects, such as reduced DM intake and productivity, and formation of *trans* and hydroxyl FAs. In fact, the transfer rates of EPA and DHA were higher when fish oil was offered in the protected form (7.62 and 5.05% for EPA and DHA, respectively) than when offered without any protection (2.50 and 3.53% for EPA and DHA, respectively) (Kitessa *et al.*, 2001).

Green herbage in the ration is a source of fat substances capable of typically characterizing milk fat. For example, it is well known that green herbage increases *cis*-9,*trans*-11 CLA in ruminant milk, due to the large amount of LNA present in the green plant, compared with hay. This is because the oxidation processes following the cutting, drying and storage of hay significantly reduce its fraction of PUFAs. Nevertheless, the increment rate of CLA in grazing animals varies with ruminant species and pasture quality. Differences in milk CLA content have been observed among animals of different species fed the same pasture (Nudda *et al.*, 2003) and among animals of the same species fed pastures with different botanical composition (Mele *et al.*, 2007b).

Conclusions

The control of milk fat yield and composition is critical for the improvement of the technological and nutritional quality of milk from dairy goats.

Similar to what is observed in dairy ewes, the energy balance largely affects milk fat concentration in goats. Lipid supplementation can improve milk fat secretion, especially at the beginning of lactation, with beneficial effects on the yield efficiency of goat's cheese.

Peculiarities of goat's milk FA composition play an important role in the nutritional quality of milk. Dietary lipid supplementation can improve the milk fat content of some FAs with nutraceutical effects.

Further research is needed on genetic, physiological and nutritional factors regulating lipid metabolism, in order to better understand some peculiar aspects of milk fat composition of dairy goats.

References

Alonso, L., Fontheca, J., Lozada, L., Fraga, M.J. and Juárez, M. (1999) Fatty acid composition of caprine milk: major, branched-chain, and *trans* fatty acids. *Journal of Dairy Science* 82, 878–884.

Andrade, P.V.D. and Schmidely, P. (2006a) Effect of duodenal infusion of *trans*-10, *cis*-12 CLA on milk performance and milk fatty acid profile in dairy goats fed high or low concentrate diet in combination with rolled canola seed. *Reproduction, Nutrition, Development* 46, 31–48.

Andrade, P.V.D. and Schmidely, P. (2006b) Influence of percentage of concentrate in combination with rolled canola seeds on performance, rumen fermentation and milk fatty acid composition in dairy goats. *Livestock Science* 104, 77–90.

Antongiovanni, M., Buccioni, A., Petacchi, F., Secchiari, P., Mele, M. and Serra, A. (2003) Upgrading the lipid fraction of foods of animal origin by dietary means: rumen activity and presence of *trans* fatty acids and CLA in milk and meat. *Italian Journal of Animal Science* 2, 3–28.

Baldwin, R.L., Smith, N.E., Taylor, J. and Sharp, M. (1980) Manipulating metabolic parameters to improve growth rate and milk secretion. *Journal of Animal Science* 51, 1416–1428.

Barber, M.C., Clegg, R.A., Travers, M.T. and Vernon, R.G. (1997) Lipid metabolism in the lactating mammary gland. *Biochimica et Biophysica Acta* 1347, 101–126.

Bauchart, D., Legay-Carmier, F., Doreau, M. and Gaillard, B. (1990) Lipid metabolism of liquid-associated and solid-adherent bacteria in rumen contents of dairy cows offered lipid-supplemented diets. *British Journal of Nutrition* 63, 563–578.

Bauman, D.E. and Griinari, J.M. (2001) Regulation and nutritional manipulation of milk fat: low fat syndrome. *Livestock Production Science* 70, 15–29.

Baumgard, L.H., Corl, B.A., Dwyer, D.A., Saebo, A. and Bauman, D.E. (2000) Identification of the conjugated linoleic acid isomer that inhibits milk fat synthesis. *American Journal of Physiology. Regulatory, Integrative and Comparative Physiology* 278, R179–R184.

Bernard, L., Leroux, C., Hayes, H., Gautier, M., Chilliard, Y. and Martin, P. (2001) Characterization of the caprine stearoyl-CoA desaturase gene and its mRNA showing an unusually long 3′-UTR sequence arising from a single exon. *Gene* 281, 53–61.

Bernard, L., Rouel, J., Leroux, C., Ferlay, A., Faulconnier, Y., Legrand, P. and Chilliard, Y. (2005) Mammary lipid metabolism and milk fatty acid secretion in alpine goats fed vegetable lipids. *Journal of Dairy Science* 88, 1478–1489.

Boggs, D.L., Bergen, W.G. and Hawkins, D.R. (1987) Effects of tallow supplementation and protein withdrawal on ruminal fermentation, microbial synthesis and site of digestion. *Journal of Animal Science* 64, 970–979.

Brown-Crowder, I.E., Hart, S.P., Cameron, M., Dsahlu, T. and Goetsch, A.L. (2001) Effects of dietary tallow level on performance of Alpine does in early lactation. *Small Ruminant Research* 39, 233–241.

Cerbulis, J., Parks, O.W. and Farrell, J.R. (1982) Composition and distribution of lipid of goats' milk. *Journal of Dairy Science* 65, 2301–2307.

Chalupa, W., Rickabaugh, B., Kronfeld, D.S. and Sklan, D. (1984) Rumen fermentation *in vitro* as influenced by long chain fatty acids. *Journal of Dairy Science* 67, 1439–1448.

Chilliard, Y. and Bocquier, F. (1993) Effect of fat supplementation on milk yield and composition in dairy goats and ewes. In: *Proceedings of the 5th International Symposium 'La qualità delle produzioni dei piccoli ruminanti'*. Camera Commercio Industria Artigianato Agricoltura di Varese, Varese, Italy, pp. 61–68.

Chilliard, Y. and Ferlay, A. (2004) Dietary lipids and forages interactions on cow and goat milk fatty acid composition and sensory properties. *Reproduction, Nutrition, Development* 44, 467–492.

Chilliard, Y., Gagliostro, G., Flechet, J., Lefaivre, J. and Sebastian, I. (1991) Duodenal rapeseed oil infusion in early and mid-lactation cows. 5. Milk fatty acids and adipose tissue lipogenic activities. *Journal of Dairy Science* 74, 1844–1854.

Chilliard, Y., Ferlay, A., Mansbridge, R.M. and Doreau, M. (2000) Ruminant milk fat plasticity: nutritional control of saturated, polyunsaturated, *trans* and conjugated fatty acids. *Annales de Zootechnie* 49, 161–192.

Chilliard, Y., Ferlay, A. and Doreau, M. (2001) Effect of different types of forages, animal fat or marine oils in cow's diets on milk fat secretion and composition, especially conjugated linoleic acid (CLA) and polyunsaturated fatty acids. *Livestock Production Science* 70, 30–48.

Chilliard, Y., Chabosseau, J.M., Rouel, J., Capitan, P., Gominard, C., Gaborit, P., Juaneda, P. and Ferlay, A. (2002a) Interactions between forage nature and sunflower or linseed oil supplementation on goat milk fatty acids of interest for human nutrition. In: Durans, J.L., Emile, J.C., Huyghe, C. and Lemaire, G. (eds) *Proceedings of 19th General Meeting of the European Grassland Federation*. EGF, La Rochelle, France, pp. 548–549.

Chilliard, Y., Ferlay, A., Loor, J., Rouel, J. and Martin, B. (2002b) *Trans* and conjugated fatty acids in milk from cows and goats consuming pasture or receiving vegetable oil or seeds. *Italian Journal of Animal Science* 1, 243–254.

Chilliard, Y., Ferlay, A., Rouel, J. and Lamberet, G. (2003) A review of nutritional and physiological factors affecting goat milk lipid synthesis and lipolysis. *Journal of Dairy Science* 86, 1751–1770.

Chilliard, Y., Rouel, J., Guillouet, P., Raynal-Ljutovac, K., Leloutre, L. and Ferlay, A. (2005) Kinetics of goat milk fatty acids according to dietary forage:concentrate and/or high doses of linseed oil or sunflower oil or extruded mixture of seeds. In: *Proceedings of the 56th Annual Meeting of the European Association for Animal Production*. Wageningen Academic Publisher, Wageningen, The Netherlands, p. 268.

Chilliard, Y., Rouel, J. and Leroux, C. (2006) Goat's α-s1 casein genotype influences its milk fatty acid composition and Δ-9 desaturation ratios. *Animal Feed Science and Technology* 131, 474–487.

Corl, B.A., Chouinard, P.Y., Bauman, D.E., Dwyer, D.A., Griinari, J.M. and Nurmela, K.V. (1998) Conjugated linoleic acid in milk fat of dairy cows originate in part by endogenous synthesis from *trans*-11 octadecenoic acid. *Journal of Dairy Science* 82, 2737–2745.

Czerkawski, J.W. (1984) Microbial fermentation in the rumen. *Proceedings of the Nutrition Society* 43, 101–118.

Devillard, E., McIntosh, F.M., Newbold, C.J. and Wallace, R.J. (2006) Rumen ciliate protozoa contain high concentrations of conjugated linoleic acid and vaccenic acid, yet do not hydrogenate linoleic acid or desaturate stearic acid. *British Journal of Nutrition* 96, 697–704.

Fay, J.P., Jakober, K.D., Cheng, K.J. and Costerton, J.W. (1990) Esterase activity of pure cultures of rumen bacteria as expressed by the hydrolysis of *p*-nitrophenylpalmitate. *Canadian Journal of Microbiology* 36, 585–589.

Fontecha, J., Rios, J.J., Lozada, L., Fraga, M.J. and Juarez, M. (2000) Composition of goat's milk fat triglycerides analysed by silver ion adsorption-TLC and GC-MS. *International Dairy Journal* 10, 119–128.

Fukuda, S., Furuya, H., Suzuki, Y., Asanuma, N. and Hino, T. (2005) A new strain of *Butyrivibrio fibrisolvens* that has high ability to isomerize linoleic acid to conjugated linoleic acid. *Journal of General and Applied Microbiology* 51, 105–113.

Fukuda, S., Suzuki, Y., Asanuma, N. and Hino, T. (2006) Augmentation of vaccinate production and suppression of vaccinate biohydrogenation in cultures of mixed ruminal microbes. *Journal of Dairy Science* 89, 1043–1051.

Griinari, J.M. and Bauman, D.E. (1999) Biosynthesis of conjugated linoleic acid and its incorporation into meat and milk in ruminants. In: Yurawecz, M.P., Mossoba, M.M., Kramer, J.K.G., Pariza, M.W. and Nelson, G.J. (eds) *Advances in Conjugated Linoleic Acid Research*, Vol. 1. AOCS Press, Champaign, Illinois, pp. 180–200.

Griinari, J.M., Dwyer, D.A., McGuire, M.A., Bauman, D.E., Palmquist, D.L. and Nurmela, K.V.V. (1998) *Trans*-octadecenoic acids and milk fat depression in lactating dairy cows. *Journal of Dairy Science* 81, 1251–1261.

Gulati, S.K., Byers, E.B., Byers, Y.G., Ashes, J.R. and Scott, T.W. (1997) Effect of feeding different fat supplements on the fatty acid composition of goat milk. *Animal Feed Science and Technology* 66, 159–164.

Ha, J.K. and Lindsay, R.C. (1993) Release of volatile branched-chain and other fatty acids from ruminal milk fats by various lipases. *Journal of Dairy Science* 76, 677–690.

Hadjipanayiotou, M. (1999) Feeding ensiled crude olive cake to lactating Chios ewes, Damascus goats and Friesian cows. *Livestock Production Science* 59, 61–66.

Harfoot, C.G. (1978) Lipid metabolism in the rumen. *Progress in Lipid Research* 17, 21–54.

Harfoot, C.G. and Hazlewood, G.P. (1988) Lipid metabolism in the rumen. In: Hobson, P.N. (ed.) *The Rumen Microbiology Ecosystem*. Elsevier, London, pp. 285–322.

Harfoot, C.G., Noble, R.C. and Moore, J.H. (1973) Factors influencing the extent of biohydrogenation of linoleic acid by rumen micro-organisms *in vitro*. *Journal of the Science of Food and Agriculture* 24, 961–970.

Ikwuegbu, O.A. and Sutton, J.D. (1982) The effect of varying the amount of linseed oil supplementation on rumen metabolism in sheep. *British Journal of Nutrition* 48, 365–375.

Jahreis, G., Fritsche, J. and Kraft, J. (1999) Species-dependent, seasonal, and dietary variation of conjugated linoleic acid in milk. In: Yurawecz, M.P., Mossoba, M.M., Kramer, J.K.G., Pariza, M.W. and Nelson, G.J. (eds) *Advances in Conjugated Linoleic Acid Research*, Vol. 1. AOCS Press, Champaign, Illinois, pp. 215–225.

Jenkins, T.C. (1993) Lipid metabolism in the rumen. *Journal of Dairy Science* 76, 3851–3863.

Jenkins, T.C. (1994) Regulation of lipid metabolism in the rumen. *Journal of Nutrition* 124, 1372 S–1376 S.

Jenkins, T.C. and Palmquist, D.L. (1984) Effect of fatty acids and calcium soaps on rumen and total nutrient digestibility of dairy rations. *Journal of Dairy Science* 67, 978–984.

Jesse, B.W., Solomon, R.K. and Baldwin, R.L. (1992) Palmitate metabolism by isolated sheep rumen epithelial cells. *Journal of Animal Science* 70, 2235–2242.

Kepler, C.R., Hirons, K.I., McNeill, P.H. and Tove, S.B. (1967) Intermediates and products of biohydrogenation of linoleic acid by *Butyrivibrio fibrisolvens. Journal of Biological Chemistry* 241, 1350–1354.

Kinsella, J.E. (1972) Stearoyl-CoA as a precursor of oleic acid and glycerolipids in mammary microsomes from lactating bovine: possible regulatory step in milk triglyceride synthesis. *Lipids* 7, 349–355.

Kitessa, S.M., Gulati, S.K., Ashes, J.R., Fleck, E., Scott, T.W. and Nichols, P.D. (2001) Utilization of fish oil in ruminants II. Transfer of fish oil fatty acids into goats' milk. *Animal Feed Science and Technology* 89, 201–208.

LeDoux, M., Rouzeau, A., Bas, P. and Sauvant, D. (2002) Occurrence of trans-C18:1 fatty acid isomers in goat milk: effect of two dietary regimens. *Journal of Dairy Science* 85, 190–197.

Lock, A.L., Teles, B.M., Perfield, J.W. II, Bauman, D.E. and Sinclair, L.A. (2006) A conjugated linoleic acid supplement containing *trans*-10, *cis*-12 reduces milk fat synthesis in lactating sheep. *Journal of Dairy Science* 89, 1525–1532.

Loor, J.J., Ferlay, A., Ollier, A., Doreau, M. and Chilliard, Y. (2005) Relationship among *trans* and conjugated fatty acids and bovine milk fat yield due to dietary concentrate and linseed oil. *Journal of Dairy Science* 88, 726–740.

Lu, C.D. (1993) Implication of feeding isoenergetic diets containing animal fat on milk composition of Alpine does during early lactation. *Journal of Dairy Science* 76, 1137–1147.

McDonald, I.W. and Scott, T.W. (1977) Food of ruminant origin with elevated content of polyunsaturated fatty acids. *World Review of Nutrition and Dietetics* 26, 144–207.

Martin, B., Fedele, V., Ferlay, A., Groleir, P., Rock, E., Gruffat, D. and Chilliard, Y. (2004) Effects of grass based diets on the content of micronutrients and fatty acids in bovine and caprine dairy products. In: Luscher, A., Jeangros, B., Kessler, W., Huguenin, O., Lobsiger, M., Millar, N. and Suter D. (eds) *Proceedings of the 20th General Meeting of the European Grassland Federation*. EGF, Lucerne, Switzerland, pp. 876–886.

Massart-Leen, A.M., De Pooter, H., Decloedt, M. and Schamp, N. (1981) Composition and variability of the branched-chain fatty acid fraction in the milk of goats and cows. *Lipids* 16, 286–292.

Mele, M., Serra, A., Rafanelli, M.R., Conte, G. and Secchiari, P. (2005) Effect of forage: concentrate ratio and soybean oil supplementation on milk yield and quality from dairy goats. *Italian Journal of Animal Science* 4(Suppl. 2), 392–394.

Mele, M., Conte, G., Serra, A., Buccioni, A. and Secchiari, P. (2007a) Relationship between β-lactoglobulin polymorphism and milk fatty acid composition in milk of Massese dairy ewes. *Small Ruminant Research* 73, 37–44.

Mele, M., Macciotta, N.P.P., Serra, A., Pollicardo, A. and Secchiari, P. (2007b) Fatty acid composition of milk and cheese from sheep fed rough or cultivated pasture. In: *Proceedings of the 5th International Symposium on the Challenge to Sheep and Goats Milk Sectors*, Alghero, Italy. International Dairy Federation, Brussels, in press.

Mir, Z., Goonewardene, L.A., Okine, E., Jaegar, S. and Scheer, H.D. (1999) Effect of feeding canola oil on constituents, conjugated linoleic acid and long chain fatty acids in goat milk. *Small Ruminant Research* 33, 137–143.

Moore, J.H., Noble, R.C., Steele, W. and Czerkawski, J.W. (1969) Differences in the metabolism of esterified and unesterified linoleic acid by rumen micro-organisms. *British Journal of Nutrition* 23, 869–878.

Morand-Fehr, P., Bas, P., Sauvant, D., Hervieu, G. and Chilliard, Y. (1984a) Utilisation des matières grasses chez les ruminants. Expériences sur chèvres laitières. In: *Peut-on et comment utiliser le matières grasses dans les rations des vaches laitières.* ADEPRINA, Paris, pp. D1–D21.

Morand-Fehr, P., Sauvant, D. and Bas, P. (1984b) Effet de la fraction grasse des graines de soja sur le performances des chèvres laitières. In: *Proceedings of the 35th Annual Meeting of the European Association for Animal Production*. Wageningen Press, The Hague, The Netherlands, p. 323.

Ntambi, J.M. (1995) The regulation of stearoyl-Coa desaturase (SCD). *Progress in Lipid Research* 34, 139–150.

Nudda, A., Mele, M., Battacone, G., Usai, M.G. and Macciotta, N.P.P. (2003) Comparison of conjugated linoleic acid (CLA) content in milk of ewes and goats with the same dietary regimen. *Italian Journal of Animal Science* 2(Suppl. 1), 515–517.

Nudda, A., Battacone, G., Usai, M.G., Fancellu, S. and Pulina, G. (2006) Supplementation with extruded linseed cake affects concentrations of conjugated linoleic acid and vaccenic acid in goat milk. *Journal of Dairy Science* 89, 277–282.

Palmquist, D.L. and Jenkins, T.C. (1980) Fat in lactation rations: review. *Journal of Dairy Science* 63, 1–14.

Patton, S.C., Stemberger, B.H. and Knudson, C.M. (1977) The suppression of milk fat globule secretion by colchicine: an effect coupled to inhibition of exocytosis. *Biochimica et Biophysica Acta* 499, 404–407.

Pulina, G., Nudda, A., Battacone, G. and Cannas, A. (2006) Effects of nutrition on the content of fat, protein, somatic cells, aromatic compounds, and undesirable substances in sheep milk. *Animal Feed Science and Technology* 131, 255–291.

Ramos, M. and Juarez, M. (1986) Chromatographic, electrophoretic and immunological methods for detecting mixtures of milks from different species. In: *IDF Document 202*. International Dairy Federation, Brussels, pp. 175–187.

Rapetti, L., Provetto, G.M., Galassi, G., Sandrucci, A., Succi, G., Tamburini, A. and Battelli, G. (2002) Effect of maize, rumen-protected fat and whey permeate on energy utilization and milk fat composition in lactating goats. *Italian Journal of Animal Science* 1, 43–53.

Rovai, M., Lock, A., Gipson, T., Goetsch, A. and Bauman, D. (2007) A conjugated linoleic acid (CLA) supplement containing *trans*-10, *cis*-12 CLA reduces milk fat synthesis in lactating goats. *Italian Journal of Animal Science* 6(Suppl. 1), 629–631.

Sanz-Sampelayo, M.R., Perez, L., Martin, B., Alonso, J.J., Amigo, L. and Boza, J. (2002) Effects of concentrates with different contents of protected fat rich in PUFAs on the performance lactating Granadina goats. Part II. Milk production and composition. *Small Ruminant Research* 43, 141–148.

Sanz-Sampelayo, M.R., Chilliard, Y., Schmidely, P. and Boza, J. (2007) Influence of type of diet on the fat constituents of goat and sheep milk. *Small Ruminant Research* 68, 42–63.

Schmidely, P., Morand-Fehr, P. and Sauvant, D. (2005) Influence of extruded soybeans with or without bicarbonate on milk performance and fatty acid composition of goat milk. *Journal of Dairy Science* 88, 757–765.

Storry, J.E., Brumby, P.E., Hall, A.J. and Johnson, V.W. (1974) Responses in rumen fermentation and milk fat secretion in cows receiving low-roughage diets supplemented protected tallow. *Journal of Dairy Research* 41, 165–172.

Taniguchi, M., Utsugi, T., Oyama, K., Mannen, H., Kobayashi, M., Tanabe, Y., Ogino, A. and Tsuji, S. (2003) Genotype of stearoyl-CoA desaturase is associated with fatty acid composition in Japanese black cattle. *Mammalian Genome* 14, 142–148.

Teh, T.H., Trung, L.T., Jia, Z.H. and Gipson, T.A. (1994) Varying amounts of rumen inert fat for high producing goats in early lactation. *Journal of Dairy Science* 77, 253–258.

Ulberth, F. and Henninger, M. (1994) Quantification of *trans* fatty acids in milk fat using spectroscopic and chromatographic methods. *Journal of Dairy Research* 61, 517–527.

Viviani, R. (1970). Metabolism of long chain fatty acids in the rumen. *Advances in Lipid Research* 8, 267–273.

Vlaeminck, B., Fievez, V., Demeyer, D. and Dewhurst, R.J. (2006a) Effect of forage: concentrate ratio on fatty acid composition of rumen bacteria isolated from ruminal and duodenal digesta. *Journal of Dairy Science* 89, 2668–2678.

Vlaeminck, B., Fievez, V., Cabrita, A.R.J., Fonseca, A.J.M. and Dewhurst, R.J. (2006b) Factors affecting odd- and branched-chain fatty acids in milk: a review. *Animal Feed Science and Technology* 131, 389–417.

Ward, R.J., Travers, M.T., Richards, S.E., Vernon, R.G., Salter, A.M., Buttery, P.J. and Barber, M.C. (1998) Stearoyl-CoA desaturase mRNA is described from a single gene in the ovine genome. *Biochimica et Biophysica Acta* 1391, 145–156.

Yurawecz, M.P., Roach, J.A.G., Sehat, N., Mossoba, M.M., Kramer, J.K.G., Fritsche, J., Steinhart, H. and Ku, Y. (1998) A new conjugated linoleic acid isomer, 7-*trans*, 9-*cis*, octadecadienoic acid, in cow milk, cheese, beef and human milk and adipose tissue. *Lipids* 33, 803–809.

4 Protein Components of Goat's Milk

G.F. GREPPI[1,2], P. RONCADA[2] AND R. FORTIN[1]

[1]*Dipartimento di Scienze Cliniche Veterinarie, Facoltà di Medicina Veterinaria, Università di Milano, Via Celoria 10, 20133 Milan, Italy;*
[2]*Istituto Sperimentale Italiano 'Lazzaro Spallanzani', Località La Quercia, 260207 Rivotta D'Adda, Italy*

Introduction

Milk is a complex biological fluid produced by the mammary gland which has three main functions: (i) nutritional; (ii) immunological; and (iii) physiological. Recently, the physiological role of milk has been emphasized in research, and consumer demand for high-quality products has increased. Consequently, the transformation industry has started to market 'high-quality milk' with particular nutritional/dietary characteristics. In industrialized countries, milk proteins represent 30% of total dietary protein intake and 75% of Ca intake (Meisel, 1993a,b, 1997; Morand-Fehr, 1996; Reinert and Fabre, 1996; Spuergin *et al.*, 1997; Restani *et al.*, 1999).

Several epidemiological studies have focused on the influence of cow's milk consumption on health, especially on cancer development. However, available data are contradictory, due to the complex interactions among milk components. Nevertheless, a positive linear relationship between cancer and energy intake or fat-rich milk consumption has been suggested (Attaie and Richter, 2000).

Milk contains lipid components which are well-known anticarcinogenic agents (Parodi, 1997). According to the World Cancer Research Fund, the lipid components of milk and derived products increase the risk of two types of cancer only (kidney and prostate) (WCRF, 1997).

In an early study on the use of goat's and ass's milk, high infant mortality occurred due to the bad microbiological characteristics of the goat's milk employed (Michel, 1910). As a consequence, for more than half a century goat's milk was not recommended for human diets. However, since the 1980s several other studies and epidemiological research have highlighted the nutritional properties and the presence of minor components of pharmacological interest in goat's milk, thus recommending its usage in human feeding (Domeniconi and Balzola, 1980; Birkbeck, 1984; Parry, 1984; Darnton Hill *et al.*, 1987; Awiatlo *et al.*, 1990; Park, 1991; Houdebine, 1995; MacDonald, 1995; Albertini *et al.*, 1999).

The US Food and Drug Administration has just concluded that food from cattle, swine and goat clones and their progenies is as safe to eat as food from animals bred conventionally (Rudenko and Matheson, 2007). This decision on food safety will certainly be greatly discussed in the very near future.

Biochemical Characteristics

Milk characteristics can be specifically associated with the environment and species. For example, there is a positive linear relationship between the logarithm of milk protein quantity and the body weight of the mother, which indicates that some mammalian species may produce insufficient milk for the newborn (Fig. 4.1).

Nowadays 4027 species are classified as mammals, but the milk composition of only 200 species, 60 of these with fewer than ten samples, has been studied (Jenness, 1974, 1980). High intraspecific variations in protein percentage are commonly observed, probably due to differences in genotype and/or in environmental conditions. A high number of scientific studies have demonstrated that milk protein content of 200 mammalian species can vary from 1% up to 20%. Milk is a colloidal heterogeneous system where mineral salts, lactose and some whey proteins dissolved in water represent the continuous phase, while casein micelles represent the dispersed phase. Fat is present as tiny globules in this

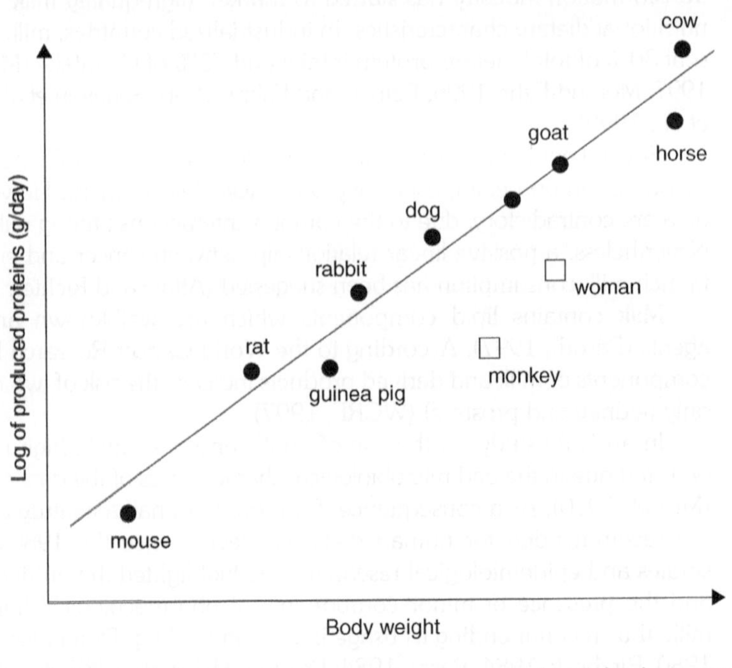

Fig. 4.1. Relationship between body weight and produced proteins in mammals (logarithmic scale).

emulsion, the membranes of which are constituted mainly by a phospholipid–protein complex. The biochemical composition and characteristics of milk vary from species to species, reflecting the newborn's nutritional needs (Jenness and Sloan, 1970; Jenness, 1979; Davis and Collier, 1983; Targowski, 1983).

For a long time, the use of milk in the diet of adults and, in particular, of infants has been studied. In comparison to the milk of other species, goat's milk has been considered a good source of proteins and amino acids (Tables 4.1–4.4).

In recent decades, milk has been considered as a message carrier to nursing infants (Teschemacher *et al.*, 1977; Teschemacher and Brantl, 1994; Teschemacher, 1995) and adults as well. In the past, very few studies indicated the presence of neurotransmitters or hormonal substances in milk (Blanc, 1982). On the other hand, in the last few years, hundreds of works have shown the presence of prolactin, melatonin, oxytocin, growth hormone, luteinizing hormone-releasing hormone, thyroid-stimulating hormone, vasoactive intestinal peptide, calcitonin, neurotensin and cholecystochinin in milk, indicating that milk can be considered an exogenous endocrine system for the newborn (Mackle and Bauman, 1998).

Table 4.1. Milk composition of different species (g/l).

	Proteins	Lipids	Lactose
Primates			
Human	9–15	38–41	70–72
Simian	16	40	70
Rodents			
Rat	81	88	38
Ruminants			
Goat	29–31	35–45	41–44
Cow	32–34	37–39	48
Cetaceans			
Dolphin	68	330	11
Whale	109	423	13

Table 4.2. Composition of human, cow's and goat's milk.

	Human milk	Cow's milk	Goat's milk
Water (%)	87.6	87.3	87.5
Dry residue (g/l)	11.7	12.5	13.6
Dust (g/l)	2.0	8.0	8.9
Total proteins (g/l)	10	34	33
Carbohydrates (g/l)	70	48	51
Total lipids (g/l)	38	37	29
Non-protein N (g/l)	3.2	2.5	3.2
Ca (mg/l)	33	125	124
P (mg/l)	15	96	105

Table 4.3. Composition of the protein fraction of human, cow's and goat's milk (g/l).

	Human milk	Cow's milk	Goat's milk
Total proteins	9–15	32–34	28–32
Caseins	2.0–2.5	26–37	22–28
αS_1-Casein	–	11–15	10
αS_2-Casein	–	3–4	3
β-Casein	1.5	9–11	11
κ-Casein	0.5	2–4	4
Whey proteins	6.3	5.8–6.5	5.5–6.5
α-Lactoalbumin	1.9–2.6	0.6–1.5	1.2
β-Lactoglobulin	–	3–4	3.1
Serum albumin	0.4	0.4	0.5
Immunoglobulin	1.1	1.0	1.0
Lactoferrin	1.7–2	0.1	0.02–0.2
Lysozyme	0.04–0.2	–	–

Table 4.4. Protein amino acid composition of human, cow's and goat's milk.

	Human milk	Cow's milk	Goat's milk
Total proteins (g/l)	9–15	32–34	28–32
Essential amino acids (mg/g protein)			
Cysteine	20	9	9
Phenylalanine	37	52	47
Isoleucine	53	64	48
Histidine	23	28	26
Leucine	104	100	96
Lysine	71	83	80
Methionine	16	27	25
Tyrosine	46	53	38
Threonine	44	51	49
Tryptophan	17	14	–
Valine	51	68	61
Non-essential amino acids (mg/g protein)			
Aspartate	86	79	75
Glutamate	190	208	209
Alanine	40	35	34
Arginine	36	37	29
Glycine	22	21	18
Proline	95	101	106
Serine	61	56	49

These hormonal substances should be considered as passengers that use milk to reach their target organ (Choick, 1998).

Knowledge of milk proteins has also increased enormously. For example, the functional meaning of some caseins and whey proteins has been identified and evidence for the importance of biopeptides (i.e. amino acid sequences in alimentary proteins) has been found. These peptides are inactive when they are present in their native protein, but they can be released in their active form after proteolytic digestion (i.e. during gastric digestion or simply during any technological treatment). Since the 1980s, many authors have described active biopeptides derived from milk proteins which, in contrast with endogenous biopeptides, have many biological functions (Yoshikawa *et al.*, 1988). Since then, the nutritional importance of milk, such as in the diet of athletes, has increased. Athletes need to consume proteins with high biological value, in order to: (i) maintain a positive N balance during physical activities; (ii) allow an increase of muscular mass; and, occasionally, (iii) repair muscular lesions after training (Fig. 4.2). In these cases, since protein needs increase from twofold to threefold with respect to normal needs, kidneys may suffer due to high protein intake. The consequences from intense physical activity can be: (i) nutritional deficiencies; (ii) formation of free radicals that could also lead to immunological deficiencies; and (iii) a lack of Fe, during aerobic stress, with consequent limited performance. Iron integration into the diet may have undesirable effects. Therefore, in all of these cases, utilization of goat's milk could be helpful (Domeniconi and Balzola, 1980).

Milk has two main protein fractions: caseins, which are predominant, and whey proteins (Law and Brown, 1994). The qualitative and quantitative content of each protein fraction depends on many factors: (i) physiological (lactation stage, lactation order); (ii) environmental (climate, hygiene); (iii) genetic (breed, genealogy); and (iv) nutritional (Polidori *et al.*, 1991). More precisely, six main proteins are present in milk: four caseins (αS_1, αS_2, β and κ) and two whey proteins (α-lactoalbumin and β-lactoglobulin) (Perez and Calvo, 1995). The protein fraction of goat's milk, like that of other domestic ruminants, is constituted mainly by caseins, which account for 80% of total proteins (Ambrosoli *et al.*, 1988).

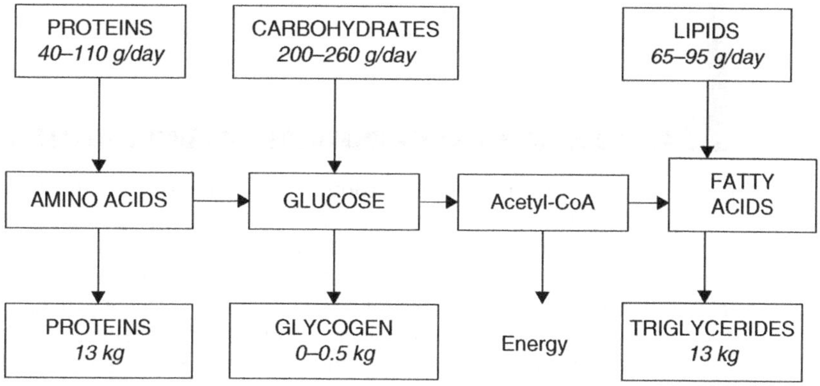

Fig. 4.2. Energy production and protein metabolism in the athlete.

It has been recently demonstrated that mRNA levels for the four caseins produced in the mammary epithelial cells during lactation are almost equally abundant (~25% each), while mature protein levels in milk differ greatly (Bevilacqua *et al.*, 2006). In particular, translational efficiency for αS_1- and β-casein is about four times higher than that for αS_2- and κ-casein, thus demonstrating that regulation of casein biosynthesis is also controlled at translational level.

The most abundant milk proteins are the three caseins (αS_1, αS_2 and β), which are sensitive to Ca. Casein biosynthesis is performed at cellular level under the influence of external agents such as hormones (Figs 4.3 and 4.4). Casein is a globular protein, present in milk as a colloidal suspension, having the aspect of micelles. It is characterized by a hydrophobic core and a charged polar hydrophilic part. The anionic regions of the polar part are responsible for sensitivity to Ca^{2+} and for some physicochemical properties of this protein (Jaubert *et al.*, 1999). Moreover, caseins are relatively hydrophobic, and under ionic conditions, such as those present in milk, they tend to associate tightly into their typical colloidal form (micelles). Micelles are constituted by a protein fraction and by a mineral component (Ca and phosphate). Caseins are a group of specific phosphoproteins in milk which precipitate: (i) when milk is acidified to a pH

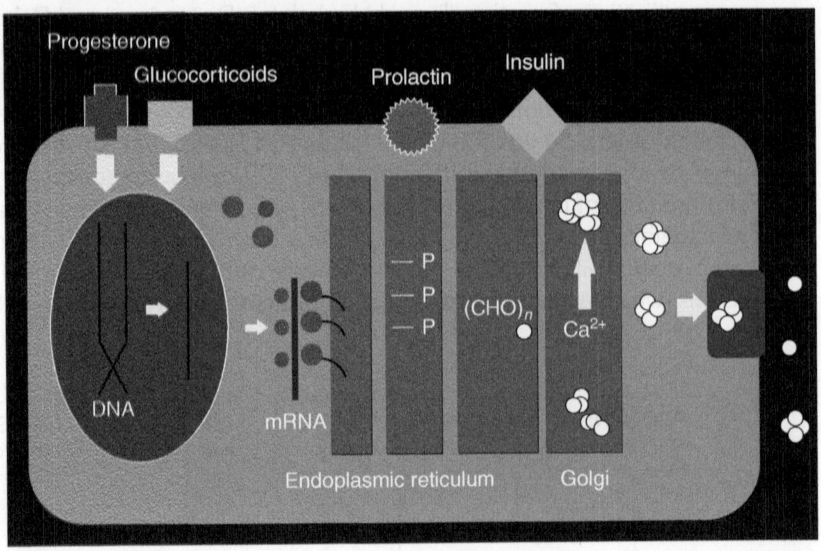

Fig. 4.3. Casein biosynthesis. Following external signalling (i.e. hormones) and signal transduction inside the nucleus, casein genes are transcribed into mRNA molecules, which undergo post-transcriptional processing once inside the cytosol. Mature mRNA molecules are translated into single casein molecules which are transferred inside the endoplasmic reticulum for post-translational modification. Casein micelle formation begins in the cis-Golgi with condensation of casein molecules; addition of Ca^{2+}, possibly in the secretory vesicle, leads to maturation of the casein micelles into particles sufficiently dense to be seen in the electron microscope.

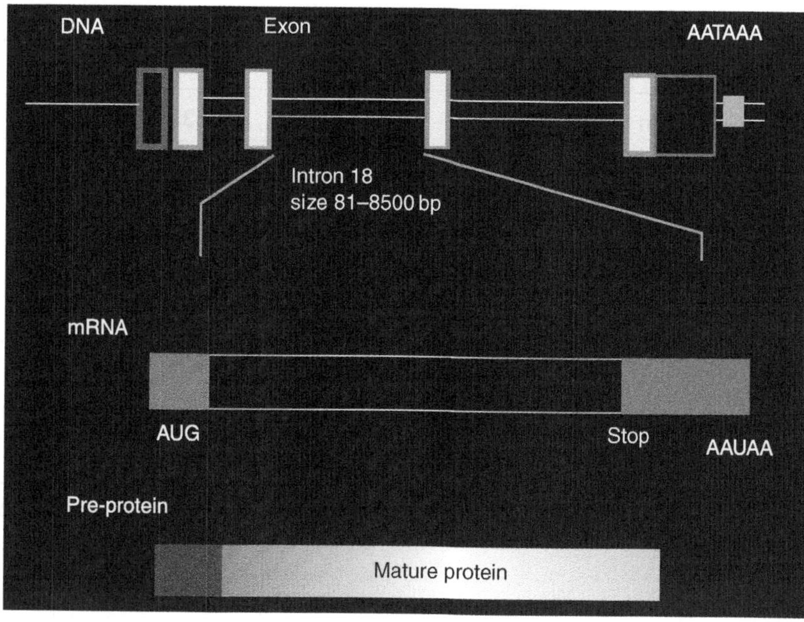

Fig. 4.4. Molecular aspects of biosynthesis.

of 4.6–4.7 at 20°C; (ii) by enzymatic treatment (chymosin); or (iii) by ultracentrifugation. Caseins are rich in proline, glutamate and glutamine, and poor in glycine and aspartate. A high amount of Ca and phosphate is required to make casein soluble in milk in a stable form, so that it becomes a nutritional source of amino acids.

A very important milk feature, especially in goat's milk, is related to its content of complex oligosaccharides, either in free form or conjugated to proteins. In fact, beneficial anti-inflammatory effects of such oligosaccharides, especially on inflammatory bowel disease, have been demonstrated (Daddaoua *et al.*, 2006).

Milk Protein Polymorphisms

Aschaffenburg and Drewry (1955) reported, for the first time, the presence of two variants of β-lactoglobulin in cow's milk. Since then, many other polymorphisms have been identified both in caseins and in whey proteins. In the past decade, molecular and biochemical characterization of protein polymorphisms has been performed and applied to the genetic improvement of animals and the improvement of milk characteristics, as described in detail in the reviews of Ng-Kwai-Hang (1997, 1998). Many protein variants were correlated with differences in casein or whey protein composition as well as in milk physicochemical properties related to caseification (Martin *et al.*, 2002).

In the last 20 years, electrophoretic methods have been widely used to identify genetic variants of proteins (Russo *et al.*, 1981, 1986; Martin, 1993; Martin *et al.*, 1999). The molecular origin of milk protein polymorphisms is the substitution

Table 4.5. Milk protein genetic variants.

Protein	Variant	Amino acid position			
αS_1-Casein		14–26	53	59	192
	A	Misses			
	B		Ala	Gln	Glu
	C			Gly	
	D		Thr		
	E		Lys	Gly	
αS_2-Casein		33–47	50–58	130	
	A	Glu	Ala	Thr	
	B				
	C	Gly	Thr	Ile	
	D	Misses			

Ala, alanine; Gln, glutamine; Glu, glutamic acid; Gly, glycine; Ile, isoleucine; Lys, lysine; Thr, threonine.

or deletion of amino acids in the protein chain (Table 4.5). Alleles associated with reduced protein synthesis have been reported by many authors (see the review of Valentine, 1998, for molecular aspects). These modifications result in changes of milk physicochemical properties such as charge, superficial hydrophobicity and molecular size and shape (Russo *et al.*, 1981, 1986; Addeo *et al.*, 1987, 1989; Grosclaude *et al.*, 1987; Dall'Olio *et al.*, 1988, 1989; Brignon *et al.*, 1989; Di Gregorio *et al.*, 1989). Such modifications have been identified through the use of electrophoretic and chromatographic techniques (Addeo *et al.*, 1988). Each protein fraction shows different genetic variants (Mercier and Vilotte, 1993). These differences in the primary structure of a protein can deeply modify its molecular characteristics. The ratios between various protein fractions can vary and are regulated by structural genes (Parma *et al.*, 1999a,b; Rando *et al.*, 2000). Since each one of the four types of casein is polymorphic, different genetic variants can be identified. Polymorphisms can be attributed to two main factors: (i) genetic variants; and (ii) post-translational modifications, due to a different localization of phosphorylation or glycosylation sites in the amino acidic sequence of the primary structure (Bouniol *et al.*, 1993, 1994; Ramunno *et al.*, 1994; Klose, 1999). Intrinsic electrophoretic heterogeneity of casein fractions is due to the presence of incompletely phosphorylated or glycosylated sites in these proteins (Bevilacqua *et al.*, 2002; Martin, 1993; Ramunno *et al.*, 2004).

The αS_1- and αS_2-caseins differ in their content of phosphoric groups, while κ-casein differs in its content of glucidic and phosphoric groups (Boulanger *et al.*, 1984; Leroux *et al.*, 1992, 2003; Iametti *et al.*, 1996). β-Casein heterogeneity derives from incomplete saturation of its phosphorylation sites (Chianese *et al.*, 1993).

The most studied goat's milk polymorphism has been regarding αS_1-casein (Bouniol *et al.*, 1993; Remeuf, 1993; Jansa-Pérez *et al.*, 1994; Remeuf *et al.*,

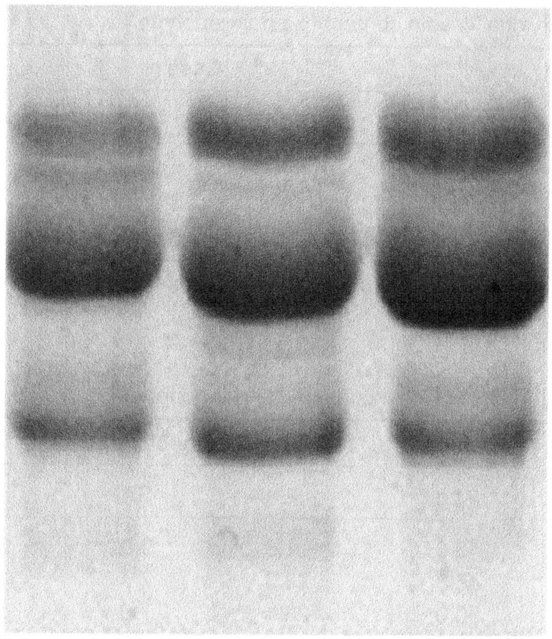

Fig. 4.5. SDS-PAGE of milk caseins. Milk proteins can be separated and identified by molecular mass (molecular weight; in kDa). One method of separating proteins is by polyacrylamide gel electrophoresis. This method essentially separates the proteins by molecular mass, with the largest proteins migrating more slowly in the gel and remaining nearer the top and the smaller proteins migrating more rapidly towards the bottom of the gel. The relative size of the caseins is ~25–35 kDa.

1995; Feligini *et al.*, 1995, 1996, 1998; Chianese *et al.*, 1996; Pena *et al.*, 1998; Martin *et al.*, 1999; Iametti *et al.*, 2000). The gene encoding this protein is a major-effect gene, due to the presence of alleles which are responsible for high differences in the amount of such protein in milk (Parson and Heflick, 1997). At chromosomal level, seven alleles are associated with the codification of three different quantities of αS_1-casein in milk: (i) αS_1-Cn A, αS_1-Cn B and αS_1-Cn C, so-called 'strong alleles', are associated with a production of 3.6 g/l per allele; (ii) αS_1-Cn E, so-called 'medium allele', is associated with a production of 1.6 g/l per allele; (iii) αS_1-Cn D and αS_1-Cn F, so-called 'weak alleles', are associated with a production of 0.6 g/l per allele; and (iv) αS_1-Cn 0, which seems to be a 'null allele', whose homozygous form does not synthesize any kind of casein (Brignon *et al.*, 1990; Pierre *et al.*, 1998a,b). Therefore, milk produced by homozygous individuals AA, EE, FF and 00 would have 7.2, 3.2, 1.2 and 0 g αS_1-casein/l, respectively (Tables 4.6 and 4.7). Some examples of the main electrophoretic techniques used in the analysis of the protein profile of goat's milk are reported in Figs 4.6–4.9 (Jaubert and Martin, 1992; Cattaneo *et al.*, 1996; Recio *et al.*, 1997; Roncada *et al.*, 1997, 2002a,b, 2003a,b,c, 2004; Murakami *et al.*, 1998; Bini and Roncada, 1999; Roncada and Greppi, 1999).

Table 4.6. Individual casein content (%) per genotype.

	Casein				
Genotype	αS_1	αS_2	β	κ	Other
Medium	28.35[a]	0.62[a]	53.74[b]	15.76	1.62[a]
Weak	5.20[b]	21.16[b]	49.01[a]	19.15	3.92[b]
Null	1.61[c]	23.98[b]	52.87[b]	17.07	4.46[b]

[a,b,c]Values in a column with different superscript letters were significantly different ($P \leq 0.05$).

Table 4.7. Major whey protein content (%) per genotype.

	Whey protein				
Genotype	β-Lg	α-La	SA	Lf	Other
Medium	43.8	21.8	9.0	2.6	22.8
Weak	45.8	23.3	7.7	1.5	21.7
Null	50.0	23.1	7.0	1.5	18.4

β-Lg, β-lactoglobulin; α-La, α-lactoalbumin; SA, serum albumin; Lf, lactoferrin.

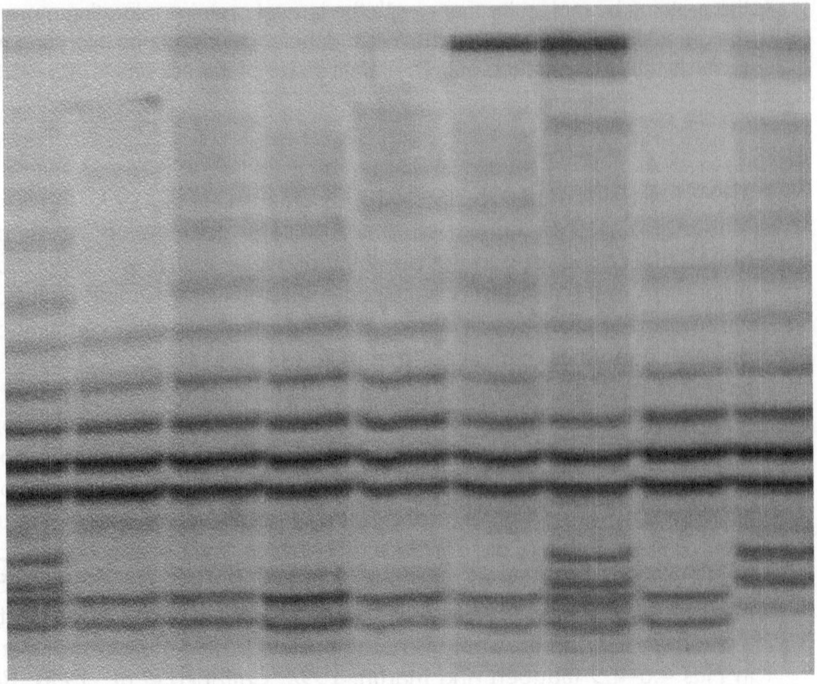

Fig. 4.6. Isoelectric focusing analysis of casein samples of goat's milk (pH 7–4 from top to bottom) (in the low part, αS_1 variants).

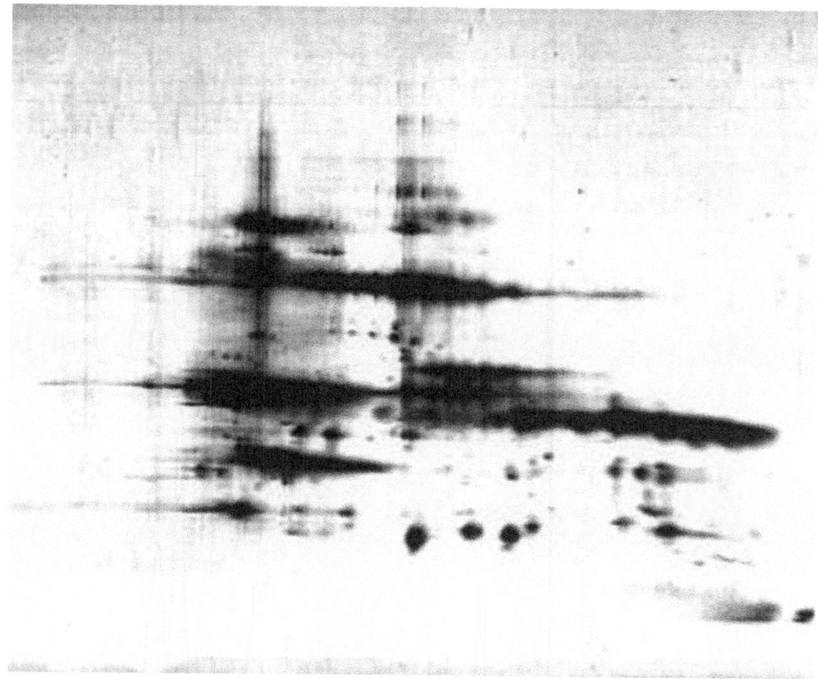

Fig. 4.7. Electrophoretogram of goat's milk in two-dimensional electrophoresis (pH 4–8 from left to right, molecular weight 220 kDa to 10 kDa from top to bottom). Protein spots have been identified by matrix-assisted laser desorption/ionization–time-of-flight mass spectrometry.

Compared with the other genotypes, milk with strong genotype at the αS_1-casein locus is well suited for dairy transformation due to the following properties: (i) higher Ca content, therefore higher propensity for enzymatic coagulation; (ii) lower casein micelles diameter, therefore higher coagula consistency; (iii) lower activation period of coagulation process, i.e. the time between curdle addition and coagula hardening; and (iv) faster formation of coagula.

Even if quantitative differences between genetic variants of single proteins are almost always small (Di Luccia *et al.*, 1990; Grousclaude *et al.*, 1994; Noè, 1995; Lamberet *et al.*, 1996; Enne *et al.*, 1997; Bramanti *et al.*, 2003; Noè and Greppi, 2003), they might directly or indirectly influence milk technological and nutritional properties (Ambrosoli *et al.*, 1988; Addeo *et al.*, 1989; Heil and Dumond, 1993; Remeuf, 1993; Pirisi *et al.*, 1994; Vassal *et al.*, 1994; Delacroix-Buchet *et al.*, 1996; Buchin *et al.*, 1998; Trujillo *et al.*, 1998; Clark and Sherbon, 2000). The main studies on goat casein polymorphisms are reported in Table 4.8.

β-Casein is the most abundant of the caseins on SDS-PAGE (Fig. 4.5), giving rise to two bands, b_1 and b_2. These two isoforms (b_1 and b_2) differ from each other due to the presence of six and five phosphoric groups, respectively (Galliano *et al.*, 2004).

Some recent studies have demonstrated that the milk of Italian goat breeds has a β-null allele (β-Cn 0) associated with the αS_1-Cn A allele (Ramunno *et al.*,

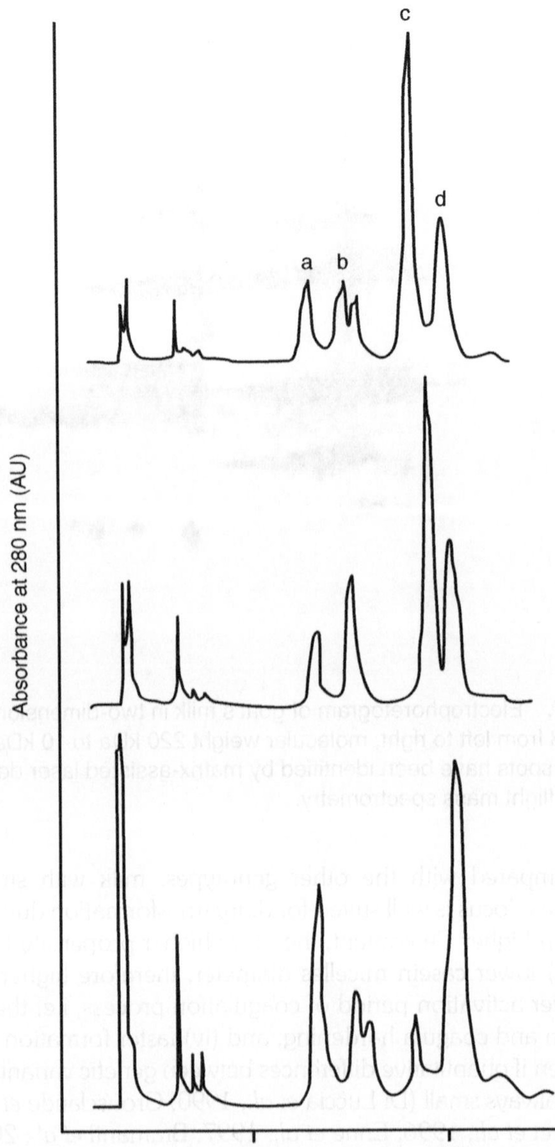

Fig. 4.8. Reversed-phase high-performance liquid chromatograms of different variants of αS_1-casein (A, B and F); a = κ-casein, b = αS_2-casein, c = β-casein.

1996; Cosenza *et al.*, 2003). This β-null allele reduces the content of β-casein, but it is also associated with an increase of αS_1- and total casein. Moreover, compared with heterozygous individuals, milk produced by homozygous β-Cn 0/0 individuals has: (i) coagulation time three times higher; (ii) coagula with reduced consistence; and (iii) lower yield of caciotta cheese (Chianese *et al.*, 1993; Pena *et al.*, 1998).

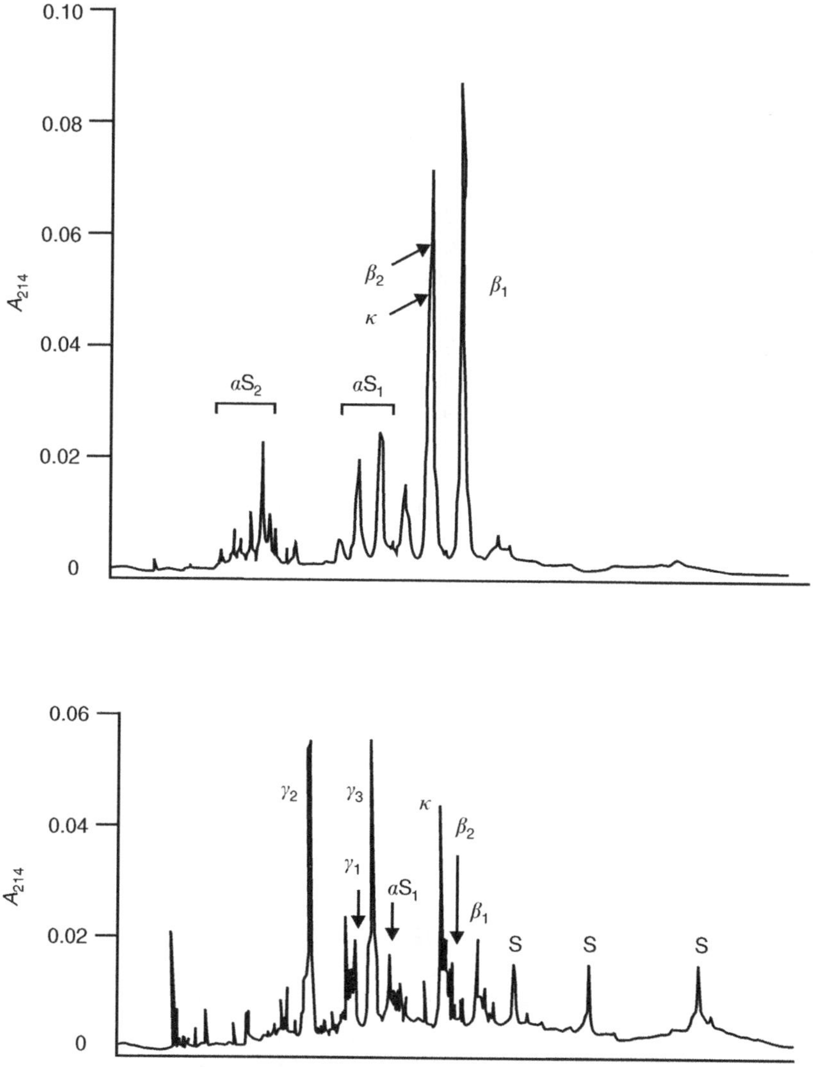

Fig. 4.9 Capillary electrophoresis of goat's milk (upper) and plasmin-treated goat's milk (lower); S indicates casein peptides.

Studies on αS_2-casein are not as numerous as those on αS_1-casein. αS_2-Casein has three known genetic variants, A, B and C (Bouniol *et al.*, 1994), which have only recently been identified through electrophoretic techniques such as isoelectric focusing and SDS-PAGE. Alleles differ from each other only in a single amino acid substitution which has ties with a phosphorylation site. The link between αS_1-Cn and κ-Cn does not seem to be as strong as that between αS_1-Cn and β-Cn (Chang *et al.*, 1993).

A genetic polymorphism at the κ-Cn locus was also found in κ-casein (Di Luccia *et al.*, 1990; Angiolillo *et al.*, 2002; Yahyaoui *et al.*, 2003; Prinzenberg *et al.*, 2005). This casein has two known variants, A and B, which probably differ

Table 4.8. Allelic frequencies at αS_1-Cn in different goat breeds.

Breed	Country	Obs.	Year	A	B	C	D	E	F	0
Alpine[a]	F	213	1994	14	5	1		34	41	5
Alpine[b]	I	38	1994	3	26	1	5	42	10	8
Alpine[c]	I	40	1993	–	–	–	–	–	–	23
Alpine[d]	I	80	1991	–	–	–	–	35	59	6
Alpine[e]	I	26	1995	13	15	–	–	9	40	21
Alpine[f]	I	113	1995	11	29	–	–	16	14	30
Alpine[g]	I	88	2005	43	20	0.6	0.6	11	23	1
Blonde A[e]	I	85	1996	1	3	–	–	5	33	58
Canarian[h]	S	74	1991	28	32	–	–	20	–	20
Corse[i]	F	106	1989	6	13	–	–	14	59	8
Frisian[e]	I	36	1995	–	3	–	–	5	37	55
Frisian[c]	I	70	2006	13	0.7	–	–	20	56	11
Garganic[b]	I	48	1994	39	41	5	4	–	–	–
Garganic[d]	I	54	1991	61	–	37	–	–	2	–
Garganic[h]	S	77	1991	8	25	–	–	62	5	–
Malaguenian[h]	S	56	1991	–	25	–	–	70	5	–
×Maltesian[e]	I	177	1995	21	38	–	–	4	11	27
Maltesian[d]	I	81	1991	33	–	28	–	–	11	27
Maltesian[b]	I	372	1994	46	16	1	–	5	14	3
×Sardinian[e]	I	181	1995	20	53	–	–	5	14	8
Ionic[b]	I	77	1994	39	27	8	–	5	1	4
Marroche[e]	M	82	1995	21	52	4	–	9	7	6
Orobic[f]	I	70	1995	4	–	–	–	2	47	45
Orobic[c]	I	66	2006	–	0.8	–	–	0.8	96	2
Payoya[h]	S	39	1991	4	14	–	–	82	–	–
Poitevinian[i]	F	209	1989	5	35	–	–	45	14	–
Poitevinian[j]	F	302	1995	4	36	–	–	–	43	9
Rove[i]	F	147	1989	12	5	0	0	62	10	11
Saanen[a]	F	159	1994	7	6	0.3	–	49	46	–
Saanen[d]	I	70	1991	5	35	–	–	45	14	–
Saanen[e]	I	50	1995	10	8	–	–	18	10	54
Sardinian[e]	I	115	1995	11	59	–	–	–	7	14
Sirian[b]	I	241	1994	50	17	–	–	–	6	16
Tunisian[e]	T	86	1995	31	41	1	–	5	5	17
Verzaschese[e]	I	57	1995	4	34	–	–	–	7	48
Verzaschese[k]	I	54	1999	3	30	–	2	5	49	11
Verzaschese[c]	I	67	2006	–	4	–	–	20	75	0.7

F, France; I, Italy; M, Morocco; S, Spain; T, Tunisia.
Data source: [a]Grosclaude *et al.* (1994); [b]Chianese *et al.* (1994, 1996); [c]Caroli *et al.* (1993, 2006); [d]Ramunno *et al.* (1991); [e]Feligini *et al.* (1998); [f]Meggiolaro *et al.* (1995); [g]Budelli *et al.* (2005); [h]Jordana *et al.* (1991); [i]Mahé and Grosclaude (1989); [j]Ricordeau *et al.* (1995); [k]Iametti *et al.* (1999).

in just a single amino acid residue. The physicochemical properties of κ-casein are characterized by solubility in the presence of Ca^{2+} and sensitivity to chymosin. Its content in milk influences coagulation, since this phase starts only when all κ-casein has been hydrolysed. Even if a strong association between αS_1- and β-casein has not yet been demonstrated, it has been hypothesized that many correlations exist between all casein loci for goats, as already found for cattle (Parma *et al.*, 1999a,b, 2003a,b).

Allergic Reaction and Milk Intolerance

Alimentary intolerance is an abnormal reaction by the organism that occurs after food ingestion, regardless of the pathogenetic mechanism involved. Allergy is defined as the sum of clinical symptoms (e.g. rhinitis, asthma and eczema) which appear after contact between an 'antigen' (e.g. dust mites, pollen, cow's milk, pig's meat, barley) and a specific antibody of the 'epsilon' (IgE) class of immunoglobulins. The IgEs are commonly attached to the surface of different cell types such as mastocytes and polymorphonucleate basophiles. Contact between an antigen and its specific IgE leads to the production of mediators, such as histamine, which cause an excessive immune response, from which derives the term hypersensitivity. Free histamine is present in considerable quantities in tuna, mackerel, other dark meat fish and in seasoned cheese. Other foods, such as strawberries, tomato and albumen, cause the release of histamine through particular mechanisms which are also valid for lecithins contained in fruit, cereals and many vegetables (Carbonaro *et al.*, 1996; Bellioni-Businco *et al.*, 1999; Cartoni *et al.*, 1999).

If the above-mentioned and many other biologically active agents (e.g. antioxidants, antimicrobials, toxins, microorganisms) are present in great quantities, they can potentially induce a great variety of symptoms, whose manifestation depends on the sensitivity of each individual. For example, subjects with hyperactive gastrointestinal and respiratory mucus membranes or with irritable skin have a low threshold and thus also react in the presence of low amounts of antigen (biologically active agent in food), showing symptomatology typical of alimentary allergy.

Cow's milk contains at least 20 different proteins, but only five are related to allergy: (i) serum albumin, γ-globulin and α-lactoalbumin, which are thermolabile and thus easily deactivated; and (ii) β-lactoglobulin and casein, which are thermostable and resistant to the proteolytic action of the digestive apparatus. In the case of infants, the structure of any milk protein ingested would remain unaltered, i.e. amino acids are not removed from its peptide chain, due to insufficient enzymatic activity of proteases or lack of a specific enzyme in the intestine. Therefore, milk protein would be recognized as a foreign body (antigen). The quantity of absorbed antigenic material depends on anatomic barriers (mucus, epithelium) and/or its removal by immunological barriers, as a consequence of the combination of antigens with secretory immunoglobulins (IgA) on the surface of the epithelium. This is an important protective action which reduces the allergenic penetration, since antigens (milk proteins) do not have access to IgE and IgG. If these barriers are weakened, by inflammatory processes or by lack of IgA, the penetration of high quantities of antigens can cause sensitivity and allergy in atopic individuals. In general, the newborn is not intolerant to maternal milk, excluding those affected by serious metabolic diseases (phenylketonuria, galactosaemia) or by diseases that may require interruption of milking (icterus, diarrhoea, constipation). However, in some cases cow's milk is not tolerated by breast-fed children as some of its thermostable protein components (β-lactoglobulin and casein) are absorbed almost intact and recognized as 'foreign' proteins by hypersensitive subjects. Cow's milk intolerance is conventionally defined as cow's milk protein allergy (CMPA) or cow's milk protein intolerance, and can affect 2–4% of the population. In paediatrics, the estimated frequency of cow's milk intolerance ranges from 1 to 8%, while that of atopic predisposition in the entire population varies between 15 and 20%. Cow's milk allergy is the first cause of digestive allergies in industrialized countries.

The CMPA can be of immunologic or non-immunologic origin. Among non-immunologic intolerances lactose intolerance is the best known, being caused by a lack of intestinal lactase. However, not all problems and symptomatologies found in sensitive individuals following ingestion of certain foods are caused by alimentary allergy (i.e. proved or likely to be of immunologic aetiology). In fact, in many cases such disturbances are related to alimentary intolerance, pseudo-allergy or idiosyncrasy.

CMPA encompasses different types of allergies: (i) type I, with immediate hypersensitivity reaction which can lead to anaphylactic shock, nettle rash, asthma crisis and serious digestive problems; (ii) type III, with complex immunologic reactions characterized by enteropathies and internal haemorrhages; and (iii) type IV, with a retarded hypersensitivity which causes chronic diarrhoea. Clinical symptoms of CMPA are very different. They can be acute or delayed, and digestive or extradigestive. Even if the risk of anaphylactic shock is rare, it can affect 10% of CMPA individuals. Digestive problems can be acute or chronic, while described extradigestive symptoms are mainly related to the skin (dermatitis, nettle rash) or the respiratory tract (asthma). Nevertheless, renal problems, blood alterations, and central nervous system or neurological alterations may also occur, at a lower frequency.

The fact that the clinical symptomatology of CMPA is protein-related has resulted in a shortage of specific diagnostic tools. For example, total and specific IgE is a diagnostic tool only for type-I CMPA, while intestinal biopsy and permeability test are difficult to perform and are suitable only for enteropathic forms. There are predisposing factors that can favour the appearance of CMPA symptoms, which are correlated with: (i) some degree of hereditability, i.e. allergic relatives; (ii) early feeding with cow's milk (fresh or powdered), during the very first days of a newborn's life; and (iii) development of gastroenteritis within the first 3 months of life, after which individuals are considered to be at high risk. As already noted, an inflammatory intestinal disease increases epithelial permeability and, as a consequence, the risk of reaction due to intolerance or allergy. More than a quarter of newborns who suffered gastroenteritis in their first months of life showed CMPA symptomatology afterwards. In addition, a relevant percentage of children who suffered from gastroenteritis within the first year and up to

3 years of life also show CMPA symptoms later on. However, whereas milk is essential in the first few months of life, it can be replaced by other food products as children get older. The few studies conducted on the use of goat's milk in human feeding are encouraging (Dore *et al.*, 1993). Moreover, goat's milk has been reported to be one of the best alternatives for the production of hypo-allergenic formulas for infants in case of cow's milk allergy (D'Auria *et al.*, 2005). Very recent studies in this field have demonstrated that goat's and sheep's milk allergy differs from cow's milk allergy because it affects older children and appears later. Cow's milk products do not cause any clinical manifestation in patients allergic to goat's and sheep's milk, whereas goat's and sheep's milk cause cross-reaction in patients allergic to cow's milk (Ah-Leung *et al.*, 2006).

References

Addeo, F., Masi, P., Perna, A.M., Ciafarone, N., Moio, L. and Tripaldi, C. (1987) Proprietà funzionale dei latti caprini selezionati in base al tipo di variante genetica di caseina αs1. *Il Vergaro* 7, 36–40.

Addeo, F., Mauriello, L. and Di Luccia, A. (1988) A gel electrophoretic study of caprine casein. *Journal of Dairy Research* 55, 413–421.

Addeo, F., Masi, P. and Rubino, R. (1989) The aptitude of caprine milk to cheese-making. Relationship between αs1 casein content and rheological properties of curd and cheese. In: *Agriculture. Programme de recherche Agrimed. L'évaluation des ovins et des caprins. Méditerranéens.* Récueils des communications. Commission des Communautés Européennes, Luxemburg, pp. 291–311.

Ah-Leung, S., Bernard, H., Bidat, E., Paty, E., Rance, F., Scheinmann, P. and Wal, J.M. (2006) Allergy to goat and sheep milk without allergy to cow's milk. *Allergy* 61, 1358–1365.

Albertini, A., Mentasti, T., Moretti, V.M., Bellagamba, F., Luzzana, U. and Valfre, F. (1999) Occurrence of free dscp-amino acids in milk from food-producing animals. *Milchwissenschaft* 51, 127–129.

Ambrosoli, R., Di Stasio, L. and Mazzocco, P. (1988) Content of αs1-casein and coagulation properties in goat milk. *Journal of Dairy Science* 71, 24–28.

Angiolillo, A., Yahyaoui, M.H., Sanchez, A., Pilla, F. and Folch, J.M. (2002) Characterization of a new genetic variant in the caprine κ-casein gene. *Journal of Dairy Science* 85, 2679–2680.

Aschaffenburg, R. and Drewry, J. (1955) Occurrence of different β-lactoglobulins in cow's milk. *Nature* 176, 218–219.

Attaie, R. and Richter, R.L. (2000) Size distribution of fat globules in goat milk. *Journal of Dairy Science* 83, 940–944.

Awiatlo, N., O'Connors, D.L., Andrews, J. and Picciano, M.F. (1990) Relative folate bioavailability from diets containing human, bovine and goat milk. *Journal of Nutrition* 120, 172–177.

Bellioni-Businco, B., Paganelli, R., Lucenti, P., Giampietro, P.G., Perborn, H. and Businco, L. (1999) Allergenicity of goat's milk in children with cow's milk allergy. *Journal of Allergy and Clinical Immunology* 103, 1191–1194.

Bevilacqua, C., Ferranti, P., Garro, G., Veltri, C., Lagonigro, R., Leroux, C., Pietrola, E., Addeo, F., Pilla, F., Chianese, L. and Martin, P. (2002) Interallelic recombination is probably responsible for the occurrence of a new α(s1)-casein variant found in the goat species. *European Journal of Biochemistry* 269, 1293–1303.

Bevilacqua, C., Helbling, J.C., Miranda, G. and Martin, P. (2006) Translational efficiency of casein transcripts in the mammary tissue of lactating ruminants. *Reproduction, Nutrition, Development* 46, 567–578.

Bini, L. and Roncada, P. (1999) Il progetto Proteoma. In: *Proceedings of the 34° Simposio Internazionale di Zootecnia*. CNR, Milan, Italy, pp. 57–101.

Birkbeck, J.A. (1984) Goat's milk in infant nutrition. *New Zealand Medical Journal* 97, 413–414.

Blanc, D. (1982) Les protéines du lait à activité enzymatique et hormonale. *Lait* 62, 350–395.

Boulanger, A., Grosclaude, F. and Mahé, M.F. (1984) Polymorphisme des caséines αs1 et αs2 de la chèvre, *Capra hircus*. *Génétique Sélection Evolution* 16, 157–176.

Bouniol, C., Brignon, G., Mahé, M.F. and Printz, C. (1993) Characterisation of goat αs2-casein A and B; further evidence of the phosphorylation code of caseins. *Protein Sequences & Data Analysis* 5, 213–218.

Bouniol, C., Brignon, G., Mahé, M.F. and Printz, C. (1994) Biochemical and genetic analysis of variant C of caprine αs2-casein (*Capra hircus*). *Animal Genetics* 25, 173–177.

Bramanti, E., Sortino, C., Onor, M., Beni, F. and Raspi, G. (2003) Separation and determination of denatured α(s1)-, α(s2)-, β- and κ-caseins by hydrophobic interaction chromatography in cows', ewes' and goats' milk, milk mixtures and cheeses. *Journal of Chromatography A* 994, 59–74.

Brignon, G., Mahé, M.F., Grosclaude, F. and Ribadeau Dumas, B. (1989) Sequence of caprine αs1-casein and characterization of those of its genetic variants which are synthesized at a high level, αs1-Cn A, B and C. *Protein Sequences & Data Analysis* 2, 181–188.

Brignon, G., Mahé, M.F., Ribadeau Dumas, B., Mercier, J.C. and Grosclaude, F. (1990) Two of the three genetic variants of goat αs1-casein which are synthesized at a reduced level have an internal deletion possibly due to altered RNA splicing. *European Journal of Biochemistry* 193, 237–241.

Buchin, S., Duboz, G., Le Quéré, J.L. and Grappin, R. (1998) Développement des caractéristiques biochimiques et sensorielles des fromages de chèvre. Etude interspèce par échange de la matière grasse et du lait écrémé de laits de vache et de chèvre. *Lait* 78, 673–687.

Budelli, E., Chessa, S., Chiatti, F., Bolla, P. and Caroli, A. (2005) Variabilità degli aplotipi caseinici nella razza caprina Camosciata. *Scienza e Tecnologia Lattiero-Casearia* 56, 205–217.

Carbonaro, M., Bonomi, F., Iametti, S., Cappelloni, M. and Carnovale, E. (1996) Aggregation of whey proteins in raw and heat-treated milks: possible nutritional consequences. *Italian Biochemical Society Transactions* 7, 266.

Caroli, A., Bolla, P., Maccarini, G. and Bruni, G. (1993) Polimorfismo genetico dell'αs1 – caseina ed effetto su caratteri quantitativi nella razza caprina Camosciata. In: *Proceedings of the XXVIII Simposio Internazionale di Zootecnia*, Milan, Italy. Società Italiana per il Progresso della Zootecnia, Milan, Italy, pp. 209–214.

Caroli, A., Chiatti, F., Chessa, S., Rignanese, D., Bolla, P. and Pagnacco, G. (2006) Focusing on the goat casein complex. *Journal of Dairy Science* 89, 3178–3187.

Cartoni, G., Coccioli, F., Jasionowska, R. and Masci, M. (1999) Determination of cows' milk in goats' milk and cheese by capillary electrophoresis of the whey protein fractions. *Journal of Chromatography A* 846, 135–141.

Cattaneo, T.M.P., Nigro, F. and Greppi, G.F. (1996) Analysis of cow, goat and ewe milk mixtures by capillary zone electrophoresis (CZE): preliminary approach. *Milchwissenschaft* 51, 616–619.

Chang, C.S., Sinclair, R., Khalid, S., Yamazaki, I., Nakamura, S. and Powers, L. (1993) An extended X-ray absorption fine structure investigation of the structure of the active site of lactoperoxidase. *Biochemistry* 32, 2780–2786.

Chianese, L., Garro, G., Nicolai, M.A., Mauriello, R., Ferranti, P., Rizzano, R., Cappuccio, U., Laezza, P. and Addeo, F. (1993) The nature of β casein heterogeneity in caprine milk. *Lait* 73, 533–547.

Chianese, L., Mauriello, R., Garro, G., D'Auria, R., Nicolai, M.A., Addeo, F., Cogliandro, E., Fedele, V. and Rubino, R. (1994) Correlazione tra polimorfismo genetico dell'αs1-caseina caprina e attitudine del latte di capra alla coagulazione. In: *Proceedings of the Giornata Scientifica 'Miglioramento dell'Efficienza Riproduttiva degli Ovini e dei Caprini'*, Bella (Potenza), Italy. Istituto Sperimentale per la Zootecnia, Potenza, Italy, pp. 4.1–4.18.

Chianese, L., Garro, G., Ferranti, P., Malorni, A., Laezza, P., Ferranti, P. and Addeo, F. (1996) Occurence of five αs1-casein variant. *Journal of Dairy Research* 63, 49–59.

Choick, W.S. (1998) Role of the insulin-like growth factors and their binding proteins in lactation. *Journal of Dairy Science* 81, 1769–1777.

Clark, S. and Sherbon, J.W. (2000) Genetic variants of αs1-CN in goat milk: breed distribution and associations with milk composition and coagulation properties. *Small Ruminant Research* 38, 135–143.

Cosenza, G., Illario, R., Rando, A., di Gregorio, P., Masina, P. and Ramunno, L. (2003) Molecular characterization of the goat CSN1S1(01) allele. *Journal of Dairy Research* 70, 237–240.

Daddaoua, A., Puerta, V., Requena, P., Martinez-Ferez, A., Guadix, E., de Medina, F.S., Zarzuelo, A., Suarez, M.D., Boza, J.J. and Martinez-Augustin, O. (2006) Goat milk oligosaccharides are anti-inflammatory in rats with hapten-induced colitis. *Journal of Nutrition* 136, 672–676.

Dall'Olio, S., Davoli, R. and Tedeschi, M. (1988) Contributo allo studio della caseina caprina mediante cromatografia. *Scienza e Tecnologia Lattiero-Casearia* 39, 167–178.

Dall'Olio, S., Davoli, R. and Russo, V. (1989) Una nuova variante di caseina caprina. *Scienza e Tecnologia Lattiero-Casearia* 40, 24–28.

Darnton Hill, I., Coveney, J. and Davery, G.R. (1987) Goat milk – nutritional and public health aspects: a review. *Food Technology in Australia* 39, 568–572.

D'Auria, E., Agostoni, C., Giovannini, M., Riva, E., Zetterstrom, R., Fortin, R., Greppi, G.F., Bonizzi, L. and Roncada, P. (2005) Proteomic evaluation of milk from different mammalian species as a substitute for breast milk. *Acta Paediatrica* 94, 1708–1713.

Davis, S.R. and Collier, R.J. (1983) Mammary blood flow and regulation of substrate supply for milk synthesis. *Journal of Dairy Science* 68, 1041–1058.

Delacroix-Buchet, A., Degas, C., Lamberet, G. and Vassal, L. (1996) Influence des variants AA e FF de la caséine αs1 caprine sur le rendement fromager et les caractéristiques sensorielles des fromages. *Lait* 76, 217–241.

Di Gregorio, P., Rando, A., Ramunno, L., Masina, P. and Pieragostini, E. (1989) Polimorfismi nelle regioni del DNA che contengono i geni delle caseine di pecora e capra. In: *Proceedings of the XXXIV Simposio Internazionale di Zootecnia*, Milan, Italy. Società Italiana per il Progresso della Zootecnia, Milan, Italy, pp. 275–282.

Di Luccia, A., Mauriello, R., Chianese, L., Moio, L. and Addeo, F. (1990) κ-Casein polymorphism in caprine milk. *Scienza e Tecnologia Lattiero-Casearia* 41, 305–314.

Domeniconi, D. and Balzola, F. (1980) Considerazioni sull'impiego del latte di capra nell'alimentazione umana. *Rivista della Società Italiana di Scienza dell'Alimentazione* 1, 33–36.

Dore, A., Enne, G. and Oggiano, R. (1993) Il latte di capra nella dietoterapia dei lattanti affetti da IPLV (Intolleranza alle proteine del latte vaccino). In: *Proceedings of the V Simposio Internazionale 'La Qualità nelle Produzioni dei Piccoli Ruminanti'*, Varese, Italy. La Tecnografica, Varese, Italy, pp. 79–82.

Enne, G., Feligini, M., Greppi, G.F., Iametti, S. and Pagani, S. (1997) Gene frequencies of caprine αs1-casein polymorphism in dairy goat. In: *Proceedings of the IDF Seminar 'Milk Protein Polymorphism'*, Palmerston North, New Zealand. International Dairy Federation, Brussels, pp. 275–279.

Feligini, M., Civardi, G., Pasquini, M., Greppi, G.F. and Enne, G. (1995) Caratterizzazione della popolazione caprina verzaschese al locus αs1-caseina. In: *Proceedings of the Convegno Nazionale 'Parliamo di Produzione e Trasformazione del Latte'*, Fossano, Italy. L'artistica Savigliano, Cuneo, Italy, pp. 211–214.

Feligini, M., Pasquini, M., Greppi, G.F. and Enne, G. (1996) Polymorphism at the locus αs1 casein in some autochthonous goats. In: Enne, G. and Greppi, G.F. (eds) *Food and Health. Role of Animal Products*. Elsevier, London, pp. 125–128.

Feligini, M., Parma, P., Aleandri, R., Greppi, G.F. and Enne, G. (1998) PCR-RFLP test for direct determination of β lactoglobulin genotype in Sarda sheep. *Animal Genetics* 29, 473–474.

Galliano, F., Saletti, R., Cunsolo, V., Foti, S., Marletta, D., Bordonaro, S. and D'Urso, G. (2004) Identification and characterization of a new β-casein variant in goat milk by high-performance liquid chromatography with electrospray ionization mass spectrometry and matrix-assisted laser desorption/ionization mass spectrometry. *Rapid Communications in Mass Spectrometry* 18, 1972–1982.

Grosclaude, F., Mahé, M.F., Brignon, G., Di Stasio, L. and Jeunet, R. (1987) A Mendelian polymorphism underlying quantitative variations of goat αs1-casein. *Génétique Sélection Evolution* 19, 399–412.

Grosclaude, F., Martin, P., Ricordeau, G., Remeuf, F., Vassal, L. and Bouillon, J. (1994) Du gène au fromage: le polymorphisme de la caséine αs1 caprine, ses effets, son évolution. *INRA Productions Animales* 7, 3–19.

Heil, F. and Dumont, J.P. (1993) Caractéristiques organoleptiques du fromage de chévre fabriqué à partir de laits contenant des variants génétiques différents de la caseine α-s1. *Lait* 73, 559–565.

Houdebine, L.M. (1995) The production of pharmaceutical proteins from the milk of transgenic animals. *Reproduction, Nutrition, Development* 35, 609–617.

Iametti, S., Testa, D., Pagani, S., Greppi, G.F., Feligini, M. and Enne, G. (1996) IEF e RP-HPLC per la caratterizzazione del latte caprino. In: *Proceedings of the XIII Congresso Nazionale SIPAOC*, Varese, Italy. Società Italiana di Patologia e di Allevamento degli Ovini e dei Caprini, Sassari, Italy, pp. 249–252.

Iametti, S., Sessa, F., Feligini, M., Greppi, G.F., Enne, G. and Pagani, S. (1999) Frequency αs1-casein variants in Italian goat breed Sarda and Verzaschese. *Italian Journal of Food Science* 3, 249–255.

Iametti, S., Feligini, M., Dimartino, S., Sessa, F. and Enne, G. (2000) Frequency of αs2-casein variants in the italian goat breed Verzaschese. In: *Proceedings of the XXXV Simposio Internazionale di Zootecnia 'Produzioni Animali di Qualità ed Impatto Ambientale nel Sistema Mediterraneo'*, Ragusa, Italy. MG Scientific Pub., Bergamo, Italy, pp. 229–236.

Jansa-Pérez, M., Leroux, C., Sanchez Bonastre, A. and Martin, P. (1994) Occurrence of a line sequence in the 3′ UTR of the goat αs1-casein E allele associated with a reduced protein synthesis. *Gene* 147, 179–187.

Jaubert, A. and Martin, P. (1992) Reverse-phase HPLC analysis of goat caseins. Identification of αs1 and αs2 genetic variants. *Lait* 72, 235–247.

Jaubert, A., Durier, C., Kobilinsky, A. and Martin, P. (1999) Structural organization of the goat casein micelle: effect of the physico-chemical environment (pH, temperature, ionic strength) on its mineral and protein composition. *International Dairy Journal* 9, 369–370.

Jenness, R. (1974) The composition of milk. In: Larson, B.L. and Smith, V.R. (eds) *Lactation*, Vol. 3. Academic Press, New York, pp. 3–107.

Jenness, R. (1979) Comparative aspects of milk proteins. *Journal of Dairy Research* 46, 197–210.

Jenness, R. (1980) Composition and characteristics of goat milk: review 1968–1979. *Journal of Dairy Science* 63, 1605–1630.

Jenness, R. and Sloan, R.E. (1970) The composition of milk of various species: a review. *Dairy Science Abstracts* 32, 599–612.

Jordana, J., Sanchez, A., Jansa, M., Mahé, M.F. and Grosclaude, F. (1991) Estudio comparativo de razas caprinas españolas en relacion a las variantes de la caseina αs1. *Información Técnica Económica Agraria* 11, 598–600.

Klose, J. (1999) Genotype and phenotypes. *Electrophoresis* 20, 643–652.

Lamberet, G., Degas, C., Delacroix-Buchet, A. and Vassal, L. (1996) Effect of characters linked to A and F caprine αs1-casein alleles on goat flavours: cheesemaking with protein–fat exchange. *Lait* 76, 349–361.

Law, A.J.R. and Brown, J.R. (1994) Compositional changes in caprine whey proteins. *Milchwissenschaft* 49, 674–677.

Leroux, C., Mazure, N. and Martin, P. (1992) Mutations away from splice site recognition sequences might *cis*-modulate alternative splicing of goat αs1-casein transcripts. Structural organization of the relevant gene. *Journal of Biological Chemistry* 267, 6147–6157.

Leroux, C., Le Provost, F., Petit, E., Bernard, L., Chilliard, Y. and Martin, P. (2003) Real-time RT-PCR and cDNA macroarray to study the impact of the genetic polymorphism at the αs1-casein locus on the expression of genes in the goat mammary gland during lactation. *Reproduction, Nutrition, Development* 43, 459–469.

MacDonald, A. (1995) Which formula in cow's milk protein intolerance? The dietitian's dilemma. *European Journal of Clinical Nutrition* 49, 556–563.

Mackle, T.R. and Bauman, D.E. (1998) Recent developments in the regulation of milk protein production. In: *Proceedings of the 1998 Cornell Nutrition Conference for Feed Manufacturers*, Ithaca, New York. Cornell University, Rochester, New York, pp. 104–112.

Mahé, M.F. and Grosclaude, F. (1989) αs1-CnD, another allele associated with a decreased synthesis rate at the caprine αs1-casein locus. *Génétique Sélection Evolution* 21, 127–129.

Martin, P. (1993) Polymorphisme génétique des lactoprotéines caprines. *Lait* 73, 511–532.

Martin, P., Ollivier-Bousquet, M. and Grosclaude, F. (1999) Genetic polymorphism of caseins: a tool to investigate casein micelle organization. *International Dairy Journal* 9, 163–171.

Martin, P., Szymanowska, M., Zwierzchowski, L. and Leroux, C. (2002) The impact of genetic polymorphisms on the protein composition of ruminant milks. *Reproduction, Nutrition, Development* 42, 433–459.

Meggiolaro, D., Leone, P., Basilico, F. and Succi, G. (1995) Polymorphism of αs1 casein fraction in Orobic goats. In: *Proceedings of the XI Congresso Nazionale ASPA*, Grado, Italy. Avenue Media, Bologna, Italy, pp. 183–184.

Meisel, H. (1993a) *Food Proteins: Structure and Functionality*. VCH Weinheim, New York.

Meisel, H. (1993b) *New Perspective in Infant Nutrition*. Thiemme Stuggart, New York.

Meisel, H. (1997) Biochemical properties of regulatory peptides derived from milk proteins. *Biopolymers* 43, 119–128.

Mercier, J.C. and Vilotte, J.L. (1993) Structure and function of milk protein genes. *Journal of Dairy Science* 76, 3079–3098.

Michel, C. (1910) Lait de chèvre et lait d'ânesse. *Revue Générale du Lait* 8, 68–69.

Morand-Fehr, P. (1996) Il latte di capra per l'alimentazione umana, le ultime novità. *Caseus* 4, 53–56.

Murakami, K., Lagarde, L. and Yuki, Y. (1998) Identification of human colostrum and mature milk by two-dimensional electrophoresis. *Electrophoresis* 19, 2521–2527.

Ng-Kwai-Hang, K.F. (1997) A review of the relationship between milk protein polymorphism and milk composition/milk production. In: *Proceedings of the IDF Seminar 'Milk Protein Polymorphism'*, Palmerston North, New Zealand. International Dairy Federation, Brussels, pp. 22–37.

Ng-Kwai-Hang, K.F. (1998) Genetic polymorphism of milk proteins: relationships with production traits, milk composition and technological properties. *Canadian Journal of Animal Science* 78, 131–147.

Noè, L. (1995) Caratteristiche fenotipiche della razza Nera di Verzasca. *Rosato's Advertising – Printed.* Mimeograph, Casargo (Lecco), Italy.

Noè, L. and Greppi, G.F. (2003) Estensivo o intensivo, ma sempre di qualità. *Informatore Zootecnico* 14, 60–64.

Park, Y.W. (1991) Hypo-allergenic and therapeutic significance of goat milk. *Small Ruminant Research* 14, 151–159.

Parma, P., Feligini, M., Noè, L., Aleandri, R., Greppi, G.F. and Enne, G. (1999a) IEF characterization and mRNA sequence of a rare αs1-casein variant in Nera Verzasca goat breed. *Animal Genetics* 29, 54.

Parma, P., Feligini, M., Noé, L., Iametti, S., Greppi, G.F. and Enne, G. (1999b) Detection of the goat αs2-casein variants by ASA-PCR. *Animal Genetics* 30, 231.

Parma, P., Feligini, M., Greppi, G.F. and Enne, G. (2003a) The complete nucleotide sequence of goat (*Capra hircus*) mitochondrial genome. *DNA Sequence* 14, 199–203.

Parma, P., Pirri, G., Damin, F., Chiari, M. and Greppi, G.F. (2003b) Application of DNA microarray technology in *Bos taurus* genomic analysis: preliminary results. *Italian Journal of Animal Science* 2(Suppl. 1), 61–63.

Parodi, P.W. (1997) Cow's milk fat components as potential anticarcinogenic agents. *Journal of Nutrition* 127, 1055–1060.

Parry, T.E. (1984) Goat's milk in infant and children. *British Medical Journal* 288, 863–864.

Parson, B.L. and Heflick, R.H. (1997) Genotypic selection methods for the direct analysis of point mutations. *Mutation Research* 387, 97–121.

Pena, R.N., Folch, J.M., Sanchez, A. and Whitelaw, C.B. (1998) Chromatin structures of goat and sheep β-lactoglobulin gene differ. *Biochemical and Biophysical Research Communications* 252, 649–653.

Perez, M.D. and Calvo, M. (1995) Interaction of β-lactoglobulin with retinol and fatty acids and its role as a possible biological function for this protein, a review. *Journal of Dairy Science* 78, 978–988.

Pierre, A., Le Quéré, J.L., Famelart, M.H., Riaublanc, A. and Rousseau, F. (1998a) Composition, yield, texture and aroma compounds of goat cheeses as related to the A and O variants of αs1-casein in milk. *Lait* 78, 291–301.

Pierre, A., Le Quéré, J.L., Riaublanc, A., Le Graët, Y., Demaizières, D. and Michel, F. (1998b) Composition and physico-chemical characteristics of goat milks containing the A or O αs1-casein variants. *Lait* 78, 191–202.

Pirisi, A., Colin, P., Laurent, F., Scher, J. and Parmentier, M. (1994) Comparison of milk composition, cheesemaking properties and textural characteristics of the cheese from two groups of goats with a high or low rate of αs1-casein synthesis. *International Dairy Journal* 4, 329–345.

Polidori, F., Baldi, A., Cheli, F. and Pulina, G. (1991) Alimentazione e qualità del latte caprino. In: *Proceedings of the III Simposio Internazionale 'Qualità del Latte Ovino-*

Caprino', Varese, Italy. Istituto Sperimentale Italiano L. Spallanzani, Milan, Italy, pp. 105–134.

Prinzenberg, E.M., Gutscher, K., Chessa, S., Caroli, A. and Erhardt, G. (2005) Caprine κ-casein (CSN3) polymorphism: new developments in molecular knowledge. *Journal of Dairy Science* 88, 1490–1498.

Ramunno, L., Rando, A., Di Gregorio, P., Massari, M. and Masina, P. (1991) Struttura genetica di alcune popolazioni caprine allevate in Italia al locus della caseina αs1. In: *Proceedings of the IX Congresso Nazionale ASPA*, Rome. Avenue Media, Bologna, Italy, pp. 579–589.

Ramunno, L., Rando, A., Pappalardo, M., Fiorella, A., Di Gregorio, P., Capuano, M. and Masina, P. (1994) Molecular analyses on quantitative alleles at goat β-Cn and cow αs1-Cn loci. In: *Proceedings of the XXIX Simposio Internazionale di Zootecnia*, Milan, Italy. Società Italiana per il Progresso della Zootecnia, Milano, Italy, pp. 233–240.

Ramunno, L., Cosenza, G., Rando, L., Pappalardo, E., Mariani, P. and Masina, P. (1996) Alcune capre producono latte apparentemente privo di caseina αs2. In: *Proceedings of the XII Congresso Nazionale SIPAOC*, Varese, Italy. Società Italiana di Patologia e di Allevamento degli Ovini e dei Caprini, Sassari, Italy, pp. 439–442.

Ramunno, L., Cosenza, G., Rando, A., Illario, R., Gallo, D., Di Berardino, D. and Masina, P. (2004) The goat αs1-casein gene: gene structure and promoter analysis. *Gene* 334, 105–111.

Rando, A., Ramunno, L. and Masina, P. (2000) Mutations in casein genes. *Zootecnica e Nutrizione Animale* 26, 105–114.

Recio, I., Perez-Rodriguez, M.L., Ramos, M. and Amigo, L. (1997) Capillary electrophoretic analysis of genetic variants of milk proteins from different species. *Journal of Chromatography A* 768, 47–56.

Reinert, P. and Fabre, A. (1996) Utilisation du lait de chèvre chez l'enfant. Expériences de Créteil. In: Freund, G. (ed.) *Intérêts Nutritionnel et Diététique du Lait de Chèvre*. INRA Editions, Versailles, France, pp. 119–121.

Remeuf, F. (1993) Influence du polymorphisme génétique de la caséine αs1 caprine sur les caractéristiques physico-chimiques et technologiques du lait. *Lait* 73, 549–557.

Remeuf, F., Verdalet-Guzman, I. and Lenoir, J. (1995) Technological adjustment of goat milk containing low synthesis-rate α-s1-casein variants. *International Dairy Journal* 5, 381–392.

Restani, P., Gaiaschi, A., Plebani, A., Beretta, B., Cavagni, G., Fiocchi, A., Poiesi, C., Velina, T., Ugazio, A.G. and Galli, C.L. (1999) Cross-reactivity between milk proteins from different animal species. *Clinical and Experimental Allergy* 29, 997–1004.

Ricordeau, G., Mahé, M.F., Amigues, Y., Grosclaude, F. and Manfredi, E. (1995). La caséine αs1 en race Poitevine. *La Chévre* 206, 34–37.

Roncada, P. and Greppi, G.F. (1999) Comparison of different milks using two-dimensional electrophoresis. In: *Proceedings of Proteine '99*, Rome. Biomedia, Milan, Italy, pp. 139.

Roncada, P., Parma, P. and Greppi, G.F. (1997) Lactoferrin in mare and donkey milk: glycosylation forms detected by immunoblotting techniques. In: *Proceedings of the 3rd International Conference on Lactoferrin*, Le Touquet, France. Lactoferrin Le Touquet-Paris Plage abst., Paris, p. 45.

Roncada, P., Gaviraghi, A. and Greppi, G.F. (2002a) Identification of lactoferrin in serum goat milk with proteomics technique. In: *Proceedings of the 26th IDF World Dairy Congress*, Paris. Società Italiana per il Progresso della Zootecnia, Milan, Italy, pp. 71–80.

Roncada, P., Gaviraghi, A., Liberatori, S., Canas, B., Bini, L. and Greppi, G.F. (2002b) Identification of casein in goat milk. *Proteomics* 2, 723–726.

Roncada, P., Carta, F. and Greppi, G.F. (2003a) Goat milk: building a proteomic map. In: *Proceedings of the 38th Simposio Internazionale di Zootecnia on Milk & Research*, Lodi, Italy. Società Italiana per il Progresso della Zootecnia, Milan, Italy, pp. 285–292.

Roncada, P., Carta, F. and Greppi, G.F. (2003b) Identification of minor proteins of goat colostrum and milk by two dimensional electrophoresis. In: *Proceedings of the 54th Annual Meeting of the European Association for Animal Production*, Rome. Wageningen Academic, Wageningen, The Netherlands, p. 22.

Roncada, P., Carta, F. and Greppi, G.F. (2003c) Identification of proteins in goat milk. International Meeting of Proteome Analysis. In: *Proceedings of Proteomic Forum 2003*, Munich, Germany. Wiley-VCH, Weinheim, Germany, p. 234.

Roncada, P., Cuccuru, M.A., Malagutti, L., Zannotti, M., Lauzi, S. and Greppi, G.F. (2004) Proteomic analysis of goats milk experimental feeded with aflatoxin B1. In: *Proceedings of the 58th Convegno Nazionale SisVET*, Grado, Italy. Springer, Dordrecht, The Netherlands, p. 89.

Rudenko, L. and Matheson, J.C. (2007). The US FDA and animal cloning: risk and regulatory approach. *Theriogenology* 67, 198–206.

Russo, V., Davoli, R. and Migliori, L. (1981) Polimorfismo genetico delle proteine del latte nelle pecore di razza Sarda e Massese. *Zootecnica e Nutrizione Animale* 7, 421–428.

Russo, V., Davoli, R., Dall'Olio, S. and Tedeschi, M. (1986) Ricerche sul polimorfismo del latte caprino. *Zootecnica e Nutrizione Animale* 12, 55–62.

Spuergin, P., Walter, M., Schiltz, E., Deichmann, R., Forster, J. and Mueller, H. (1997) Allergenicity of α-casein from cow, sheep and goat. *Allergy* 52, 293–298.

Targowski, S.P. (1983) Role of immune factors in protection of mammary gland. *Journal of Dairy Science* 66, 1781–1789.

Teschemacher, H. (1995) *Handbook of Experimental Pharmacology*. Springer, Berlin.

Teschemacher, H. and Brantl, V. (1994) *β-Casomorphin and Related Peptides: Recent Developments*. VCH Weinheim, New York.

Teschemacher, H., Koch, G. and Brantl, V. (1977) Milk protein-derived opioid receptor ligands. *Biopolymers* 43, 99–117.

Trujillo, A.J., Jordana, J., Guamis, B., Serradilla, J.M. and Amills, M. (1998) Polymorphism of the caprine αs1-casein gene and its effect on the production, composition and technological properties of milk and on cheese making and ripening. *Food Science & Technology International* 4, 217–235.

Valentine, C.R. (1998) The association of non sense codons with exons skipping. *Mutation Research* 411, 87–117.

Vassal, L., Delacroix-Buchet, A. and Bouillon, J. (1994) Influence des génotypes AA, EE et FF de la caséine αs1 sur le rendement fromager et les caractéristiques sensorielles des fromages traditionnels, premières observations. *Lait* 74, 89–103.

WCRF (1997) *Food, Nutrition and the Prevention of Cancer: a Global Perspective*. American Institute for Cancer Research, Washington, DC.

Yahyaoui, M.H., Angiolillo, A., Pilla, F., Sanchez, A. and Folch, J.M. (2003) Characterization and genotyping of the caprine κ-casein variants. *Journal of Dairy Science* 86, 2715–2720.

Yoshikawa, M., Tani, F. and Chiba, H. (1988) *Peptide Chemistry*. Protein Research Foundation, Osaka, Japan.

5 Aromatic and Nutritional Quality of Goat's Milk

V. Fedele

CRA – Unità di Ricerca per la Zootecnia Estensiva, 85054 Bella Scalo, Potenza, Italy

Introduction

In human nutrition, milk has always been considered important for its balanced supply of proteins, fat, minerals and vitamins. Once milk turns into cheese, it is appreciated for many other features, such as taste and odour.

Numerous studies on the taste and smell of milk and cheese have been carried out in the past throughout the world, especially in Europe and New Zealand. Such research has clearly shown that what our senses perceive is based on a multitude of molecules, each one being able to characterize a product. Substances such as alcohols, esters, ketones and terpenes (and many others) are the basis of many of the odours that we perceive in a dairy product. Research on nutritional quality has deeply changed in recent decades. The components taken into account are no longer only fat and protein, but also molecules previously almost unknown, such as the conjugated linoleic acids (CLA), *n*–3 fatty acids (FAs), β-carotenes, tocopherols and phenols, which are compounds with defence functions in the human organism. These substances are present in the herbage grazed by animals and can subsequently be found in dairy products.

From popular beliefs to science

In Italy, in the past it was common for cultured men to praise the qualities of a milk or cheese in this way:

> They are famous for being delicious, for being only second to the ones of Pollino [mountain]. Their good quality is due to sweet herbs forming the animal pasture . . . the milk of the province . . . with no difficulty it is a high quality milk. Rich and fertile pastures, made of sweet plants, make milk sweet-smelling and savoury.
>
> (Demarco, 1988)

(eds A. Cannas and G. Pulina)

The fact that the taste and flavour of milk or cheese are a direct consequence of what the animal eats has always been rooted in the mind of people who, being careful observers, used to take note of the repeating phenomena and their causes. Thanks to this spirit of observation, they were able to notice that the tastes and flavours of a cheese changed with seasons. Based on that, the best time to produce cheese could be chosen:

> cheese should not be made in every season, but only in March, April, August and September. The worst cheese is made in May, that is why the good cheese is called 'formaggio', i.e. 'fuori di maggio', which meant it was made 'out of May'. . . In May herbs are unripe, their juice is nasty and not yet elaborated and savory.
>
> (Demarco, 1988)

Therefore, the kind of pasture and the season were the elements that made the difference in milk and thus cheese quality. As indoor intensive systems spread, the traditional pastoral civilization and, consequently, the perception and the importance of 'quality' faded. Old evaluation parameters have been replaced gradually by high production, hygiene, dairy yield, technology and fat content parameters, without taking into consideration what a careful and sensitive writer such as Italo Calvino had already understood: 'Behind each cheese, there is a different pasture of a different green under a different sky' (Italo Calvino – from the poetry Palomar).

Only in the last decades have these issues been studied scientifically, demonstrating that each kind of pasture, or plant grouping, can represent a factor of qualitative differentiation of a product.

Gas chromatographic techniques have allowed the characterization of the compounds responsible for smell and taste. They are specific molecules or molecule groupings that are often defined as secondary metabolites. Only recently has the origin or the formation pathway of some of these secondary metabolites been understood. For many others no information is available today.

There have been many hypotheses regarding the origin of the compounds which produce taste and smell. For years, smells and tastes were considered the main expression, if not the only one, of microbiological activity (Urbach, 1977; Keen and Wilson, 1993). Later, it was supposed that, at least to a certain extent, enzymatic activity and some oxidation processes of substances were involved (Forss, 1979; Molinard and Spinnler, 1996). Only recently the importance of molecules transferred from the feed eaten by the animals to the milk was acknowledged (Mariaca *et al.*, 1997; Viallon *et al.*, 1999; Fedele *et al.*, 2000).

Sensorial and instrumental assessment of taste and smell

In the assessment of the aromatic qualities of a product, the borderline between certain and uncertain is extremely narrow, as there is no clear discriminator between cause and effect or, better, between what is perceived and the cause of that perception.

Most assessments are made either at the sensory level or at the instrument level, for various reasons.

The sensorial assessment is purely subjective and gives information about: (i) the perception of the difference between two or more products (e.g. ISO-5495, 2005); (ii) the classification and arrangement of products on the basis of a quality attribute (ISO-6658, 2005); and (iii) the differences between two or more products in terms of well-defined sensory parameters. Therefore, it is not based on objective measures and neglects the study of causes and mechanisms.

The instrumental assessment is based on measurements made by a device whose reliability is mostly based on the validity of the analytical methods used. Instruments provide data that do not always correspond to the actual differences in taste perceived by humans. This happens because a single product can contain many different molecules in variable concentrations and each compound can affect sensory perception at different concentration thresholds. Some molecules are able to transmit their characteristics at infinitesimal concentrations, while others, on the contrary, are not able to influence sensory perceptions even at very high concentrations. In addition, each aromatic note is the result of a combination of several flavours. For all of these reasons, the instrumental assessment does not represent a complete means for evaluating the sensorial quality of a product.

The combined assessment based on instrumental and sensorial techniques is undoubtedly best, because it enables one to: (i) study similarities and differences between instrumental and sensorial results; and (ii) understand causes, effects and mechanisms involved in the aromatic qualities of a dairy product. Despite the limited research in this field, some indications on the aromatic quality of dairy products are given in this chapter.

Aromatic quality

Flavour is the sensory impression (smell and taste combined) that can be attributed to all substances present in food and feed of plant (herbs, preserved forage, grains) or animal (milk, cheese, meat) origin and that can be pleasant or unpleasant to the consumer. Since the mid-1970s, much research has been carried out on flavours. Many studies showed that what was perceived by the senses, by smelling and/or tasting a milk or cheese, was the direct or indirect consequence of what the animal ate (Urbach *et al.*, 1972; Keen and Wilson, 1993; Bosset *et al.*, 1997; Coulon *et al.*, 2000). With the advance of studies, it became clear that at least part of the secondary metabolites contained in food are not subjected to any change. O'Sullivan (1967) suggested that when the feed reaches the animal's mouth, the most volatile metabolites (many terpenes, alcohols and ketones) volatilize and, through the breathed-in air, reach the lungs, where they are absorbed into the blood, and then transferred to the milk. The least volatile of these metabolites (short-chain terpenes) are freed only after a partial digestion of the feed and, through belched gases, reach the lungs following the same pathway just described.

All of these aspects have been extensively studied. Here only those closely linked to the feeding system, rather than those associated with each specific feed, are discussed.

Pasture and flavours

In everyday language, aromatic plants are associated with those used for cooking or for flavouring alcoholic drinks (e.g. thyme, mint, marjoram, oregano and rhubarb). However, all plants contain aromatic substances, even though only some of them supply flavours that can be easily perceived by our senses. In a pasture, some spontaneous plants can predominate over the others, depending on the environment and season, thus characterizing the flavour of dairy products. Many natural chemical compounds present in feeds are able to increase these kinds of perceptions. Among them, terpenes, alcohols and ketones are the most studied in milk and cheese.

Major Chemical Compounds Affecting the Flavour of Dairy Products

Terpenes

Terpenes are a class of unsaturated hydrocarbons present in the form of odorous and volatile liquids in leaves, flowers and fruit. The best-known terpenes belong to two groups: (i) monoterpenes ($C_{10}H_{16}$), made up of two structural units of isoprene; and (ii) sesquiterpenes ($C_{15}H_{24}$), made up of three units of isoprene.

Most studies have been carried out on this class of compounds, perhaps because it is considered to characterize, more than the others, the flavour of dairy products.

In many traditional Mediterranean goat feeding systems, the forage basis is mostly made up of natural pastures directly grazed for a variable time, depending on the environment and season. For example, in summer and winter, hay from swards or natural pastures totally or partially replaces the pasture.

According to many goat breeders and cheese makers, the flavour of cheese is affected by changes in the type of forage used. An experiment was carried out to test this hypothesis, by comparing three groups of goats fed: (i) hay from clover–Italian ryegrass stands; (ii) hay from natural pasture; and (iii) a natural pasture grazed at the phenological stage in which hay is normally made (Fedele *et al.*, 2005a). Instrumental analysis showed that milk from the three groups did not differ in the content of monoterpenes, but differed highly in the content of sesquiterpenes (Fig. 5.1). Changing the forage basis from sward hay to natural pasture hay and grazed grass increased sesquiterpene levels by about three- and fourfold, respectively. This experiment showed that the feeding of natural pasture herbage can be a useful factor for the aromatic diversification of dairy products. This can also be observed when animals are supplied with a feeding integration based on concentrates (Fedele *et al.*, 2004a). Indeed, the sesquiterpene concentration in the milk from supplemented goats is even higher than that of goats eating only herbage (12,000 versus 8000 ng/l) (Fig. 5.2), while it is lowest in the milk from housed animals fed hay plus concentrates (1000 ng/l). A direct effect of concentrates was excluded since concentrates were the same for grazing and

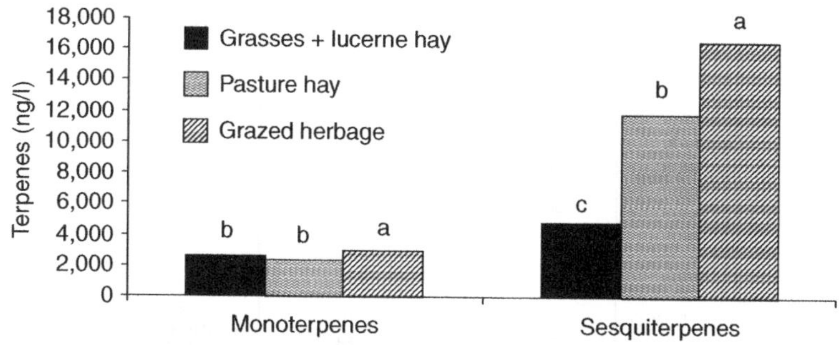

Fig. 5.1. Milk terpenes content in relation to the forage source fed to goats. a,b,c, mean values with different superscript letters were significantly different at $P < 0.05$. (Adapted from Fedele *et al.*, 2005b.)

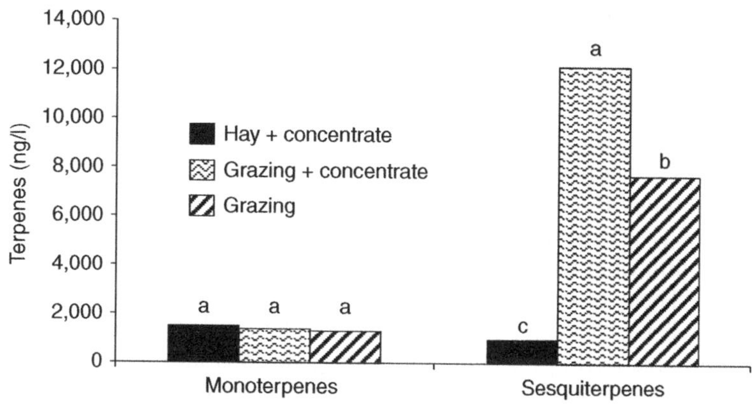

Fig. 5.2. Milk terpenes content (annual average) in relation to different feeding systems. a,b,c, mean values with different superscript letters were significantly different at $P < 0.05$. (Adapted from Fedele *et al.*, 2004a.)

zero-grazing animals. What was the cause, then, of these variations? The phenomenon could be explained by disaggregating the data by season.

Terpenes and season

The increase in the yearly average concentration of terpenes of the grazing group supplemented with concentrates depended mostly on the high content of sesquiterpenes in the milk produced in summer (about 32,000 ng/l), almost twice as high as that of the milk from unsupplemented grazing goats (16,500 ng/l) (Fig. 5.3). In fact, in summer, the Mediterranean pastures of mountain areas show the maximum diffusion of the so-called 'weeds', plants belonging to several botanical families, such as *Geranium*, *Cichorium*, *Rumex*, mint, *Gallium* and *Asperula*. Some of these plants, as shown below, are particularly rich in terpenes and are mostly browsed when goats are supplemented, especially with protein concentrate. Observations carried out on pasture intake showed that in summer the presence of the above-mentioned plants was 7–11% higher in the

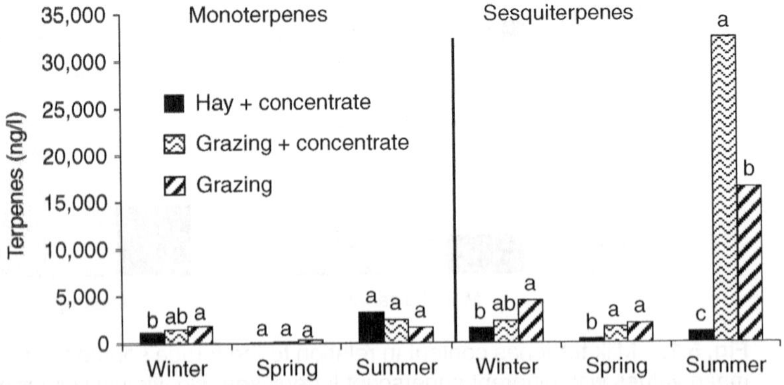

Fig. 5.3. Seasonal variation in milk terpenes content in relation to different feeding systems. a,b,c, mean values with different superscript letters were significantly different at $P < 0.05$. (Adapted from Claps *et al.*, 2005.)

diet of supplemented goats than in that of unsupplemented grazing goats. In the other two seasons, the milk was much less rich in terpenes, with the lowest concentrations during spring. Therefore, it seems aromatic quality is expressed differently in each season, being more marked in summer and less intense in spring, at least under the environmental conditions of the trials. This seems to confirm what was observed in the past and reported by Demarco (1988).

These observations were confirmed later by another experiment in which the diets were enriched by infusion of plants selected from those most browsed by goats during the different grazing seasons into the rumen through rumen cannulae (Fedele *et al.*, 2004a). In winter, milk sesquiterpenes were decreased by diets enriched with *Lolium perenne* and, to a greater extent, by *Dactylis glomerata*, compared with the non-enriched pasture diet (Fig. 5.4). By contrast, in spring milk sesquiterpenes were much higher when diets were enriched with *Asperula odorosa* than when they were rich in *Geranium molle*, and the latter diet led to concentrations of sesquiterpenes not as high as those observed in summer, when the diet was enriched with *Galium verum* or *Cichorium intybus*.

The following conclusions can be drawn from the results presented above: (i) the milk produced in spring was poorer in sesquiterpenes, and remained such even when the animals' diet was enriched with plants that are potentially aromatic, such as *G. molle* and *A. odorosa*; (ii) the milk produced in the summer had sesquiterpene concentrations eight times higher than in spring and four times higher than in winter (16,500, 5500 and 2020 ng of sesquiterpenes per litre in summer, winter and spring, respectively); and (iii) some plants (*G. verum* and *C. intybus*) can enrich milk with aromatic components more than others (*L. perenne* and *D. glomerata*), confirming the results of other studies (Bosset *et al.*, 1997; Mariaca *et al.*, 1997). Similarly, in cattle, Bugaud *et al.* (2001) found that the quantity of monoterpenes and of some sesquiterpenes identified in milk was significantly correlated with that determined in the pasture herbage.

These results did not show any remarkable influence of the digestive process on the terpenic components present in herbage. The main results on the effects

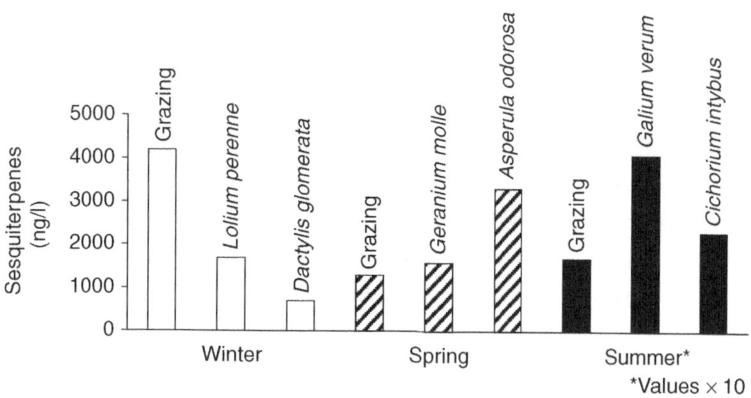

Fig. 5.4. Effect of forage species on sesquiterpenes content in milk (Fedele *et al.*, 2004a).

of herbage on specific single compounds of milk, using a very simple approach (i.e. profile comparison), are described in the following subsection.

Aromatic profile

Data from an experiment on the monoterpene profile of the herbage eaten by the animals in two seasons (Fedele *et al.*, 2005a) were compared with those of a study on the cheese produced in the same seasons (Fedele *et al.*, 2005c) (Fig. 5.5). The monoterpene profile of grass corresponded to that of cheese, while concentrations of single molecules showed an irregular variability. The dilution effect, the different volatility of molecules and the cheese-making process probably influenced the concentration of the single components. Unlike in winter, when a better correspondence between grass and cheese profiles occurred, in spring the correlation between them was poor. Since the concentration of terpenes is remarkably lower in spring than in other seasons, as mentioned above, thermal treatment during cheese-making might have impoverished milk in those components that are more volatile or present in a lower concentration. The aromatic note of certain aromatic compounds detected in cheese is known. For example, α- and β-phellandrene, camphene and α-pinene give characteristic notes of herbaceous, mint, camphor and resin, respectively.

In the same experiments (Fedele *et al.*, 2005a,c), what was discovered about monoterpenes was further confirmed with ketones, organic compounds whose name derives from acetone, which is the simplest compound in the series. They are characterized by a carbonyl (C=O) functional group, directly linked to two alkyl groups. Ketones are present in many plants' essential oils, and some of them transfer particular odours (e.g. carvone, mint, lavender and walnut) to products. In the grass grazed by goats, ketones can reach concentrations much higher than those of terpenes. In winter grass, their concentrations can be up to three times higher than those of terpenes, while in summer they can be one- to twofold higher. This higher concentration was probably responsible for the almost identical ketone profiles of cheese and herbage (Fig. 5.6). In cheese, ketones

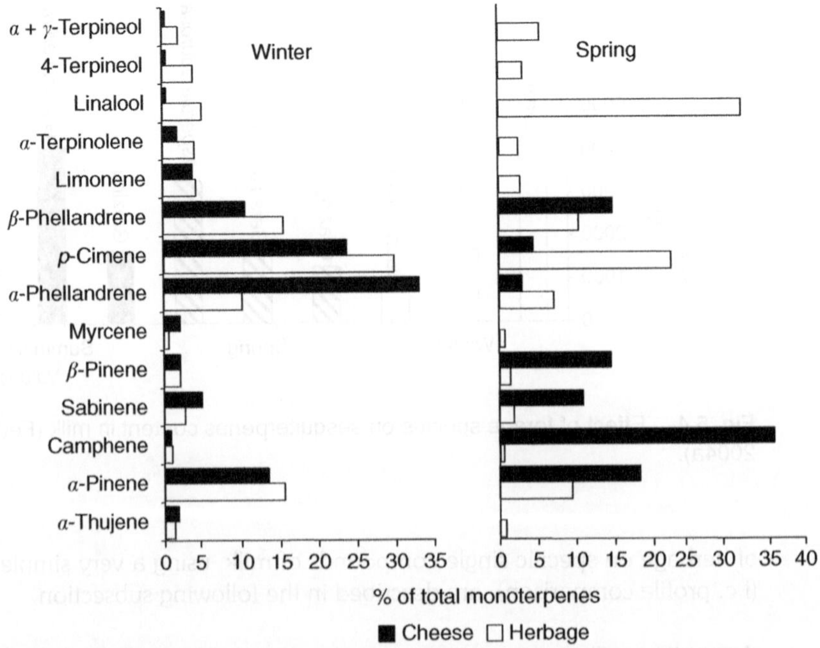

Fig. 5.5. Comparison of monoterpene profiles between grazed herbage (Fedele *et al.*, 2005a) and the corresponding cheese in two seasons (Fedele *et al.*, 2005c).

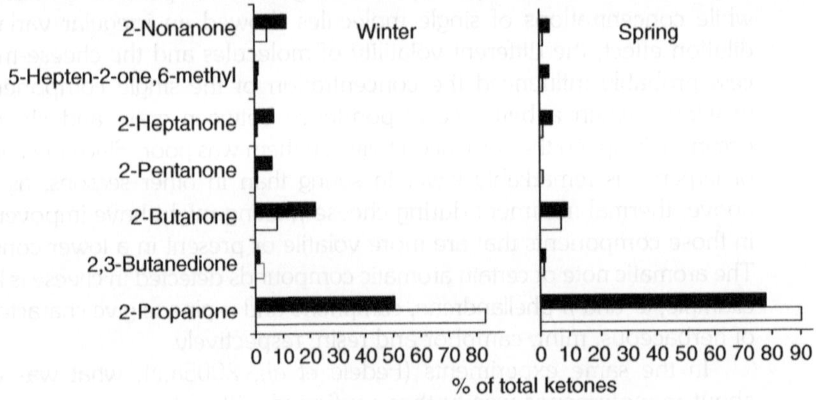

Fig. 5.6. Comparison of ketone profiles between grazed herbage and the corresponding cheese in two seasons. (Adapted from Fedele *et al.*, 2003.)

such as 2-propanone, 2-butanone and 2-nonanone, which can give characteristic notes of hay, butter and milk, respectively, were found. These results on ketones and terpenes show that there is no clear evidence of the influence of the digestive process on the aromatic profile of milk and cheese in goats.

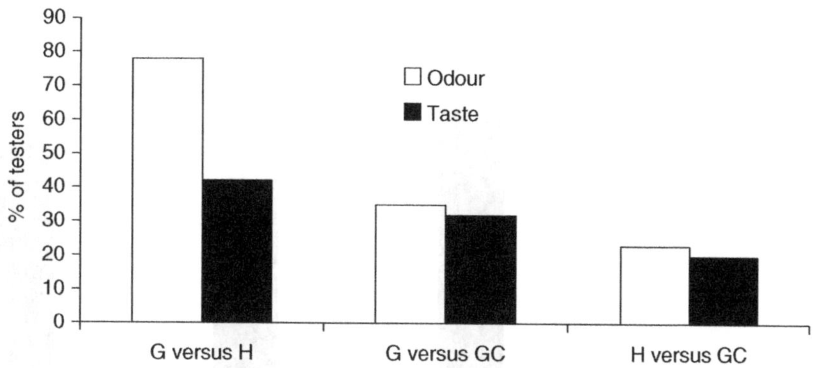

Fig. 5.7. Sensory evaluation expressed as percentage of testers able to distinguish milk obtained from different feeding systems (G = grazing; GC = G + concentrate; H = hay + concentrate). (Adapted from Claps *et al.*, 2001.)

Sensory evaluation

In order to verify if analytical differences can be associated with sensory perceptions, different goat's milk and caciotta cheeses, ripened for 1 and 20 days, were submitted to a sensory evaluation by a trained panel. Milk and cheese, which were the expression of one feeding system based on hay and concentrate (H) and two based on pasture (grazing (G) and grazing plus concentrate (GC)), were assessed through a triangular test: each panellist was given three milk samples and three cheeses, of which two were the same. The results showed that 72% of panellists were able to distinguish by odour and 40% by taste the milk produced by unsupplemented grazing goats (G) from that of zero-grazing goats (H) (Fig. 5.7). These percentages increased to 82 and 68%, respectively, for 1-day cheese (Fig. 5.8) and to 100% for both parameters for the cheese ripened for 20 days (Fig. 5.9). When the rations had an alimentary factor in common, such as herbage in the case of the two grazing groups (G and GC) and concentrate in the case of the integrated pasture (GC) and indoor animals fed hay and concentrate (H), the tasters' ability to distinguish different products decreased markedly. The fact that only one feed in common creates trouble in discriminating among products is a further confirmation that feeding influences to some extent, and with unknown mechanisms, the chemical, aromatic and sensory characterization of products.

Milk Nutritional Quality

In human nutrition, milk has been considered for centuries as a source of high-quality proteins, fat, minerals and vitamins. The assessment of milk nutritional quality based on macro-compounds (fat, protein, lactose) is losing importance with the increase of knowledge about micronutrients (e.g. vitamins, *trans* FAs, CLA and *n*–3 FAs). Therefore, the current approach on milk quality evaluation has been improved by this new perspective.

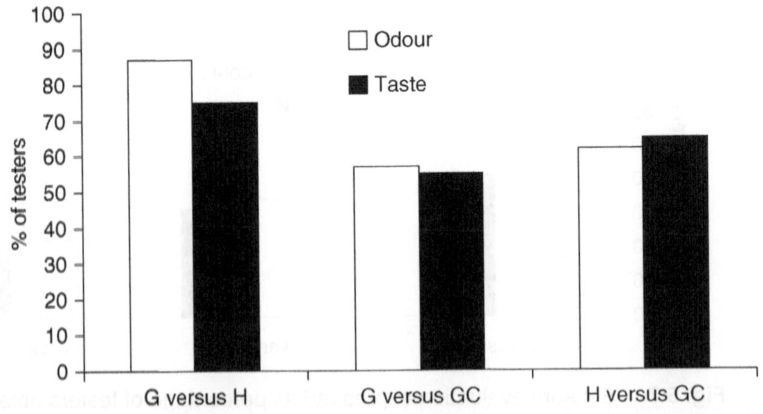

Fig. 5.8. Sensory evaluation expressed as percentage of testers able to distinguish 1-day-ripened cheese obtained from different feeding systems (G = grazing; GC = G + concentrate; H = hay + concentrate). (Adapted from Claps *et al.*, 2001.)

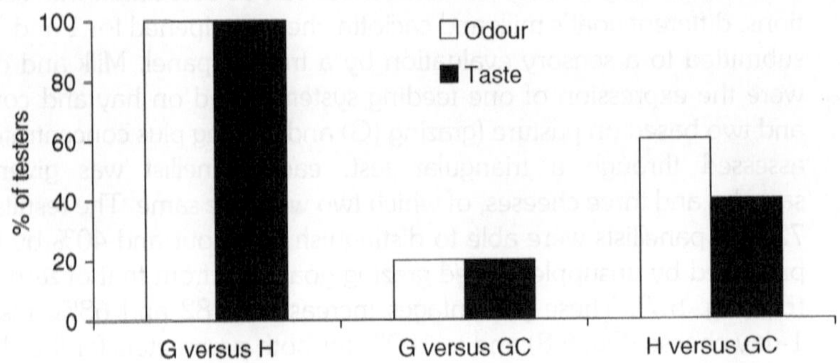

Fig. 5.9. Sensory evaluation expressed as percentage of testers able to distinguish 20-day-ripened cheese obtained from different feeding systems (G = grazing; GC = G + concentrate; H = hay + concentrate). (Adapted from Claps *et al.*, 2001.)

Vitamins

These nutrients are present to a great extent in the green tissues of plants. Leaves are especially rich in vitamins, the youngest leaves being much richer than senescent ones. Since ruminants, and especially goats, browse preferably on leaves and young buds (Fedele *et al.*, 1993), their diet is particularly rich in vitamins. Goats fed hay and high quantities of concentrate were likely to show a mineral and vitamin deficiency (especially of vitamin E) (Fedele *et al.*, 1994), while those grazing pasture did not show any deficiency (Claps *et al.*, 1994). A large number of the vitamins contained in the green tissues of plants are destroyed by heat and light; that is why pasture hays and senescent grass have a vitamin content that can be up to 90% lower than fresh herbage (McDonald *et al.*, 1992).

Retinol (vitamin A)

ORIGIN IN THE MILK. Vitamin A is known chemically as retinol. It can be supplied as such in animal products (e.g. cod-liver oil and liver), or indirectly in vegetable products (particularly carrots and spinach). The latter do not contain vitamin A, but pro-vitamins such as α- and β-carotene that are transformed into retinol at the level of intestine and liver. Milk and cheese are a remarkable source of vitamin A for consumers, especially for newborns (Derby, 2001). In ruminants, vitamin A mainly comes from the conversion of carotenoids contained in fodder. These substances are pigments whose primary function is to protect plants from photo-oxidation and to contribute, together with chlorophyll, to converting sunlight into energy. Vitamin A has an antioxidant activity, being able to neutralize the oxidative activity of oxygen and eliminate free radicals. Milk's vitamin A derives from blood through filtration at the level of the mammary gland. The yellow colour of cow's fat and milk is mainly linked to the content of β-carotene (Prache *et al.*, 2002). In goat's milk, vitamin A is present as such and not as β-carotene; that is why this kind of milk has a white colour.

EFFECTS OF FEEDING. Green vegetables contain good quantities of retinol, but when they are exposed to light and heat they partly lose it in proportion to the time and temperature of exposure (Park *et al.*, 1983). Milk contains *trans* and *cis* isomers of retinol, which have a different A-vitamin activity. A lower vitamin activity is ascribed to *cis*-retinol (Scott and Rodriquez-Amaya, 2000). In cow's milk, the forage of the diet represents an element of nutritional diversification of production. The highest concentrations of vitamin A were found in the milk produced from cows fed on pasture, when grass is young, or fed ensiled ryegrass, while the lowest values in the milk were from cows fed hay or diets rich in concentrates (Martin *et al.*, 2002). Goat's milk also shows remarkable variations in this vitamin, depending on the type of forage used. The presence of natural pastures in the ration greatly enriches milk in vitamin A compared with hay-based rations. In fact, with the same quantity of concentrate in the diet, milk produced by grazing goats (GMB and GBC) contained 20% more *trans*-retinol than that produced by housed goats (I); this difference increased up to 31% when unsupplemented grazing goats (G) were compared with housed goats (I) (Fedele *et al.*, 2004b) (Fig. 5.10). The content of *trans*-retinol in the milk of grazing goats seemed to be directly correlated to pasture intake: going from the zero-grazing group (I) to groups GBC and GMB, the fresh herbage dry matter (DM) intake increased by 381.7 ± 96.5 g/day per head, reaching 568.7 ± 69.4 g/day per head in group G.

Not always and not under all conditions has pasture herbage the same effect on retinol. For example, the grazing season, especially in terms of vegetation phenological stage, seems to affect the retinol levels of the milk. If the ratio of total retinol in the milk to total milk production (TR/TMP), a useful parameter to standardize these comparisons, is considered instead of total milk retinol content, it is possible to observe that this ratio increased at different rates for the different feeding treatments, as milk production decreased (Fig. 5.11). The hay and concentrate treatment (I) was the least effective of all and the least variable among seasons; grazed herbage (G, GBC and GMB) had different effects on this

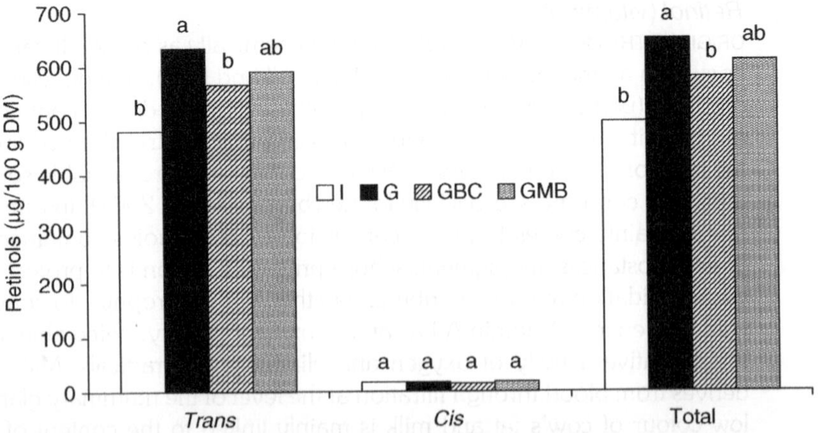

Fig. 5.10. Effect of feeding treatments (I = hay + commercial concentrate 600 g/day; G = grazing; GBC = G + mixed barley and chickpea grain 600 g/day; GMB = G + mixed maize and broad bean grain 600 g/day) on milk content of retinols (on dry matter, DM, basis). a,b, mean values with different superscript letters were significantly different at $P < 0.05$. (Adapted from Fedele *et al.*, 2004b.)

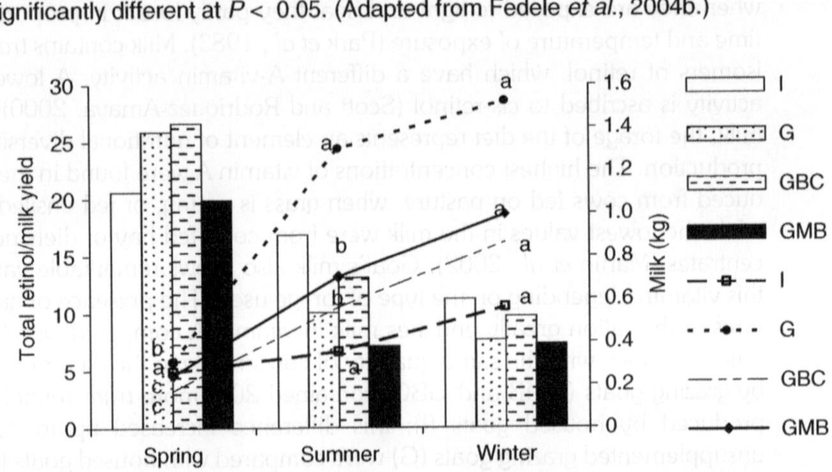

Fig. 5.11. Seasonal variation of the ratio of total retinol to total milk production (lines) in relation to milk production (bars) obtained with different feeding treatments (I = hay + concentrate; G = grazing; GBC = G + mixed barley and chickpea grain; GMB = G + mixed maize and broad bean grain). a,b,c, line means with different superscript letters were significantly different at $P < 0.05$. (Adapted from Fedele *et al.*, 2004b.)

ratio depending on the quantity of herbage eaten by goats and the grazing season. The concentrate decreased grass intake (–28% for GMB treatment and –32% for GBC) compared with G and, as a consequence, decreased TR/TMP. The variation in herbage intake did not always explain the variation in the TR/TMP ratio. In fact, in summer and winter TR/TMP ratio was higher for all grazing treatments, despite the lower herbage intake (an average 29%), in comparison to spring.

A higher TR/TMP ratio was also observed in winter compared with summer, despite their quite similar herbage intake. The following hypotheses most likely explain this phenomenon.

1. In spring, most plants are blooming and goats graze above all on flowers and floral buds; during this phenological stage, it seems that plants synthesize mainly secondary metabolites with protective function (Strauss *et al.*, 2004), so it is likely that the retinol content in the grazed grass is lower than in other seasons.
2. In summer, herbage is mostly dry or semi-dry and, as a consequence, the diets of goats contain dry vegetation (Fedele *et al.*, 1993) that, for the previously explained reasons, can contain up to 90% less vitamins than fresh herbage.

α-Tocopherol (vitamin E)

ORIGIN IN MILK AND CHEESE. Vitamin E is one of the most common vitamins in green forage, especially in leaves, where it reaches concentrations 20% higher than in stems (McDonald *et al.*, 1992). In nature, vitamin E originates from α-, β-, γ- and δ-tocopherols; among them, the α form is the most biologically active and the one with the highest antioxidant power.

EFFECTS OF FEEDING. The diet is the main source of variation of vitamin E content in milk. Grass consumed directly by cows enriches milk more than dry forage supplied in the stable (Thompson *et al.*, 1964; Kanno *et al.*, 1968), since ensiled ryegrass is better than hay or rations rich in concentrates (Martin *et al.*, 2002). Similarly, in goats grazing different types of pasture and supplemented with variable doses of concentrates, pasture enriched milk and cheese in α-tocopherol more than hay and concentrates (Fedele *et al.*, 2002) (Fig. 5.12). With the exception of cheese produced with milk from grazing goats receiving concentrate 900 g/day per head, all other samples of milk and cheese had much higher vitamin E contents than samples from housed goats fed hay and concentrate (I). The supplementation of grazing goats with concentrate at 600 g/day per head decreased the content in α-tocopherol, but never down to the levels of milk and cheese of housed animals.

By distributing 600 g concentrate/day, the average decrease of vitamin E compared with the group grazing on valley pastures reached 25 μg/100 g DM; by increasing the integration to 900 g/day, this decrease reached 323 μg/100 g DM.

The grazing environment (hill or mountain) did not seem to affect the content of vitamin E.

The fact that fresh herbage guarantees the production of milk richer in tocopherol than dry diets is confirmed by the regression line shown in Fig. 5.13 (Pizzoferrato, 2001). As grass intake increased in the diet from zero to 1150 g/day per head, the content of tocopherol in the milk increased almost threefold.

Cholesterol

This component of milk and milk products is affected by many factors, such as the animal species (Popjipac *et al.*, 1951; Cowie *et al.*, 1971; Mills *et al.*, 1974), breed

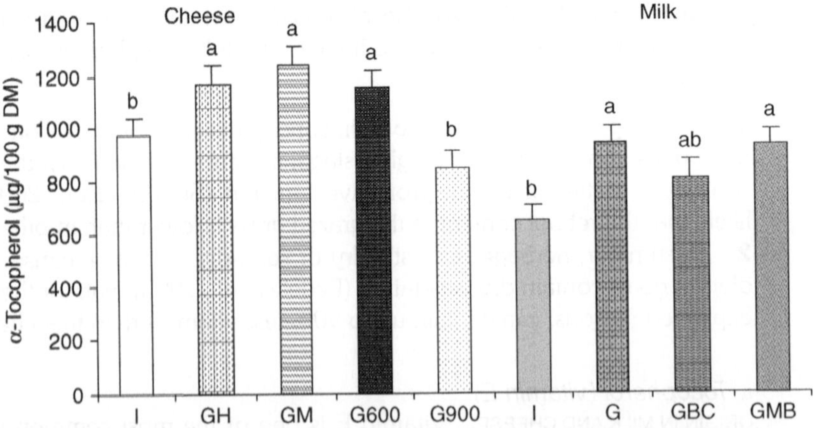

Fig. 5.12. Effect of different feeding treatments (I = hay + concentrate; GH = grazing hill pasture; GM = grazing mountain pasture; G600 = GH + concentrate 600 g/day; G900 = GH + concentrate 900 g/day; G = grazing valley pasture; GBC = G + mixed barley and chickpea grain; GMB = G + mixed maize and broad bean grain) on milk and cheese α-tocopherol content (on dry matter, DM, basis). a,b, mean values with different superscript letters were significantly different at $P < 0.05$. (Adapted from Fedele *et al.*, 2002.)

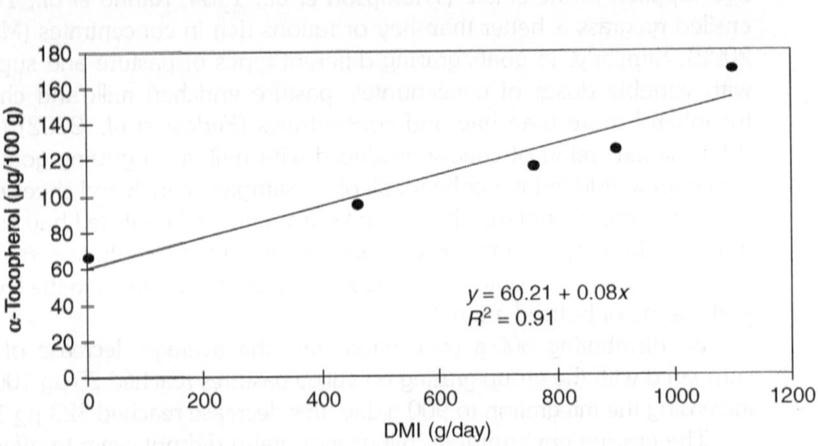

Fig. 5.13. Relationship between daily herbage dry matter intake (DMI) and milk α-tocopherol content. (Adapted from Pizzoferrato, 2001.)

(Di Trana *et al.*, 2001), interval from delivery (Di Trana *et al.*, 2001) and feeding system (Storry, 1988; Marsico *et al.*, 1993; Di Trana *et al.*, 2001; De Jong *et al.*, 2003). In ruminants, cholesterol originates in two main ways (Long, 1979).

1. Absorption by the mammary gland from the plasma lipoproteins, which contain either exogenous cholesterol of alimentary origin or cholesterol synthesized in the liver and small gut by endogenous pathways.

2. Cholesterol neo-synthesis of the mammary gland.

In goats, it has been calculated that 50–60% of cholesterol in milk comes from blood plasma (Mills *et al.*, 1974).

Feeding does not seem to affect this milk organic component in any remarkable way, probably because the interaction with other factors has greater importance.

Research on the subject often gives contradictory results, as shown by some results illustrated here. Goats fed on swards of *Festuca arundinacea* produced milk with about 20% less cholesterol than milk from goats fed hay and concentrates (Fig. 5.14) (Di Trana *et al.*, 2001). The goats of the pasture system received a supplementation of hay and concentrate based on their energy requirements, in order to reduce the effect of lipomobilization of body fat on milk cholesterol levels (Fedele *et al.*, 1997). The feeding treatment gave different results according to the breed used. Maltese goats tended to synthesize less cholesterol than Derivata di Siria goats (Di Trana *et al.*, 2001), as observed in Alpine, Beetal and Alpine × Beetal goats (Arora *et al.*, 1976).

The results were partly different when relating to cholesterol content of the same samples where tocopherol was measured (Fig. 5.15). In cheese, cholesterol was lower when animals were fed pasture grass only or pasture plus a supplementation of concentrate at 600 g/day per head. In milk, feeding solely pasture herbage did not cause the same effect, but the milk cholesterol content in these animals was almost similar to that of housed animals fed hay and concentrate at 600 g/day per head. In the latter case, the integration of grazed herbage with concentrates, rapidly (GBC) or less rapidly (GMB) degradable, decreased cholesterol levels in milk. These results indicate that the effect of pasture herbage on cholesterol is not clear at all, or becomes clear only if herbage can satisfy the animals' needs. Indeed, when grazing animals received the same quantity of concentrate as the housed group (660 g/day per head), cholesterol in milk decreased (Fig. 5.15).

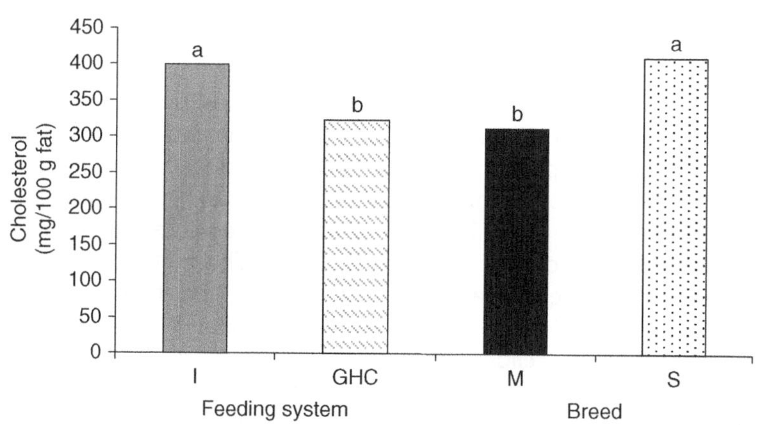

Fig. 5.14. Effect of feeding system (I = hay + concentrate; GHC = grazing + hay and concentrate) and breed (M = Maltese goat; S = Siriana goat) on milk cholesterol content. a,b, mean values with different superscript letters were significantly different at $P < 0.05$. (Adapted from Di Trana *et al.*, 2001.)

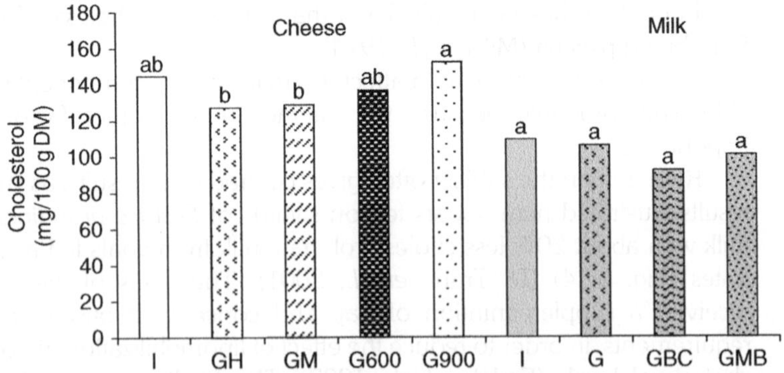

Fig. 5.15. Effect of feeding treatments (I = hay + concentrate; GH = grazing hill pasture; GM = grazing mountain pasture; G600 = GH + concentrate 600 g/day; G900 = GH + concentrate 900 g/day; G = grazing valley pasture; GBC = G + mixed barley and chickpea grain; GMB = G + mixed maize and broad bean grain) on milk and cheese cholesterol content (on dry matter, DM, basis). a,b, mean values with different superscript letters were significantly different at $P < 0.05$. (Adapted from Fedele *et al.*, 2002.)

Antioxidant protection

Molecules are considered to be antioxidant if they can neutralize, or slow down, the oxidative processes that affect a substance. α-Tocopherol (antioxidant) is able to slow down the oxidation reactions that can affect cholesterol (substance to be protected), and also FAs, especially the unsaturated ones. Many studies showed that the products of cholesterol oxidation ingested with diet are those mostly involved in the aetiology of coronary heart diseases and atherosclerosis (e.g. Kumar and Singhall, 1991); that is why the total content of cholesterol in a food is less important than the level of protection from oxidative processes it can benefit from.

In order to improve the nutritional and health characteristics of dairy products, the best possibility would be to lower the cholesterol content and increase its antioxidant protection, even if achieving only the second goal would be an important result. The antioxidant protection can be measured in several organic materials (Miller *et al.*, 1993). In milk and cheese it can be calculated by using the method of Pizzoferrato and Manzi (1999). This method defines the molar relationship between the antioxidant compound and the one to be protected, e.g. the molar relationship between α-tocopherol and cholesterol, as the degree of antioxidant protection (DAP).

The DAP assessment on milk and cheese, analysed according to their α-tocopherol and cholesterol content, highlighted that all products coming from feeding treatments based on hay, or on pasture where concentrate dominates over the grass, showed much less protected cholesterol than treatments based on grass (Fig. 5.16) (Pizzoferrato *et al.*, 2000). The fact that herbage guarantees increased antioxidant protection is also demonstrated by the

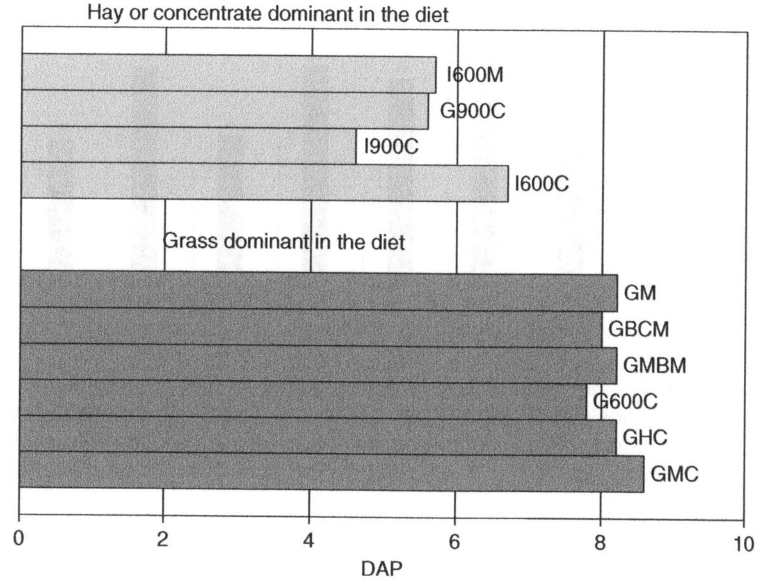

Fig. 5.16. Effect of feeding system (I600M = hay + 600 g concentrate, in milk; G900C = grazing + 900 g concentrate, in cheese; I900C = hay + 900 g concentrate, in cheese; I600C = hay + 600 g concentrate, in cheese; GM = grazing, in milk; GBCM = grazing + barley + chickpea, in milk; GMBM = grazing + maize + broad beans, in milk; G600C = grazing + 600 g concentrate, in cheese; GHC = grazing hill, in cheese; GMC = grazing mountain, in cheese) on the degree of antioxidant protection (DAP) of cholesterol in milk and cheese. (Adapted from Pizzoferrato *et al.*, 2000.)

observed marked reduction of DAP in cholesterol of cheese from grazing goats (grass-based diet) compared with hay-fed goats, both supplemented with concentrates at 900 g/day.

Conjugated linoleic acids and *n*–3 fatty acids

Although this aspect of milk quality is dealt with in more detail by Mele *et al.* (Chapter 3, this volume), here the effects of natural pasture in general, and the grazing season in particular, on CLA and *n*–3 FAs concentration in milk are described. Milk from unsupplemented grazing goats had almost twice the CLA and *n*–3 FAs content as that from goats fed indoors with hay and 600 g concentrate/day per head (Fig. 5.17) (Di Trana *et al.*, 2004). The grazing season, associated with different botanical composition and phenological stages of the plants, modifies the milk content in CLA. In winter, when plants were young and at vegetative rest, and pasture was dominated by grasses (e.g. *L. perenne*, *D. glomerata* and *Bromus ordeaceus*), milk had higher content of CLA and *n*–3 FAs. These variations were correlated to the linoleic acid content in feeds.

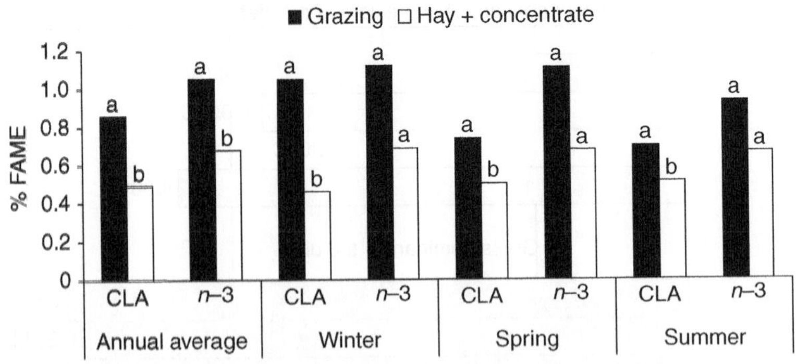

Fig. 5.17. Effect of feeding system and season on the content of conjugated linoleic acids (CLA) and *n*–3 fatty acids, expressed as percentage of total fatty acid methyl esters (FAME), in goat's milk. a,b, mean values with different letters were significantly different at $P < 0.05$. (Adapted from Di Trana *et al.*, 2004.)

This fatty acid went from an average of about 50% of fatty acid methyl esters in grass to an average of about 6% in the diet of housed goats, and from an average of 40% of the total fatty acid methyl esters in spring- and summer-grazed grass to about 68% in winter-grazed grass (Di Trana *et al.*, 2004). As grass develops (i.e. at more advanced phenological stages), it loses its effect of enrichment of milk in CLA and *n*–3 FAs, simply because it becomes poorer in linoleic acid.

Molecules of pharmacological interest

Many of the plants growing in a natural pasture are used by mankind for medicinal purposes to treat several diseases. The grazing goat browses a high number of plants, of which a quite high number are classified as medicinal (e.g. borage, hawthorn, mallow, chicory, fennel and sow thistle). The animal, through its selective action, can transfer some pharmacologically interesting molecules from grass to milk (Fig. 5.18) (De Feo *et al.*, 2006) and, consequently, to humans. In that study, goats fed borage (*Borago officinalis*) and hawthorn (*Crataegus oxyacantha*) produced milk with quite different profiles. Milk coming from goats fed the borage-based diet was the only one to contain β-sitosterol and 5,7,4-trihydroxyflavonol and lacked rutin, which was found, instead, in the milk from animals fed hawthorn. The difference between these two milks and the one from the control group was much more interesting. In the latter group, whose diet was based on natural pasture grass lacking the two studied plants, small traces of flavone and no traces of other molecules were found. Flavones have a certain antioxidant activity. Some of them are also acknowledged for other interesting activities: (i) rutin has been acknowledged as having a role in cancer prevention and an antioxidant activity (Aliaga and Lissi, 2004); (ii) β-sitosterol inhibits the growth of tumour cells (Awad *et al.*, 2005) and raises the pain threshold; and (iii) β-amirine has an anti-inflammatory action (Otuki *et al.*, 2005).

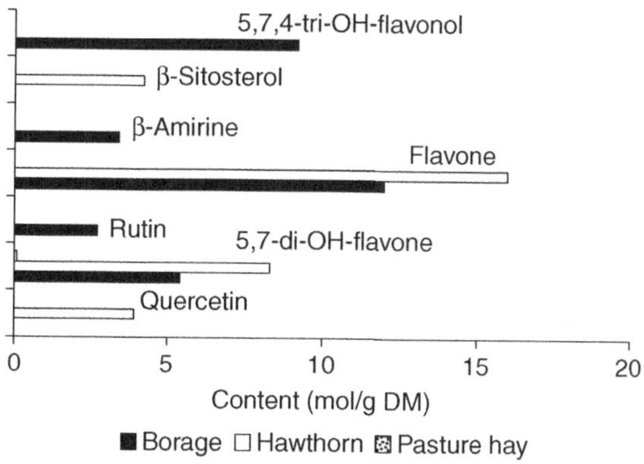

Fig. 5.18. Effect of botanical species or pasture hay on milk content (on dry matter, DM, basis) of compounds of pharmacological importance. (Adapted from De Feo *et al.*, 2006.)

Conclusions and Future Perspectives

Even if only some aspects of the aromatic and nutritional qualities of milk and cheese have been examined in this chapter, the information reported here indicates that feeding systems based on natural pastures represent a factor of quality diversification of milk and milk products. Nevertheless, many questions remain unanswered because of the lack of specific research on this topic, especially on goats. For example, it is not known if grazing systems on cultivated forage resources (swards and meadows) could guarantee the production of milk and cheese with characteristics similar to those expressed by natural pastures. The limited information available suggests that the vegetal species considered play an important role in these aspects. Grass species are poorer in terpenes compared with forbs (Mariaca *et al.*, 1997; Viallon *et al.*, 2000; Bugaud *et al.*, 2001), suggesting that swards and meadows cultivated with these species should be excluded from the diet, to produce milk and cheeses rich in these volatile components. Unfortunately, it is inconceivable or not feasible to cultivate species belonging to the botanical families richest in aromatic components, such as *Apiaceae*, asters and plantains (Mariaca *et al.*, 1997; Bugaud *et al.*, 2001). Ketones, alcohols and esters, highly present in vegetation and in milk, are not as well known as terpenes, and little or no information is available about their role in the aromatic characterization of milk and cheese. Milk and cheese nutritional quality is better understood. Preservation treatments, whatever they are, impoverish animal feeds, above all in vitamins. Even if future research does confirm, or consolidate, the value of natural pasture as an element of enrichment of milk in aromatic components appreciated by consumers, and in nutritional molecules useful for humans, will pasture systems be able to survive in industrialized society? Different answers could be given, but it is likely that only tomorrow's consumer will give the right answer. On the basis of present data on dairy products consumption, it seems that the consumption of food with an 'organic' brand or with a 'natural' image has increased by more than 300% over the last few years, despite a trend towards a rise in price of these products.

References

Aliaga, C. and Lissi, E.A. (2004) Comparison of free radical scavenger activities of quercetin and rutin – an experimental and theoretical study. *Canadian Journal of Chemistry* 82, 1558–1673.

Arora, R.L., Bindal, M.L. and Jain, M.K. (1976) Variation in fat unsaponifiable matter and cholesterol contents of goats' milk. *Indian Journal of Dairy Science* 29, 191–196.

Awad, A.B., Burr, A.T. and Fink, C.S. (2005) Effect of resveratrol and β-sitosterol in combination of reactive oxygen species and prostaglandin release by PC-3 cells. *Prostaglandins, Leukotrienes, and Essential Fatty Acids* 72, 219–226.

Bosset, O.J., Berger, T., Bühler-Moor, U., Bütikofer, U., Collomb, M., Gauch, R., Lavanchy, P., Mariaca, R., Sieber, R., Jeangros, B., Scecovic, J., Troxeler, J. and Dafflon, O. (1997) Comparison of some highland and lowland Gruyere-type cheese of Switzerland: a study of their potential PDO/AOC/AOP characteristics. In: Amado, R. and Battaglia, R. (eds) *Proceedings of Euro Food Chemistry*. Swiss Society of Food and Environmental Chemistry (SGLUG), Interlaken, Switzerland, pp. 395–400.

Bugaud, C., Buchin, S., Coulon, J.B., Hauwuy, A. and Dupo, D. (2001) Influence of the nature of alpine pastures on plasmin activity, fatty acid and volatile compound composition of milk. *Lait* 81, 401–414.

Claps, S., Rubino, R., Cogliandro, E. and Pilla, A.M. (1994) Fabbisogni alimentari e nutritivi, comportamento alimentare e produzione di latte di capre allevate al pascolo con differenti integrazioni alimentari. In: Rubino, R. (ed.) *Miglioramento dell'efficienza produttiva degli ovini e dei caprini*. Istituto Sperimentale per la Zootecnia, Potenza, Italy, pp. 2.0–2.16.

Claps, S., Sepe, L., Morone, G. and Fedele, V. (2001) Differenziazione sensoriale del latte e della caciotta caprina in rapporto al contenuto d'erba della razione. In: *Proceedings of I formaggi d'alpeggio e loro tracciabilità*. Agenzia Lucana per lo Sviluppo–Associazione Nazionale Formaggi Sotto il Cielo, Potenza, Italy, pp. 191–199.

Claps, S., Rubino, R., Fedele, V., Morone, G. and Di Trana, A. (2005) Effect of concentrate supplementation on milk production, chemical features and milk volatile compounds in grazing goats. *Options Méditerranéennes, Serie A* 67, 201–204.

Coulon, J.B., Martin, B., Verdier-Metz, I., Buchin, S. and Viallon, C. (2000) Etude du lien entre terroir et produit dans le cas de fromages AOC: influence de la composition floristique des fourrages sur les caractéristiques chimiques et sensorielles des fromages affinés. *Rencontres Recherches Ruminants* 7, 304–307.

Cowie, A.T., Ducombe, W.G., Folley, S.J. and Fran Easter, D.J. (1971) The turnover time of dietary cholesterol in the lactating rat. *Lipids* 6, 645.

Derby, G. (2001) *Lait, nutrition et santé*. Lavoisier, Paris.

De Feo, V., Quaranta, E., Fedele, V., Claps, S., Rubino, R. and Pizza, C. (2006) Flavonoids and terpenoids in goat milk in relation to forage intake. *Italian Journal of Food Science* 1, 85–92.

De Jong, A., Plat, J. and Mensink, R.P. (2003) Metabolic effect of plant sterols and stanols. *Journal of Nutritional Biochemistry* 14, 262–369.

Demarco, D. (1988) *La Statistica del Regno di Napoli*. Accademia Nazionale dei Lincei, Rome.

Di Trana, A., Cifuni, G.F., Claps, S. and Brughieri, A. (2001) L'influenza del sistema alimentare e della razza sul contenuto di colesterolo del latte di capre. In: *Proceedings of I formaggi d'alpeggio e loro tracciabilità.* Agenzia Lucana per lo Sviluppo–Associazione Nazionale Formaggi Sotto il Cielo, Potenza, Italy, pp. 200–206.

Di Trana, A., Cifuni, G.F., Fedele, V., Brughieri, A., Claps, S. and Rubino, R. (2004) Il sistema alimentare e la stagione influenzano il contenuto di CLA, omega-3 e acidi grassi trans nel latte di capra. *Progress in Nutrition* 6, 109–115.

Fedele, V., Pizzillo, M., Claps, S., Morand-Fehr, P. and Rubino, R. (1993) Grazing behaviour and diet selection of goats on native pasture in Southern Italy. *Small Ruminant Research* 11, 305–322.

Fedele, V., Claps, S., Rubino, R., Cogliandro, E. and Pilla, A.M. (1994) Fabbisogni alimentari e nutritivi, comportamento alimentare e produzione di latte di capre allevate in stalla in differenti sistemi alimentari. In: Rubino, R. (ed.) *Miglioramento dell'efficienza produttiva degli ovini e dei caprini.* Istituto Sperimentale per la Zootecnia, Potenza, Italy, pp. 1.0–1.21.

Fedele, V., Rubino, R. and Claps, S. (1997) Produzione foraggera, utilizzazione dell'erba e 'performances' produttive di capre allevate con differenti sistemi alimentari. *Rivista di Agronomia* 31, 333–337.

Fedele, V., Signorelli, F., Brancaleoni, E., Ciccioli, P. and Claps, S. (2000) Effect of concentrate grain source and herbage intake on physical-chemical features and milk aroma of grazing goats. In: Gruner, L. and Chabert, Y. (eds) *Proceedings of the 7th International Conference on Goats*, Vol. I. Institut National de la Recherche Agronomique, Paris, pp. 167–170.

Fedele, V., Sepe, L., Pizzoferrato, L., Manzi, P. and De Feo, E. (2002) Influenza del pascolo sulla qualità nutrizionale e di tipo farmacologico del latte e dei formaggi. Paper presented at *Comité Plénier Fromages 'Special fromages de montagne'*, Saint-Vincent, Val d'Aoste, Italy, 21–22 November.

Fedele, V., Rubino, R., Claps, S. and Pizzillo, M. (2003) Composés volatils: de l'herbe au fromage. *Abstracts of Talks of the 5th International Meeting on Mountain Cheese*, Arèches, France, 11–12 September, p. 18.

Fedele, V., Claps, S., Rubino, R., Sepe, L. and Cifuni, G.F. (2004a) Variation in terpenes content and profile in milk in relation to the dominant plants in the diet of grazing goats. *South African Journal of Animal Science* 34(Suppl. 1), 202–204.

Fedele, V., Rubino, R., Claps, S., Manzi, P., Marconi, S. and Pizzoferrato, L. (2004b) Seasonal retinol variation in goat milk associated with grazing compared to indoor feeding. *South African Journal of Animal Science* 34, 165–167.

Fedele, V., Claps, S., Rubino, R., Sepe, L. and Cifuni, G.F (2005a) Volatile compounds in herbage intake by goats in two different grazing seasons. *Options Méditerranéennes* 67, 261–267.

Fedele, V., Pizzillo, M., Claps, S. and Cifuni, G.F. (2005b) Effect of types of forage on terpenes content and profile in goat milk. 11th Seminar of Sub-Network FAO-CIHEAM on Sheep and Goat Nutrition, Catania, Italy. *Options Méditerranéennes*, in press.

Fedele, V., Rubino, R., Claps, S., Sepe, L. and Morone, G. (2005c) Seasonal evolution of volatile compound content and aromatic profile in milk and cheese from grazing goats. *Small Ruminant Research* 59, 273–279.

Forss, D.A. (1979) Review of the progress of dairy science: mechanism of formation of aroma compound in milk and milk products. *Journal of Dairy Research* 46, 671–706.

ISO-5495 (2005) *International Standard of Paired Comparison Test of Sensory Analysis Methodology*. International Organization for Standardization, Geneva, Switzerland.

ISO-6658 (2005) *General Guidance of Sensory Analysis*. International Organization for Standardization, Geneva, Switzerland.

Kanno, C., Yamauchi, K. and Tsugo, T. (1968) Occurrence of γ-tocopherol and variation of α and γ-tocopherol in bovine milk fat. *Journal of Dairy Science* 51, 1713–1719.

Keen, A.R. and Wilson, R.D. (1993) Effect of breed on colour and flavour. In: *Milk Fat Flavour*. New Zealand Dairy Research Institute, Palmerston North, New Zealand, pp. 50–54.

Kumar, N. and Singhall, O.P. (1991) Cholesterol oxides and arteriosclerosis: a review. *Journal of the Science of Food and Agriculture* 55, 497–510.

Long, C.A. (1979) Investigation of the origin of milk cholesterol in the goat. PhD thesis, Pennsylvania State University. UMI Dissertation Service, Ann Arbor, Michigan, pp. 1–67.

McDonald, P., Edwards, R.A. and Greenhalgh, J.F.D. (1992) Vitamine. In: *Nutrizione Animale*. Tecniche Nuove, Milan, Italy, pp. 58–87.

Mariaca, R., Berger, T., Gauch, R., Imhof, M., Jeangros, B. and Bosset, O.J. (1997) Occurrence of volatile mono- and sesquiterpenoids in highland and lowland plant species as possible precursors for flavor compounds in milk and dairy products. *Journal of Agricultural Chemistry* 45, 4423–4434.

Marsico, G., Di Trana, A., La Notte, E. and Muscio, A. (1993) Influence of breeding sub-system on the chemical and acid composition of milk fat in Podolica cows. *Archivio Veterinario Italiano* 44, 108–117.

Martin, B., Ferlay, A., Pradel, P., Rock, E., Grolier, P., Dupont, D., Gruffat, D., Besle, J.M., Ballot, N., Chilliard, Y. and Coulon, J.B. (2002) Variabilité de la teneur des laits en constituants d'intérêt nutritionnel selon la nature des fourrages consommés par les vaches laitières. *Rencontres Recherche Ruminants* 9, 347–350.

Miller, N.J., Rice-Evans, C., Davies, M.J., Gopinathan, V. and Milner, A. (1993) A novel method for measuring antioxidant capacity and its application to monitoring the antioxidant status in premature neonates. *Clinical Science* 84, 407–412.

Mills, C.S., Scott, T.W., Cook, L.J. and Nestel, P.J. (1974) Cholesterol metabolism in lactating ruminants. *Proceedings of the Australian Biochemical Society* 7, 35.

Molinard, P. and Spinnler, H.E. (1996) Review: compounds involved in the flavour of surface mold-ripened cheeses: origins and properties. *Journal of Dairy Science* 79, 169–184.

O'Sullivan, A.C. (1967) *Milk Flavour and Off-Flavour*. National Dairying Research Centre, Fermoy, Ireland.

Otuki, M.F., Vieri-Lima, F., Malheiros, A., Yunes, R.A. and Calixto, J.B. (2005) Topical anti-inflammatory effect of the ether extract from *Protium kleinii* and α-amyrin pentacyclic triterpene. *European Journal of Pharmacology* 507, 253–259.

Park, Y.W., Anderson, M.J., Walters, J.L. and Mahoney, A.W. (1983) Effect of processing methods and agronomic variables on carotene contents in forages and predicting carotene in alfalfa hay with near-infrared-reflectance spectroscopy. *Journal of Dairy Science* 66, 235–245.

Pizzoferrato, L. (2001) La ricerca della qualità. *Caseus* 6, 66.

Pizzoferrato, L. and Manzi, P. (1999) Effetto del pascolo sulla 'potenzialità' antiossidante di formaggi caprini: risultati preliminari. *Caseus* 4, 28–30.

Pizzoferrato, L., Manzi, P., Rubino, R., Fedele, V. and Pizzillo, M. (2000) Componenti bioattivi in prodotti lattiero-caseari caprini: influenza del sistema alimentare. In: Brandolini, V. and Menzioni, E. (eds) *Proceedings of the IV Congresso Nazionale di Chimica degli Alimenti*. Società Chimica Italiana, Ferrara, Italy, pp. 426–431.

Popják, G., French, T.H. and Folley, S.J. (1951) Utilization of acetate for milk-fat synthesis in the lactating goat. *Biochemical Journal* 48, 411–416.

Prache, S., Priolo, A., Tounadre, A., Jailler, R., Dubroeucq, H., Micol, M. and Martin, B. (2002) Traceability of grass-feeding by quantifying the signature of carotenoid pigments in herbivores meat, milk and cheese. *Grassland Science in Europe* 7, 592–593.

Scott, K.Y. and Rodriquez-Amaya, D. (2000) Pro-vitamin A carotenoids conversion factors: retinol equivalents – fact or fiction? *Food Chemistry* 69, 125–127.

Storry, J.E. (1988) The effect of dietary fat on milk composition. In: Haresign, W. and Cole, D.A.J. (eds) *Recent Development in Ruminant Nutrition 2*. Butterworths, London, p. 111.

Strauss, S.Y., Irwin, R.E. and Lambrix, V.M. (2004) Optimal defence theory and flower petal colour predict variation in the secondary chemistry of wild radish. *Journal of Ecology* 92, 132–141.

Thompson, S.Y., Henry, K.M. and Kon, S.K. (1964) Factors affecting the concentration of vitamins in milk I. Effect of breed, season and geographical location on fat-soluble vitamins. *Journal of Dairy Research* 31, 1–25.

Urbach, G. (1997) The flavour of milk and dairy products: II cheese: contribution of volatile compounds. *International Journal of Dairy Technology* 50, 79–89.

Urbach, G., Stark, W. and Forss, D.A. (1972) Volatile compounds in butter oil II. Flavour and flavour threshold of lactones, fatty acids, phenols, indoles and skatole in deodourised synthetic butter. *Journal of Dairy Research* 39, 35–47.

Viallon, C., Verdiez-Metz, I., Denoyer, C., Pradel, P. and Coulon, J.B. (1999) Desorbed terpenes and sesquiterpenes from forages and cheeses. *Journal of Dairy Research* 66, 319–326.

Viallon, C., Martin, B., Verdier-Metz, I., Pradel, P., Garel, J.P., Coulon, J.B. and Bedagué, J.L. (2000) Transfer of monoterpenes and sesquiterpenes from forages into milk fat. *Lait* 80, 635–641.

6 Energy and Protein Requirements of Goats

A. Cannas,[1] A.S. Atzori,[1] F. Boe[1] and I.A.M.A. Teixeira[2]

[1]Dipartimento di Scienze Zootecniche, Università di Sassari, Via E. De Nicola 9, 07100 Sassari, Italy; [2]Departamento de Zootecnia, Universidade Estadual Paulista 'Júlio de Mesquita Filho' UNESP, Via Prof. Paulo Donato Castellane km 5, 14884-900 Jaboticabal, Brazil

Introduction

Goats are widespread worldwide due to their nutritional and environmental adaptability. They are reared in a wide range of breeding (from extensive to highly intensive) and feeding (from grazing diets to total mixed rations) systems and geographical areas. As a result, milk production and body composition vary probably more in goats than in other ruminant species. Thus, correct estimation of their energy and protein requirements in each unique condition is necessary to optimize diet composition or to determine optimum stocking rates of goats.

Influence of Production Levels on Animal Requirements

The variability in milk production largely affects the relative importance of maintenance and productive requirements of goat flocks and thus the efficiency of utilization of dietary nutrients for milk production. In fact, considering a flock with constant number of heads and fecundity, as the mean milk production level increases, the flock individual and total maintenance, growth and pregnancy requirements vary little, while milk production requirements increase very quickly. This applies not only to energy requirements (Table 6.1), but also to any other nutritional requirement. The increase in total flock requirements is slower than the increase in production level, because of the fixed costs of maintenance. Therefore, considering energy requirements as an example, if total energy requirements (maintenance, growth, pregnancy and milk production) are divided by total milk production of the flock, the mean energy cost to produce each litre of milk can be easily estimated. As described in Table 6.1, such costs decrease rapidly as mean milk production level increases, varying from 4.16 to 1.47 Mcal of net energy of lactation (NEL) per kilogram of milk for an annual production level going from 200 kg/head (extensive systems) to 1000 kg/head (very intensive

Table 6.1. Simulation of the effect of various production levels (at constant milk fat composition of 4%) on the annual milk production and energy requirements of a herd of goats. The flock used for the simulation had 18 replacement kids, 18 primiparous does (14 lactating and four dry), 82 pluriparous does (76 lactating and six dry), and three intact males. The mean body weight for adult does was 60 kg.

Milk per goat milked (kg/year)	Total flock milk yield (kg/year)	Total NEL required by the flock (Mcal/year)	Total NEL required per kilogram of milk (Mcal/kg)	NEL for milk production (% of total requirements)	NEL for maintenance and growth requirements (% of total requirements)
200	18,000	75,143	4.16	17	83
400	36,000	89,531	2.48	29	71
600	54,000	103,919	1.92	38	62
800	72,000	118,199	1.63	46	54
1000	90,000	132,694	1.47	52	48

NEL, net energy of lactation.

systems), respectively. Even though the nutritional costs per unit of milk produced decrease drastically as production levels change from the lowest to the highest level, the optimal production level depends on the amount, quality and cost of the feeds available on the farm.

Main Feeding Systems for Goats

Most studies on the nutrition of goats have been published quite recently. A still widely used publication on nutrient requirements of goats was published by the US National Research Council (NRC, 1981b). Several years after the publication of the NRC (1981b) system, other feeding systems for goats were developed and published. In this chapter, estimates of the energy and protein requirements of goats reported in the literature, utilizing the most important and widely used feeding systems, are compared. The analysed systems are: (i) the Institut National de la Recherche Agronomique (INRA, 1988, 2007; INRA); (ii) the Agricultural and Food Research Council (AFRC, 1998; AFRC); (iii) the Estación Experimental del Zaidin, Granada (Aguilera, 2001; EEZ); and (iv) the E(Kika) of the Garza Institute for Goat Research–Langston University (Luo *et al.*, 2004a,b,c,d,e; Moore *et al.*, 2004; Sahlu *et al.*, 2004; IGR). The IGR system was almost integrally adopted by the recently published NRC (2007) system for goats. The INRA and AFRC systems deal with dairy breeds only, the EEZ deals mostly with the dairy breed Murciano-Granadina, while the IGR system reports the requirements of dairy breeds, of meat and indigenous breeds, and of the Angora breeds separately.

All feeding systems allow the prediction of nutrient requirements, feed value and feed intake. Although determination of the nutritional value of feed is often calculated independently from the animal species that will utilize it, nutrient requirements and feed intake are usually calculated with species-specific equations.

Because of the interaction existing between the estimation of animal requirements and the amount and type of ingested nutrients, the comparisons of the various systems here described do not allow determination of their accuracy. Such types of evaluation can be done only in experimental trials in which all variables of each system, including feed value, are measured.

Instead, in this chapter the comparisons of the main feeding systems to estimate energy and protein requirements of goats are aimed at: (i) highlighting differences in inputs required, aggregation level and flexibility among them, and also in relation to sheep and cattle systems; and (ii) quantifying, by simulating different scenarios, the quantitative differences in the energy and protein requirements estimated by the various systems.

Energy Requirements of Adult Goats

In all studied systems the calculation of total requirements is based on a factorial approach. Therefore, the requirement for each function is calculated separately and then added to the energy requirements for the other functions.

The INRA system uses a single energy unit for all functions (e.g. maintenance, milk production, pregnancy), the UFL, which corresponds to 1700 kcal of NEL (Table 6.2). The AFRC and EEZ systems calculate net energy (NE) requirements for each function, which are then converted to metabolizable energy (ME) using a specific conversion efficiency (k) for each physiological function, while the IGR system based its estimates either on the above mentioned approach (called Method 1) or on the direct estimation of ME requirements obtained by regressing MEI on RE (called Method 2) (Table 6.2). Only Method 2 is used by NRC (2007).

For this reason, the equations proposed by the various systems to estimate the efficiency of conversion of ME to NE could be compared only by attributing discrete values to their independent variables (Tables 6.2 and 6.3). In order to convert NE to ME, it was necessary to calculate the ratio of ME to gross energy (GE) in the diet (called metabolizability or q_m) for the AFRC and INRA systems, and the concentration of ME in the diet (MEC) for the IGR system. The MEC was calculated by multiplying q_m by GE. The latter term was considered constant and equivalent to 4.398 Mcal per kilogram of dry matter (DM; 18.4 MJ/kg DM). Both q_m and MEC were calculated at maintenance feeding level. In this chapter, efficiencies of the AFRC, INRA and IGR systems were compared using a series of q_m or MEC values, within the range normally observed in diets for goats. By contrast, since the EEZ system utilizes fixed numbers for the various efficiencies, it did not require the estimation of variables related to the diet and feeding level.

The AFRC and IGR (Method 1) systems use the same equation to estimate the efficiency of utilization of ME for maintenance (k_m), which varies from 0.64 (for low-quality diets) to 0.75 (for high-quality diets) (Table 6.3). The EEZ system uses a k_m of 0.73, which is very similar to that proposed by AFRC and IGR for high-quality diets. The k_m used by the INRA system is identical to the efficiency of ME utilization for milk production (k_l), with values varying from 0.56

Table 6.2. Efficiencies of conversion of metabolizable energy (ME) to net energy (NE) for different physiological functions.

	INRA	AFRC	EEZ	IGR
Maintenance	k_l	$k_m = 0.503 + 0.35q_m$[a]	0.73	$k_m = 0.503 + 0.019$ MEC[b]
Lactation	$k_l = 0.24q_m + 0.463$	$k_l = 0.42 + 0.35q_m$	0.67[c]	0.62/0.59[d]
Growth and fattening	k_l[e]	$k_g = 0.006 + 0.78q_m$[f]	NA	
k_g of body reserves in lactation	k_l	$0.95k_l$	$0.907k_l$	0.75
k_g of body reserves in dry periods	k_l	k_g	NA	
NE body loss → NE milk	0.80	0.84	0.84	0.84
NE body loss → NE maintenance	NA	0.80	NA	k_m
Hair growth	NA	k_g[g]	NA	0.151 (Angora breed)

NA, not available.

[a] q_m = ME/gross energy (GE) at maintenance feeding level; calculated assuming GE = 4.398 Mcal/kg dry matter.

[b] Same equation as AFRC, rewritten to use MEC instead of q_m; MEC = metabolizable energy per kilogram of dry matter of the diet (MJ). Based on IGR method 1.

[c] Obtained on pelleted diets.

[d] Two methods to measure ME_m, ME_l and k_l, are used; for k_l, the first value corresponds to the IGR method 1, which uses the same ME_m, derived from AFRC, for all breeds; the second value corresponds to the IGR method 2, in which ME_m varies between genetic groups (dairy versus indigenous breeds) and is reported only in terms of ME. In both cases k_l was estimated correcting its value to account for the cost of N excretion on maintenance requirements.

[e] For replacements.

[f] For pelleted diets, $k_g = 0.024q_m + 0.465$.

[g] Only if the NE is retained in fibre in significant amounts.

Table 6.3. Efficiencies of conversion of metabolizable energy (ME) to net energy (NE), calculated by four different systems, considering q_m in the diet ranging from 0.4 to 0.7 (from 1.76 to 3.08 Mcal of ME/kg dry matter).

	INRA	AFRC	EEZ	IGR
Maintenance	0.56–0.64	0.64–0.75	0.73	0.64–0.75
Lactation	0.56–0.64	0.56–0.67	0.67	0.62–0.59
Growth and fattening	0.56–0.64	0.32–0.55	NA	
k_g of body reserves in lactation	0.56–0.64	0.53–0.64	0.66	0.75
k_g of body reserves in dry periods	0.56–0.64	0.32–0.55	NA	
NE body loss → NE milk	0.80	0.84	0.84	0.84
NE body loss → NE maintenance	NA	0.80	NA	0.64–0.75

NA, not available.

to 0.64, depending on the q_m. Such values are much lower than the k_m values used by the previous systems.

The k_l used by INRA, AFRC and IGR are very similar among themselves, while the EEZ system proposes a k_l of 0.67, which is very similar to those suggested by INRA and AFRC for high-quality diets. Indeed, high-quality diets were utilized in the studies of EEZ. This indicated that the conversion efficiency for the different physiological functions is not species-specific (Aguilera, 2001).

The AFRC estimates with a specific equation for this function the conversion efficiency of ME to NE for growth and fattening (k_g), while INRA considers k_g equal to k_l, which is much higher than the k_g estimated by the AFRC system (Table 6.3). The EEZ system, instead, does not suggest any value for k_g. The IGR system does not use k_g because it estimated ME requirements for growth and fattening directly, by regressing MEI on average daily gain.

A relevant difference among the systems is the value of the efficiency for fat deposition during lactation, which is much higher in IGR than in the other systems (Table 6.3), probably because of an imperfect separation between k_l and k_g in IGR. In fact, if the average between the k_l and k_g values for lactation is calculated for that system, the obtained mean efficiency value for lactation is similar to that measured experimentally by EEZ for lactation and fat deposition together (Aguilera *et al.*, 1990).

Maintenance requirements

Even in high-producing flocks, maintenance costs represent the major component of total requirements (Table 6.1). For this reason their accurate estimation is of great importance.

The maintenance requirements estimated by AFRC are mainly based on the equations of the UK Agricultural Research Council (ARC, 1980), derived from calorimetric studies conducted mostly on fasting sheep and cattle. The AFRC system suggests increasing the basic maintenance requirements by 10% to account for the minimum movement performed in confinement. Such correction is already included in the values reported in Table 6.4. The INRA system is based on material balance studies on goats in the descending phase of lactation, integrated with information obtained by the other feeding systems. The requirements estimated by EEZ are based on calorimetric studies conducted by this research group on the Granadina breed (Aguilera and Prieto, 1985, 1986, 1991; Lachica *et al.*, 1995). Maintenance requirements of EEZ were estimated by regression equations between the total daily ME intake (MEI), using diets that assured a slightly positive energy balance, and the retained energy (RE), estimated by subtracting basal body heat production (i.e. maintenance energy), measured calorimetrically, from the MEI. The IGR system presented two approaches. One (Method 1 of IGR) was directly derived from the AFRC method. The other method (Method 2 of IGR) was based on the statistical analysis of data from several calorimetric and comparative slaughtering and feeding studies from the literature. The database included animals from eight different breeds in different physiological stages. Maintenance requirements were

Table 6.4. Basic requirements and corrections applied to estimate maintenance energy requirements by four different feeding systems for goats and by the Cornell Net Carbohydrate and Protein System (CNCPS) for sheep.

	Sheep	Goats			
	CNCPS[a]	INRA	AFRC	EEZ	IGR
Basic requirements					
NE (kcal/kg $BW^{0.75}$)	60.3–51.9[b]	63.3	82.8[c]	73.6	82.8[c]
k_m or k_l[d]	0.64	0.56–0.64	0.64–0.75	0.73	0.64–0.75[c]
ME (kcal/kg $BW^{0.75}$)	94.2–80.4	116.6–102.1	129.4–110.4	100.8	129.4–110.4[c] 101.0–119.8[e]
Corrections					
Breed	no	no	no	Granadina	Meat and indigenous; dairy; Angora
Gender	yes	yes	no	yes	yes
Age	yes	no	no	yes	yes
Grazing activity	yes	yes	yes	yes	yes
Cold stress	yes	no	no	no	no
Temperature	yes	no	no	no	no
Rain	yes	no	no	yes	no
Acclimatization	yes	no	no	no	yes
Hair/wool depth	yes	no	no	no	no
Body reserves	no	no	no	no	yes
Cost of urea excreted	yes	no	no	no	no
Production	0.09MEI[f]	no[g]	no[h]	no[g]	no

NE, net energy; BW, body weight; ME, metabolizable energy.
[a]Cannas *et al.* (2004).
[b]Variations due to the age of the animals.
[c]The values include a 10% increase over fasting metabolism for movement in confined systems. For IGR, Method 1 is used here.
[d]The range of k_m (AFRC and IGR) and k_l (INRA) was estimated considering variations from 0.4 to 0.7 of q_m (AFRC and INRA), and from 7.36 to 12.88 MJ/kg dry matter (DM) (IGR), assuming GE = 18.40 MJ/kg DM.
[e]The lowest value is for meat and indigenous breeds; the highest is for dairy breeds. Based on the IGR Method 2. For the Angora breed, ME_m requirements are equal to 113.0 kcal/kg $BW^{0.75}$.
[f]Multiplicative factor; MEI = ME daily intake.
[g]Total UFL (NEL) requirements (maintenance, production, pregnancy) are multiplied by the correction factor (2.5 × milk yield at 3.5% fat/BW) to account for the decease in digestibility and the associative effects that occur as feeding levels increase.
[h]Total ME requirements (maintenance, production, pregnancy) are multiplied by the correction factor [1 + 0.018 (L–1)], where L is multiple of maintenance ME requirements, to account for the decrease in digestibility that occurs at feeding levels higher than 1.

estimated directly in terms of ME with regression equations between MEI and RE. The IGR system thus estimated maintenance requirements specific for three breed groups (dairy, meat and indigenous, and Angora) and various animal categories (growing animals, dry animals and wethers, lactating goats, intact males).

In all feeding systems for goats, maintenance requirements are calculated in relation to basic requirements, proportional to metabolic body weight ($BW^{0.75}$) and corrected for various factors, as already done in the most recent sheep (Cannas, 2004) (Table 6.4) and cattle systems (Cannas, 2000).

The basic requirements, expressed as ME, are highest for AFRC and IGR Method 1 and lowest for INRA (Table 6.4). The EEZ system uses a fixed value for basic requirement obtained with high-quality diets, as mentioned before. This value is similar to that of INRA calculated for high-quality diets and much lower than the values of AFRC and IGR Method 1. Consistent variations in the basic ME maintenance requirements estimated by IGR Method 2 depend on production aptitude (meat and indigenous, dairy and Angora breeds). In all feeding systems, goats' maintenance requirements are higher than those of sheep (Cannas, 2004) (Table 6.4).

When maintenance requirements are expressed as NE, they are the highest for AFRC and IGR Method 1 intermediate for EEZ, and lowest for INRA.

All systems adjust maintenance requirements for the physical activity associated with grazing. Since such activity is particularly important for goats, it will be discussed in a separate section of this chapter.

Effects of gender and age on maintenance requirements are taken into account only by the EEZ and IGR systems, using higher values for intact males and growing animals, while INRA accounts only for the effect of gender.

In contrast to the feeding systems for sheep, none of the systems for goats estimates the effects of cold or heat stress on maintenance requirements (Table 6.4). However, IGR corrects maintenance requirements to account for acclimatization, based on the mean temperature of the previous and current months, applying the same method, proposed previously by NRC (1981a), used by the Cornell Net Carbohydrate and Protein System (CNCPS) for sheep (Cannas *et al.*, 2004). The lack of estimation of cold and heat effects on maintenance requirements is an important limit of the feeding systems for goats, especially because this species often lives in the mountains of temperate areas, where cold stress is likely, and is widespread in tropical and subtropical areas, where heat stress is common.

In contrast to the most recent systems for cattle (Fox *et al.*, 2004) and sheep (Cannas *et al.*, 2004), none of the systems for goats has equations to estimate the energetic cost related to the transformation of ruminal or endogenous ammonia in urea (Table 6.4). However, IGR adjusted maintenance requirements to take into account the trials in which a protein excess caused a urea production cost.

Since underfed animals reduce their maintenance requirements by a physiological mechanism of adaptation, IGR uses a correction factor which takes into account body reserves (body condition score < 3) and number of weeks of feed restriction prior to the moment of calculation of the requirements (Sahlu *et al.*, 2004). Instead of using this correction, the CNCPS for sheep (Cannas *et al.*, 2004) considers a direct effect of the feeding level on maintenance requirements (Table 6.4).

Physical activity requirements

Requirements for physical activity are due to the transformation of chemical energy into mechanical work. Requirements for normal activity in confinement are already included in the basic values of Table 6.4. For grazing animals, movement, related to ground roughness and the distance walked by the goats while grazing, and chewing activity are particularly important. Requirements for movement vary among species because of morphological, physiological and behavioural differences (Aguilera, 2001).

The various systems for goats express the requirements for grazing in two ways. The simplest one increases maintenance requirements by multiplying them by a percentage value, which varies with grazing conditions and biomass availability (NRC, 1981b), while the other method estimates grazing requirements as a function of more than one variable, following a factorial approach (Tables 6.4 and 6.5).

The INRA system (Table 6.5) is based on the studies of Blaxter (1962) and suggests separate requirements per unit of distance walked horizontally, uphill or downhill. In addition, it uses values, expressed as a percentage of maintenance, varying from 10% for good pastures to 30% or even 60% for pastures in marginal areas. Such values are similar to those proposed by NRC (1981b), which estimates grazing requirement values from 25 to 75% of maintenance.

The EEZ system (Table 6.5), based on experimental measurements of body heat production caused by movement, proposes an equation to estimate walking requirements as a function of soil slope (from –10 to +10%) and of the horizontal distance walked, and two other equations for ascending and descending movements (vertical component) (Lachica *et al.*, 1997c). The effects of meteorological conditions on requirements for walking are also estimated by EEZ (Lachica *et al.*, 1997b, 1999). Among the various methods proposed by this system for calculating the energy required for physical activity, there are calorimetric and

Table 6.5. Net energy costs of various physical activities of goats estimated by INRA, AFRC and EEZ.

Activity	INRA	AFRC	EEZ
Walking, horizontal	51 kcal × km[a,b]	0.836 kcal/kg BW × km	0.80 kcal/kg BW × km
Walking, vertical component uphill	374 kcal × km[a]	6.69 kcal/kg BW × km	7.58 kcal/kg BW × km
Walking downhill	170 kcal × km		3.15 kcal/kg BW × km
Standing		2.39 kcal/kg BW × day	
Changing position, down and up		0.062 kcal/kg BW	
Efficiency of ME → NE	k_l	k_m[c]	0.309

ME, metabolizable energy; NE, net energy; BW, body weight.

[a]In practice INRA suggests increasing maintenance requirements from 10 to 60% depending on the quality and type of pasture used.

[b]km = kilometres.

[c]Efficiency of maintenance from Table 6.2.

non-calorimetric techniques and a method based on the use of ^{13}C (Prieto *et al.*, 2001). The movement requirements for goats estimated by EEZ are lower than those reported in the literature for sheep and cattle.

The AFRC system (Table 6.5) calculates physical activity requirements on the basis of data reported by ARC (1980), by INRA (Morand-Fehr *et al.*, 1987) and by the research group of EEZ (Lachica *et al.*, 1997b,c; Prieto *et al.*, 1997). The AFRC values for physical activities are in accordance with the ARC, INRA and EEZ systems. Such a system proposes unitary requirements for each type of movement and total requirements according to pasture type and breeding conditions, thus considering the mean distances walked by the animals.

The physical activity requirements of IGR (Table 6.6) are based on the energy requirements estimated by all previously mentioned systems, modified to account for many other variables, following the methods of George (1984) and Rochinotti (1998, cited by Sahlu *et al.*, 2004). The equation used to estimate these requirements is reported by NRC (2007). In brief, requirements for physical activity are proportional to grazing time, are calculated as multiples of maintenance requirements, and are corrected for diet quality, walked distance and ground conditions. Diet quality is evaluated on the basis of the apparent digestible organic matter, while walked distance is measured in kilometres covered horizontally. Ground type is measured by a scale ranging from 1 to 5, in which 1 represents flat ground and 5 represents uneven and very hilly ground. The calculated final value is a coefficient to be multiplied by maintenance requirements. This method does not take into account possible interactions among the various variables considered.

Very few studies have estimated the energy requirements for muscle contraction and distension for prehension, chewing, swallowing, rumination and feed movement along the gastrointestinal tract, and the influence of the chemical–physical characteristics of feed on such processes. The chewing energy requirements for goats were estimated for the first time by EEZ on fistulated goats kept in metabolic cages. In goats, increase in body heat production was

Table 6.6. Inputs and output of the IGR system to estimate the physical activities of grazing goats. The equation used to estimate the total energetic cost of grazing activities is reported by NRC (2007).

Variable	Units
Inputs	
Time spent grazing and in movement	hours
Dietary quality as TDN concentration or OM digestibility	per cent
Distance walked	kilometres
Ground type	1–5 (qualitative scale)
Output	
Energetic cost of grazing activities	ME_m multiplicative coefficient

TDN, total digestible nutrients; OM, organic matter; ME_m, metabolizable energy for maintenance.

measured for various feeds, as a function of BW, ingestion and rumination time, and amount of feed digestible DM (Lachica *et al.*, 1997a). Such studies demonstrated that the energy cost of chewing activity is mainly due to the ingestion process, whose cost depends on digestible DM and type of feed, and varies from 0.4% of the feed ME for barley grain to 4.7% of the ME for olive branches.

Milk production requirements

Energy requirements for milk production are calculated from the milk NE cost of production, by measuring the milk energy content, using a calorimetric bomb.

Differences in ME requirements for milk production are due to variations in the conversion efficiency from ME to NE, depending on whether the energy comes from feed or body fat mobilization (Table 6.2). All systems propose a milk energy content associated with a specific milk fat content, which is the main factor influencing it, and estimate it by equations including milk fat content as predictor (Table 6.7). Estimated milk NE costs of production are similar for INRA, AFRC and IGR, while they are much lower for EEZ. Such discrepancies may be due to differences in the protein content of the milk used in the various systems.

Maintenance and milk production requirements

In order to determine the effects of the variables considered so far on total requirements of goats, a simulation was performed using goats with different weight (50 and 70 kg BW) and production level (from 0 to 4 kg of milk with 4% fat per day) (Fig. 6.1). All systems estimate similar ME maintenance requirements, except for IGR, which predicts 10% higher values than the other systems. The initial differences among IGR, AFRC and INRA systems remain constant as milk production increases, indicating that requirements for milk production are very similar (Fig. 6.1). In fact, their regression lines are parallel and slightly different at the intercept. By contrast, the EEZ system estimates the lowest total energy

Table 6.7. Estimates of milk energy content by various feeding systems.

System	Units	Estimation of milk NE	NE of milk with 4% fat
INRA	kcal NE/kg	$[0.4 + 0.0075 \times (F - 35)] \times 1700$	744
AFRC	kcal NE/kg	$(0.0406 \times F + 1.509)$[b] $\times 239$	749
EEZ	kcal NE/kg	In proportion to the energy value of milk with 4% fat	670
IGR	kcal NE/kg	$(1.4694 + 0.4025 \times F\%) \times 239$	736

NE, net energy; F%, percentage of fat in the milk; F, g fat/kg milk.

[a]NE of milk with 3.5% fat, 2.9% protein, 4.3% lactose corrected as a function of the variations in milk fat percentage (ΔF%) with respect to 3.5%.

[b]Equation developed on dairy cows (Tyrrell and Reid, 1965).

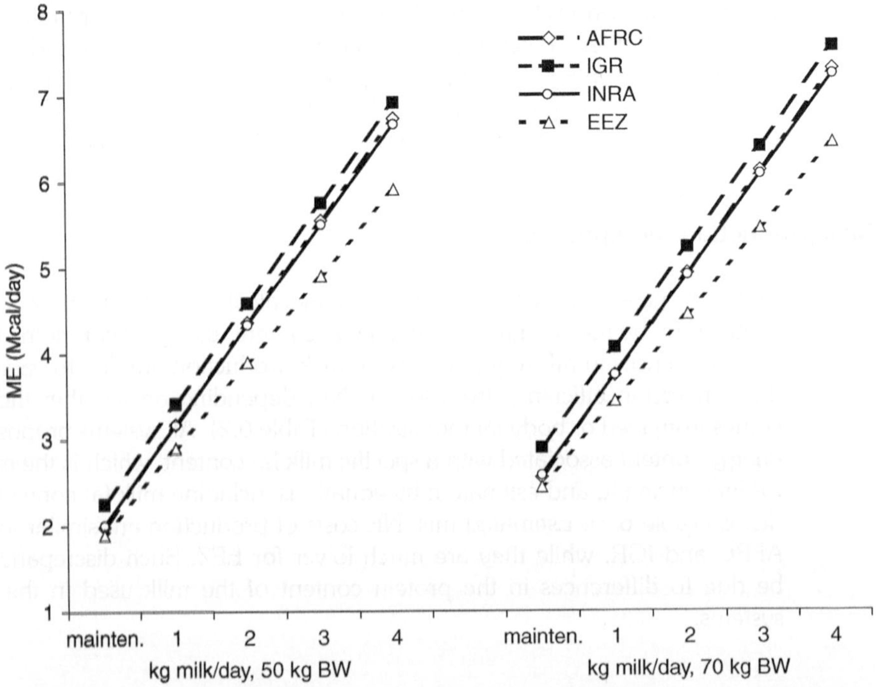

Fig. 6.1. Sum of daily maintenance (mainten.) and milk production metabolizable energy (ME) requirements by four feeding systems, assuming $q_m = 0.6$ (2.639 Mcal ME/kg dry matter) and milk production with 4% fat. For IGR, the maintenance value for dairy breeds was considered (Method 2).

costs in terms of ME, due to the lowest NE requirement for milk production (Table 6.7) and the highest k_l (Tables 6.2 and 6.3) values used. The difference between the highest estimate of ME requirement for maintenance (IGR) and the lowest one (EEZ) is above 1 Mcal ME/day (15% of total requirements) for the highest production levels.

Pregnancy requirements

All systems indicate that the increase in energy requirements during pregnancy depends on the costs associated with the growth of uterus, fetus, placenta and mammary gland. The estimated energy cost for pregnancy varies from 2 to 2.5 times the energy fixed in the uterus (INRA, 1988).

The INRA system calculates pregnancy NE requirements on the basis of total weight of kids born per birth (Morand-Fehr *et al.*, 1987; INRA, 1988). The equation used by INRA, initially proposed by Tissier and Theriez (1978; cited by INRA, 1988), seems to be wrong because it calculates very low values of NE requirements. Anyway, on the basis of this equation, INRA proposed a simplified approach to estimate pregnancy requirements, suggesting values equal to 15 and 30% of the maintenance requirements of does at the fourth and fifth month

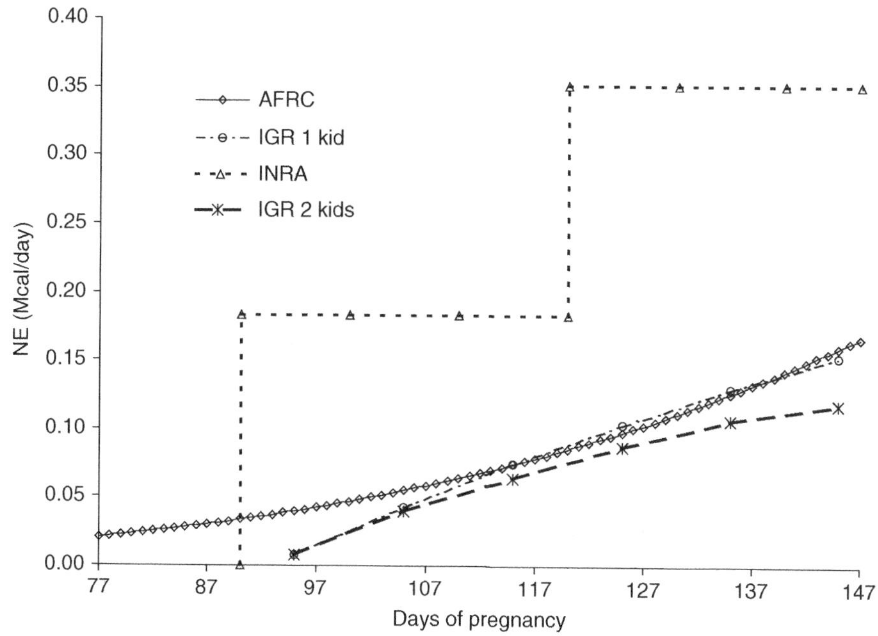

Fig. 6.2. Net energy (NE) requirements for pregnancy, assuming the delivery of one 4-kg kid (AFRC and IGR) or two 2-kg kids (IGR) at birth. Estimates by INRA were calculated assuming a goat body weight (BW) of 55 kg, because this system estimates pregnancy requirements based on mother's BW.

of pregnancy, respectively. Thus, INRA in practice does not take into account the number and the weight of kids, but it considers only the weight of the mother. The NE requirements of pregnancy proposed by INRA are much higher than those proposed by the other systems (Fig. 6.2), probably to satisfy the requirements of highly prolific goat breeds. Pregnancy requirements during the first months of pregnancy are considered negligible by INRA.

The estimates for pregnancy by AFRC are based on the assumption that fetus weight and pregnancy duration of goats are equal to those of ewes. Energy and protein requirements are calculated on the basis of retained energy (NE) by the gravid uterus during pregnancy, according to a growth curve described by the model of Gompertz (ARC, 1980).

Based on sheep data (McDonald *et al.*, 1979), adjusted for British Saanen kids' birth weights of 4.4, 3.95 and 3.65 kg for a single, twin and triplet pregnancy, respectively, AFRC determined the relationship between total uterus content and the weight and number of fetuses. Thus, the system gives a series of values of pregnancy requirements as a function of number of days from delivery (starting from the 60th day of pregnancy) and number of fetuses.

The approach used by IGR, which is very similar to that of AFRC (1998), hypothesizes kids of 2, 3, 4 and 5 kg, and divides pregnancy into 10-day intervals from the 91st day on, considering that energy requirements for pregnancy are relevant only after the third month of pregnancy. In addition, IGR considers that, for a constant total fetal weight, requirements decrease as the number of kids per birth increases. This is based on the fact that for a total weight of kids at birth, as the number of newborn kids increases, their individual weight decreases, thus increasing the proportion of bone tissue in relation to muscle and lipid tissues.

The EEZ system does not give any recommendation for pregnancy requirements.

Pregnancy requirements for the last 3 months of pregnancy were compared, assuming that a goat of 55 kg BW delivers a single 4-kg kid at birth and, only for IGR, two kids of 2 kg each also. The NE (Fig. 6.2) and ME (Fig. 6.3) requirements of pregnancy are very similar for AFRC and IGR, despite the fact that the shape of the energy requirement curve as a function of time from delivery is concave for the first system and convex for the second. The two curves of the IGR system, i.e. for one 4-kg kid or two 2-kg kids, show that, for a constant total fetal weight, as the number of fetuses increases the total energy requirement decreases, above all at the end of pregnancy. The NE requirements of pregnancy estimated by INRA are much higher than those of the other two systems (Fig. 6.2), probably because its estimates are based on does' weight instead of kids' weight. On the contrary, the ME requirements at the end of pregnancy estimated by INRA are much lower than those predicted by AFRC and IGR. The discordant INRA estimates are probably an effect of the very high efficiency of

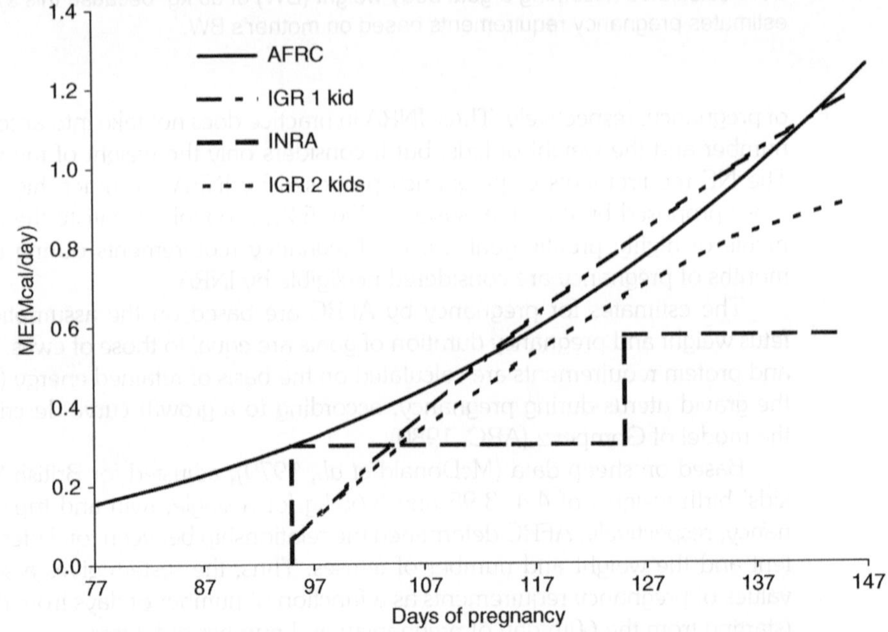

Fig. 6.3. Metabolizable energy (ME) requirements for pregnancy, assuming the delivery of one 4-kg kid (AFRC and IGR) or two 2-kg kids (IGR) at birth. Estimates by INRA were calculated assuming a goat body weight (BW) of 55 kg, because this system estimates pregnancy requirements based on mother's BW.

conversion of ME to NE adopted by INRA ($k = 0.64$) in comparison to that adopted by AFRC and IGR ($k = 0.133$), previously proposed by ARC (1980), and utilized by most feeding systems for sheep and cattle. The latter conversion efficiency is very low compared with those used for other physiological functions. However, this depends on the method of calculation used, because all energy costs of pregnancy not included in the fetus retained energy, i.e. energy for uterus and mammary gland growth, are considered dispersed energy, thus not making part of NE. A k of 0.133 for pregnancy represents the maternal efforts for reproduction, but if the whole physiological process is considered, including the energy accumulated in all organs and tissues which contribute to pregnancy, the conversion efficiency of ME to NE would certainly be much higher, becoming similar to that of other physiological processes.

Variations in body reserves

Interpreting BW changes is often difficult because of the related variations in volume and content of the digestive system.

The INRA system considers different energy variations for BW gain (6.63 Mcal NEL/kg BW gain) or BW loss (6.29 Mcal NEL/kg BW loss), using similar values to those applied for sheep and cattle by INRA (1988). Conversion efficiency of ME to NE for BW gain is equal to k_l, while k is 0.80 for tissue mobilized for milk production only, the latter value being similar to those of the other systems (Table 6.2).

The AFRC system deals with BW changes only for growing animals and lactating goats. For the second group, AFRC considers a value of 5.5 Mcal NE/kg BW gain, proposed by Dunshea *et al.* (1990), which is very similar to the 5.71 Mcal NE/kg BW gain previously proposed by ARC (1980). The conversion efficiency of NE from body reserves to milk production NE is 0.84, which is slightly higher than that of INRA, while conversion efficiency of ME to NE for BW gain of lactating goats is 0.95 k_l.

Even though EEZ uses a net energy value for each kilogram of BW change similar to that of ARC (1980), it proposes a lower value (0.907 k_l) than those of other systems for the conversion efficiency of ME to NE of BW gain. This efficiency estimate was obtained from a regression between ME intake and BW change of animals in positive energy balance, considering maintenance requirements of 95.7 kcal ME/kg $BW^{0.75}$ (Table 6.2). This maintenance value, which is lower for lactating than dry goats (Table 6.4), is different from other literature data and could be responsible for the lower efficiency in deposition of body reserves hypothesized by the EEZ system in comparison to the other systems. The adopted conversion efficiency of mobilization NE to milk production NE is equal to the 0.84 value proposed by ARC (1980).

The estimates of energy variations due to BW changes by IGR are based on regression equations between daily BW changes and ingested ME. The energy value for BW change is then obtained by subtracting the energy necessary for maintenance requirements from MEI. The IGR system proposes a series of values for total requirements, which represent the sum of maintenance requirements and daily ME for gain. An example of the requirements for maintenance and BW gain of adult goats is reported in Table 6.8. Males have higher requirements than females and wethers (15% higher maintenance requirements; NRC, 2000), while dairy breeds' requirements are higher than indigenous or meat breed ones (18% higher maintenance requirements; Nsahlai *et al.*, 2004a). The energy value of each kilogram of BW is the same as that proposed by ARC (1980). For lactating animals, BW loss is related to an energy concentration of 23.9 kJ/g (5.71 kcal/g) (AFRC, 1993) and an efficiency of use of NE from body loss to NE for milk synthesis of 0.84, as proposed by ARC (1980), while BW gain has an energy concentration similar to that considered for BW loss but lower efficiency of conversion of ME to NE (k_g = 0.75), as proposed by NRC (1981b) (Table 6.2).

Protein Requirements of Adult Goats

Protein requirement is defined as the amount of protein to be supplied by the diet to compensate for N losses and consumption and to assure high feed utilization efficiency, without influencing animal health and reproduction (INRA, 1988). Requirements are calculated on the basis of the protein composition of animal products and the efficiency of metabolic utilization of the protein absorbed by animals fed balanced diets.

The AFRC, INRA and IGR systems express protein requirements as metabolizable protein (MP), i.e. the feed and bacterial protein absorbed by the small intestine. The intestinal digestibility of feed and microbial protein is considered in terms of truly digested protein by AFRC and INRA, and in terms of apparently digested protein by IGR. The INRA system calls MP by the acronym PDI (protein digested in the intestine). The INRA system established its requirements on the

Table 6.8. Requirements for maintenance and body weight (BW) gain estimated by the IGR system, assuming a BW of 50 kg for does and intact males.

	Does and wethers		Intact males	
Item	Dairy breeds	Meat and indigenous breeds	Dairy breeds	Meat and indigenous breeds
Maintenance (Mcal ME × kg $BW^{0.75}$)	0.119	0.101	0.138	0.116
Maintenance (Mcal ME/day × 50 kg BW)	2.25	1.90	2.59	2.18
Total ME requirements (Mcal/day)				
BW gain of 20 g/day	2.39	2.04	2.73	2.32
BW gain of 40 g/day	2.53	2.17	2.86	2.46
BW gain of 60 g/day	2.66	2.31	3.00	2.59
BW gain of 80 g/day	2.80	2.44	3.14	2.73

ME, metabolizable energy.

basis of the results of feeding trials in which the animals were compared in different physiological functions (Morand-Fehr *et al.*, 1987). The AFRC and IGR estimate total protein requirements using a factorial approach, in which the protein requirements for each physiological function are calculated separately and then summed up.

The estimation of dietary MP by IGR is based on the composition of the diet and on a database of crude protein (CP) degradability and concentration of fermentable energy of the feeds. The EEZ system estimated net requirement of total N by calculating N balances of dry and lactating goats. After that, regression equations were used to estimate daily requirements of total ingested N (NI), assuming an apparent N digestibility of 0.7. The same equations were then used to calculate the different conversion efficiencies of NI to net protein (NP) for each physiological stage. The EEZ system cannot be compared with the other systems, because it converts N requirements to CP, instead of MP.

Maintenance requirements

Maintenance protein requirements are represented by the amount of the endogenous protein, derived from tissue catabolism (not from feed), which is lost from dermal tissues (scurf and wool), in urine (urinary endogenous protein, UEP) and in faeces (faecal endogenous protein, FEP). Endogenous losses are estimated by feeding N-free diets or by regression equations such as $EN = b \times NI - c$, where NI is N intake, EN is excreted N, b is the true N digestibility and c is the maintenance endogenous loss. Even though all systems are based on these principles, they estimate maintenance protein requirements using different approaches (Table 6.9).

The INRA system compared two methods to estimate protein requirements: (i) the factorial method which sums up urinary, faecal and dermal N, measured directly on lactating goats or by using the above-mentioned regression approach on wethers or pregnant goats; and (ii) the N balance approach based on feeding trials (Morand-Fehr *et al.*, 1987). Results were not influenced by gender or breed and did not differ significantly between the two methods.

The AFRC system adopted the ARC (1980, 1984) approach. Requirements for N excreted with urine (urinary endogenous N, UEN) and with faeces (faecal endogenous N, FEN) were defined as the average of measurements from the literature, including papers published by INRA and EEZ. The calculated FEN is expressed as a function of ingested DM or $BW^{0.75}$. Loss from dermal tissues is the same as that estimated for cattle by ARC (1980). Calculation of MP requirements is based on the sum of endogenous losses (total endogenous N, TEN), using a higher value for the utilization efficiency of MP in NP than that of INRA.

The EEZ system estimated maintenance protein requirements on the basis of N balances measured on dry and lactating goats, then calculating TEN with regression equations between NI and retained N. The same approach and database were used to estimate UEN. However, for wethers the obtained UEN is higher than TEN, because it is calculated at nutritional levels slightly higher than

Table 6.9 Maintenance nitrogen and protein requirements by various feeding systems.

Variable	Units	INRA	AFRC	EEZ	IGR
UEN	g/day	$0.10–0.13 \times BW^{0.75}$	$0.12 \times BW^{0.75}$	$0.119 \times BW^{0.75}$ [a] $0.218 \times BW^{0.75}$ [b]	$0.165 \times BW^{0.75}$
FEN	g/day	$0.10–0.19 \times BW^{0.75}$	$0.15–0.20 \times BW^{0.75}$		$4.27 \times DMI$
Hair and dermal	g/day	$0.02 \times BW^{0.75}$	$0.018 \times BW^{0.75}$		$0.03 \times BW^{0.60}$
TEN	g/day	$0.28 \times BW^{0.75}$	$0.35 \times BW^{0.75}$	$0.108 \times BW^{0.75}$ [a] $0.244 \times BW^{0.75}$ [b]	
Total MP	g/day	$2.1–2.3 \times BW^{0.75}$ [c]	$2.19 \times BW^{0.75}$		UEN + FEN + hair and dermal
Efficiency MP → NP		0.83	1		1
Total CP	g/day			$2.56 \times BW^{0.75}$ [a] $2.98 \times BW^{0.75}$ [b]	MP/(0.64–0.80)[d]
Efficiency CP → NP				0.51	0.64–0.80[d]

UEN, urinary endogenous N; FEN, faecal endogenous N; TEN, total endogenous N; MP, metabolizable protein; NP, net protein; CP, crude protein; BW, body weight (kg); DMI, daily dry matter intake (kg).

[a]Castrated males.

[b]Lactating does.

[c]In practice, the suggested value is 2.3 g MP × $BW^{0.75}$.

[d]Assuming that the ratio MP/CP is 0.64 and 0.80 for diets with ruminal degradability of CP from 100 to 0%, respectively.

Table 6.10. Quantitative comparison of maintenance protein requirements as calculated by three feeding systems.

	50 kg BW			70 kg BW		
Level of intake	INRA[a]	AFRC	IGR	INRA[a]	AFRC	IGR
1% of BW	44	41	35	56	53	46
3% of BW	44	41	62	56	53	84
5% of BW	44	41	88	56	53	121

[a]Estimates based on the value reported in note c of Table 6.9.

maintenance (Prieto *et al*., 1990). As already mentioned, EEZ converts TEN, and thus NP, directly to CP, hypothesizing an average conversion efficiency of CP to NP of 0.51. For this reason, requirements estimated by this system are comparable to those of the other systems only for NP but not for MP.

The IGR system estimated maintenance MP requirements separately for growing and mature goats and for meat, dairy and indigenous breeds. Both UEN (Luo *et al*., 2004c) and FEN (Moore *et al*., 2004) were estimated with the regression approach applied to a data set constituted by experimental results from the literature. The FEN was proportional to ingested DM but not to BW, as already proposed by CSIRO (1990) for sheep and cattle, and confirmed by Cannas *et al*. (2004) for sheep. Dermal loss was calculated using the estimates for meat cattle breeds by NRC (1984). Values of body composition and conversion efficiency of MP to NP adopted by IGR were similar to those adopted by AFRC (1998).

Maintenance MP requirements estimated by INRA are higher than those of AFRC (Table 6.10). The estimates by IGR, which proposes requirements that vary according to intake level, are the lowest at low intake levels, typical of dry animals, and the highest at high intake levels, typical of lactating animals (Table 6.10). In the latter case, maintenance requirements proposed by IGR are more than two times higher than those proposed by AFRC. Such large differences certainly lead to marked differences in the optimal diet composition calculated by the different systems.

Milk production requirements

In all systems, calculations of NP requirements for milk production are based on the amount of true protein contained in milk. However, the efficiencies of conversion to express requirements in terms of MP or CP vary among systems (Table 6.11).

The INRA system considers a single value of MP required for each litre of standard milk, and then provides only tolerance intervals expressed in g total MP during lactation, taking into consideration the ability of animals to mobilize part of the body protein. A tolerated daily deficit of 80–90 g for the first week, goes down to 20–30 g for the second week of lactation, and should be absent at the third week of lactation. In the INRA system, the utilization efficiency of MP which exceeds maintenance needs is 0.64, similar to that of cattle (Morand Fehr *et al*., 1987).

Table 6.11. Lactation protein requirements, in g MP/kg milk, predicted by various feeding systems.

	INRA	AFRC	EEZ[a]	IGR
MP (g/kg milk)	TP[b]/0.64	P[c] × 0.9/0.68	NA	TP/0.69
Efficiency MP → NP of milk	0.64	0.68	NA	0.69
Milk with 3% TP (g MP/kg)	47	44	NA	44

MP, metabolizable protein; NP, net protein; NA, not available.
[a]This system gives only crude protein requirements, assuming an efficiency of conversion of dietary crude protein to milk NP of 0.51.
[b]TP = milk true protein (g/kg milk).
[c]P = milk crude protein (g/kg milk).

The AFRC system considers that the content of true protein in milk is 90% of its CP, the utilization efficiency of amino acids absorbed for milk protein synthesis being equal to 0.68. According to AFRC (1998), the value of this conversion efficiency of MP to NP is lower than those found experimentally in some French studies, which are equal to 0.8 (Giger, 1987; cited by AFRC, 1998) and 0.75 (Brun Bellut, 1986; cited by AFRC, 1998) for diets rich and poor in fermentable N, respectively.

In the IGR system, the MP requirement for milk production (MP_l) is estimated with a factorial approach (Nsahlai *et al.*, 2004b), by subtracting maintenance MP (conversion efficiency of MP to NP equal to 1) from dietary MP, and correcting for BW gains (efficiency of 0.59) and for mobilization of protein reserves for milk production (efficiency of 0.69). Conversion efficiency of MP to NP for milk production is 0.69. By using a feed database containing information on the proportion of feed protein digested by the intestine, IGR calculated requirements in terms of CP.

The conversion efficiency of MP to NP of 0.68 for AFRC is very similar to the value proposed by IGR (0.69) and slightly higher than that of INRA (0.64) (Table 6.11).

The approach used by EEZ is in part similar to that adopted by NRC (1981b), but cannot be compared directly with the other systems. In fact, EEZ does not distinguish N requirements for milk production from those needed for body reserves. Nevertheless, by using the same calorimetric studies that measured energy requirements on goats, EEZ developed a regression equation of milk plus retained N on NI to estimate total N requirements and the average utilization efficiency from NI to net N, which was equal to 0.51.

Pregnancy requirements

The approaches used to calculate protein requirements for pregnancy are the same as those already reported for energy requirements. The INRA system does not report a conversion efficiency of MP to NP, while AFRC and IGR use values of 0.85 and 0.33, respectively. The latter value is also suggested by NRC (2001). The EEZ system does not deal with pregnancy protein requirements.

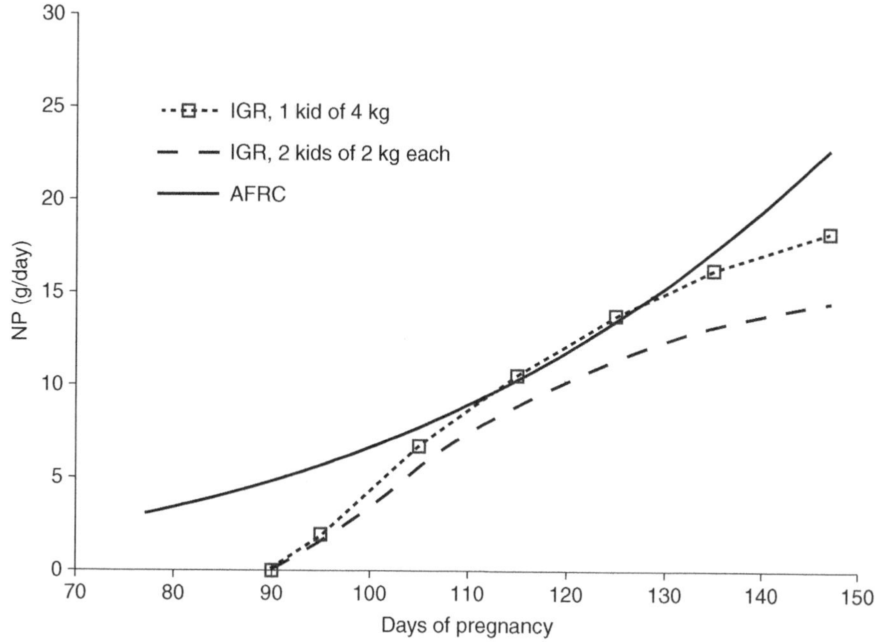

Fig. 6.4. Net protein (NP) requirements for pregnancy, assuming the delivery of one 4-kg kid (AFRC and IGR) and two 2-kg kids (IGR).

Pregnancy NP and MP requirements were compared among the various systems, assuming a goat of 55 kg delivering a 4-kg kid. The NP requirements of AFRC and IGR were quite similar and showed a trend similar to that of energy requirements (Fig. 6.4). The INRA system does not estimate pregnancy NP requirements. Pregnancy MP requirements of INRA were higher than those of AFRC, in contrast to what was observed for ME requirements, and similar to those of IGR (Fig. 6.5). Pregnancy requirements estimated by IGR are higher than those of AFRC mainly because of differences in the coefficients used to convert MP to NP and in the calculation approach used. In the INRA system, protein requirements of a goat at the fourth month of pregnancy range from 55 to 65% of maintenance requirements and at the fifth month vary from 110 to 130% of maintenance requirements. Such variation depends on the weight of the mother and of the fetuses. In the AFRC and IGR systems, instead, protein requirements depend only on the number and total weight of the kids delivered, but not on mother's weight.

Energy and Protein Requirements of Growing Goats

Many factors affect the nutritional requirements of growing goats, such as diet, level of intake, genotype, age, gender, body composition, activity, growth rate, stage of maturity, season and weather conditions. However, published feeding systems account for only a few of these variables.

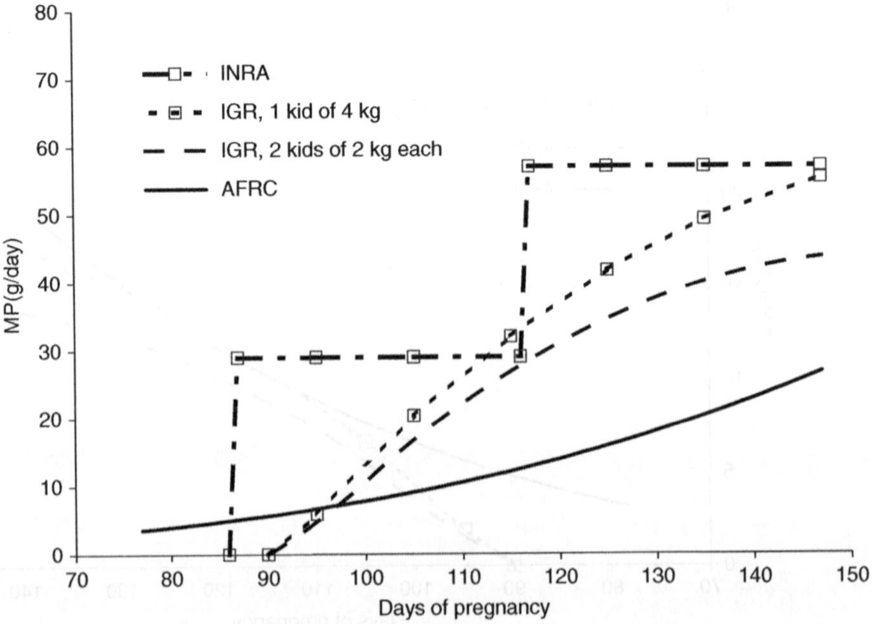

Fig. 6.5. Metabolizable protein (MP) requirements for pregnancy, assuming the delivery of one 4-kg kid (AFRC and IGR) and two 2-kg kids (IGR). Estimates by INRA were calculated assuming a goat body weight (BW) of 55 kg, because this system estimates pregnancy requirements based on mother's BW.

Among the feeding systems compared so far, the EEZ system does not provide requirements for growing goats, while the INRA system provides specific tables on them, without giving enough information on the equations used to calculate such values.

The AFRC system estimated energy and protein requirements of growing goats based on body composition studies carried out on Saanen and Toggenburg, weaned, growing, castrated males. Based on this information, prediction linear regression equations and tables relating fat, protein and energy composition of the gain to the BW of growing castrated males were developed. Energy maintenance requirements, the efficiency of conversion of NE to ME (Tables 6.2 and 6.4) and protein maintenance requirements are calculated as for adult goats, while the efficiency of use of truly absorbed amino acid for growth is considered equal to 0.59.

The IGR system created a large database from the literature and estimated ME requirements for maintenance and growth (Tables 6.12 and 6.13) by regressing ME intake against average daily gains of growing goats. This system categorized growing goats into three different biotypes, according to specific production characteristics: (i) dairy (e.g. Saanen, Alpine, Damascus, Norwegian, Swedish Landrace and dairy crossbreeds); (ii) meat (≥ 50% Boer); and (iii) indigenous (neither dairy nor meat; not including Angora) (Luo *et al.*, 2004b,d; Moore *et al.*, 2004; Sahlu *et al.*, 2004). The categorization was mainly due to differences in

Table 6.12. Recommended growth requirements according to the IGR system.

	Suckling	Growing biotypes[a]		
		Meat	Dairy	Indigenous
ME_m (kcal/kg $BW^{0.75}$)	115.9	116.9	138.6	116.9
ME_g (kcal/g ADG)	3.20	5.52	5.52	4.73
MP_m (g/kg $BW^{0.75}$)		3.07	3.07	3.07
CP_m (g/kg $BW^{0.75}$)		MP_m/ (0.64 – 0.80)	MP_m/ (0.64 – 0.80)	MP_m/ (0.64 – 0.80)
MP_g (g/g ADG)		0.404	0.290	0.290
CP_g (g/g ADG)		MP_g/ (0.64 – 0.80)	MP_g/ (0.64 – 0.80)	MP_g/ (0.64 – 0.80)

ME_m, metabolizable energy requirement for maintenance; BW, body weight; MP_g, metabolizable protein for whole body gain; ADG, average daily gain; MP_m, metabolizable protein for maintenance; CP_m, crude protein for maintenance; ME_g, metabolizable energy requirements for whole body gain; CP_g, crude protein for whole body gain.
[a]Growing, post-weaning to 18 months of age; meat, ≥ 50% Boer; dairy, Saanen, Alpine, Damascus, Norwegian, Swedish landrace and dairy crossbreeds; indigenous, neither meat nor dairy, not including Angora.

body composition among genotypes, which affected their maintenance and growth requirements (Tables 6.12 and 6.13). The efficiency of conversion of NE to ME is calculated as for adult goats (Tables 6.2 and 6.4). The IGR estimated MP requirements for maintenance and growth (Tables 6.12 and 6.13) by regressing MP intake against average daily gains of growing goats. MP intake was estimated by using published CP degradation values of feedstuffs, which were used together with predicted ruminal passage rates and intestinal digestions to calculate the MP content of the feeds (Moore *et al.*, 2004). Since the approach used did not take into consideration either the body protein content or the composition of gain, no information was reported on the efficiency of conversion of MP to NP. Maintenance MP requirements for growing goats are specific for this category and do not differ among genotypes (Tables 6.9 and 6.12).

The comparison of the NE and NP requirements for growth as predicted by different feeding systems and as estimated by different slaughtering experiments (Resende, 1989; Ribeiro, 1995; Medeiros, 2001; Ferreira, 2003; Teixeira, 2004; Fernandes *et al.*, 2007) is reported in Figs 6.6 and 6.7.

In the NRC (1981b) and IGR systems, NE requirements for growth are considered fixed and independent from BW of the growing goats, while in all other publications energy costs increase as BW increases. The fact that growth requirements increase as BW approaches maturity is a well-known phenomenon of growth biology (ARC, 1980; CSIRO, 1990).

The NP cost of growth is similar among most studies, even though NP requirements increase or decrease, depending on the study, as BW increases (Fig. 6.7). The IGR system is not included in this comparison, because it predicts MP but not NP requirements.

In contrast to all other systems and publications reported in the comparison (Figs 6.6 and 6.7), the Small Ruminant Nutrition System (described in more

Table 6.13. Total metabolizable energy (ME) and metabolizable protein (MP) requirements of growing kids assuming an average daily gain of 100 g/day (numbers in brackets refer to ME and MP for maintenance), according to the IGR system.

Biotype	Body weight (kg)	ME (Mcal/day)	MP (g/day)
Doelings and castrated males			
Dairy	20	1.76 (1.21)	58 (29)
	30	2.19 (1.64)	68 (39)
	40	2.59 (2.04)	78 (49)
Meat	20	1.57 (1.02)	69 (29)
	30	1.94 (1.38)	80 (39)
	40	2.27 (1.72)	89 (49)
Indigenous	20	1.49 (1.02)	58 (29)
	30	1.86 (1.38)	68 (39)
	40	2.19 (1.72)	78 (49)
Intact males			
Dairy	20	1.96 (1.41)	58 (29)
	30	2.46 (1.91)	68 (39)
	40	2.92 (2.37)	78 (49)
Meat	20	1.74 (1.19)	69 (29)
	30	2.17 (1.62)	80 (39)
	40	2.56 (2.00)	89 (49)
Indigenous	20	1.66 (1.19)	58 (29)
	30	2.09 (1.62)	68 (39)
	40	2.48 (2.00)	78 (49)

detail in the following section; Cannas *et al.*, 2007), which adopted the approach of CSIRO (1990), estimates energy and protein growth requirements based on the relative maturity of the kids (ratio of current weight to mature weight for a certain gender and breed) and on their feeding level. In this system kids of very different BW are considered equal in terms of fat, protein and energy content of the gain, and thus in terms of energy and protein requirements for growth, as long as their relative maturity is the same. Part of the large variability observed in the comparisons (Figs 6.6 and 6.7) could be due to the fact that for equal BW, the relative maturity of the animals used to estimate energy and protein requirements was different. In addition, differences in body composition for goats at the same relative maturity cannot be excluded. For example, higher protein concentration, and thus higher NP growth requirements, were observed in the empty body (i.e. body weight minus gastrointestinal tract) of Boer crossbred kids than in that of dairy and indigenous goats (Teixeira, 2004; Fernandes *et al.*, 2007).

The use of relative maturity also allows one to account for the effect of gender on growth requirements (Figs 6.6 and 6.7). Within a breed, males have higher mature weight than female goats. Thus, if males and females of the same breed and same BW are compared, the relative maturity and, as a consequence, growth requirements will be lower for males than for females. This aspect, fully

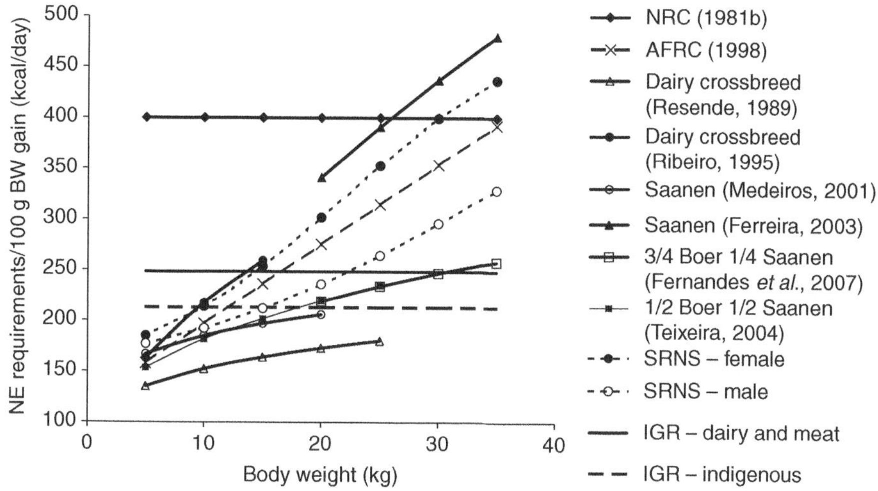

Fig. 6.6. Comparisons of net energy (NE) requirements for body weight gain (kcal/day per 100 g BW gain) across studies and feeding systems using different genotypes and ages. For the Small Ruminant Nutrition System (SRNS; see text), a mature weight of 55 kg for females and 85 kg for males was considered to estimate the relative maturity of growing goats. For the IGR system, the efficiency of conversion of metabolizable energy to NE for gain was considered equal to 0.45.

consistent with growth biology, has been neglected by the other feeding systems here considered. Part of the large variability observed in the comparison of growth requirements (Fig. 6.6) could also be due to the fact that for equal BW, requirements are different depending on the gender considered.

New Developments on Goat Requirements

A new feeding system for small ruminants, called the Small Ruminant Nutrition System (SRNS), is under development (Cannas *et al.*, 2007). The SRNS (http://nutritionmodels.tamu.edu/srns.htm) includes two sub-models, one for sheep and one for goats. The SRNS for sheep is an evolution of the CNCPS for sheep (Cannas *et al.*, 2004, 2006). In the SRNS for goats, energy requirements are predicted based on the equations developed for the SRNS for sheep, modified to account for specific requirements of goats.

The SRNS uses animal and environmental factors to predict energy, protein, Ca and P requirements. In particular, energy requirements for basal metabolism (higher for dairy goats than for other breeds) are adjusted for age, physiological stage, environmental effects, physical activity, cost of urea excretion, and acclimatization. ME requirements for milk production are estimated from milk NE, based on measured milk yield, fat and true protein. The energy available for body reserves changes depending on the energy balance after maintenance, lactation and pregnancy requirements are satisfied. New equations were developed to estimate the relationships among body condition score (BCS), BW and body

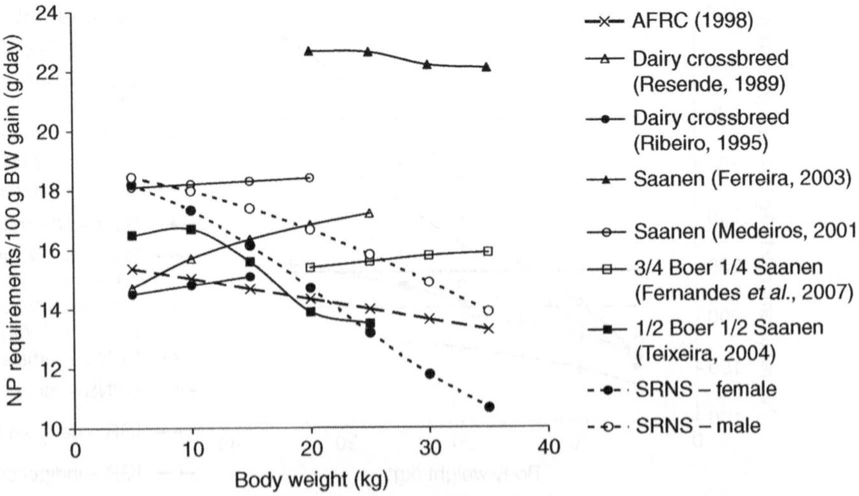

Fig. 6.7. Comparisons of net protein (NP) requirements for body weight gain (g/day per 100 g BW gain) across studies and feeding systems using different genotypes and ages. For the Small Ruminant Nutrition System (SRNS; see text), a mature weight of 55 kg for females and 85 kg for males was considered to estimate the relative maturity of growing goats.

composition in adult goats. A model that predicts the relationship between BW and BCS in various goat breeds was developed. It allows the prediction of current BW when BCS and breed mature weight at BCS 2.5 are known inputs or the prediction of BW at BCS 2.5 when current BCS and BW are known. The relationship between the proportion of fat in the empty body and BCS is also predicted.

The SRNS computes average daily gain of kids with equations based on the CSIRO (1990) model. Average daily gain depends on the energy available for gain after maintenance costs are covered and on the energy value of the gain. The latter and the fat and protein composition of the gain are estimated based on the relative maturity of the kids (ratio of current weight to mature weight for a certain gender and breed) and on their feeding level.

The SRNS uses a mechanistic ruminal model that predicts feed biological values based on carbohydrate and protein fractions and their ruminal degradation rates, forage, concentrate and liquid passage rates, microbial growth and physically effective fibre.

The evaluation of the SRNS based on literature data showed that it accurately predicted the average daily gain of kids (root of the mean squared error of prediction, RMSEP, 32.5 g/day; $r^2 = 0.85$; concordance correlation coefficient, CCC = 0.91; Cannas *et al.*, 2007). In addition, the SRNS was able to predict accurately and precisely the daily MEI (when DM intake inputs were accurate) of adult goats and wethers (RMSEP = 0.24 Mcal/day; $r^2 = 0.99$; CCC = 0.99), their milk NE (RMSEP = 0.012 Mcal/day; $r^2 = 0.99$; CCC = 0.99) and their energy balance (RMSEP = 0.20 Mcal/day; $r^2 = 0.87$; CCC = 0.90), with a systematic tendency to under-predict the energy balance slightly as the observed values increased (Cannas *et al.*, 2007).

Conclusions

Estimating energy and protein requirements is a fundamental aspect of goat nutrition. Many feeding systems have been proposed for goats, but many estimates are based on research conducted on other species.

For example, the INRA system uses an approach very similar to that developed for dairy sheep and is better suited for confined than grazing goats. The AFRC system uses a simplified approach, based mainly on the ARC (1980), deriving many of its equations from the system developed for sheep and cattle.

The EEZ system, even though not completely developed, is based on experimental measurements, often calorimetric ones, conducted on goats by this research group. For some sub-models, such as that for physical activity requirements, EEZ presented completely new data and equations.

Among the systems compared in this chapter, IGR is the most recent one and is based on the statistical analysis of databases created from the scientific literature on goats. Most of the databases were analysed by linear regressions. This is the only system which proposes specific requirements for breeds with different production aptitudes. In addition, IGR developed a unique procedure to calculate requirements for physical activity. Due to the large number of data used and the good information developed, IGR can be considered a modern and complete feeding system for goats. However, this system has the major limitation of considering constant growth requirements for animals of different BW, relative maturity and gender.

In general, the systems which predict energy and protein requirements for goats tend to be more empirical and less flexible than the most recent sheep and cattle systems. This depends partly on the scarce number of studies conducted on goats in comparison to those on sheep and cattle. In addition, the distribution of goats throughout very different environments makes the modelling of their requirements more complex. However, more mechanistic feeding systems for goats are currently under development.

References

AFRC (1993) *Energy and Protein Requirements of Ruminants*. CAB International, Wallingford, UK.

AFRC (1998) *The Nutrition of Goats*. Technical Committee on Response to Nutrients, Report No. 10. CAB International, Wallingford, UK.

Aguilera, J.F. (2001) Aportaciones al conocimiento de la nutrición energética de pequeños rumiantes, con particular referencia al ganado caprino. *Archivos de Zootecnia* 50, 565–596.

Aguilera, J.F. and Prieto, C. (1985) An open respiration unit for calorimetric studies with small animals. *Archives of Animal Nutrition* 35, 825–833.

Aguilera, J.F. and Prieto, C. (1986) Description and function of an open-circuit respiration plant for pigs and small ruminants and the techniques used to measure energy metabolism. *Archives of Animal Nutrition* 36, 1009–1018.

Aguilera, J.F. and Prieto, C. (1991) Methane production in goats given diets based on lucerne hay and barley. *Archives of Animal Nutrition* 41, 77–82.

Aguilera, J.F., Prieto, C. and Fonollá, J. (1990) Protein and energy metabolism of lactating Granadina goats. *British Journal of Nutrition* 63, 165–175.

ARC (1980) *The Nutrient Requirements of Ruminant Livestock*. Technical Revision of the Agricultural Research Council Working Party. Commonwealth Agricultural Bureaux, Farnham Royal, UK.

ARC (1984) *The Nutrient Requirements of Ruminant Livestock. Supplement N. 1*. Report of the Protein Group of the Agricultural Research Council Working Party. Commonwealth Agricultural Bureaux, Farnham Royal, UK.

Blaxter, K.L. (1962) *The Energy Metabolism of Ruminants*. Hutchinson, London.

Cannas, A. (2000) Sheep and cattle nutrient requirement systems, ruminal turnover, and adaptation of the Cornell Net Carbohydrate and Protein System to sheep. PhD dissertation, Cornell University, Ithaca, New York.

Cannas, A. (2004) Energy and protein requirements. In: Pulina, G. (ed.) *Dairy Sheep Nutrition*. CAB International, Wallingford, UK, pp. 31–49.

Cannas, A., Tedeschi, L.O., Fox, D.G., Pell, A.N. and Van Soest, P.J. (2004) A mechanistic model for predicting the nutrient requirements and feed biological values for sheep. *Journal of Animal Science* 82, 149–169.

Cannas, A., Tedeschi, L.O., Atzori, A.S. and Fox, D.G. (2006) Prediction of energy requirement for growing sheep with the Cornell Net Carbohydrate and Protein System. In: Kebreab, E., Dijkstra, J., Bannink, A., Gerrits, W.J.J. and France, J. (eds) *Nutrient Digestion and Utilization in Farm Animals: Modelling Approaches*. CAB International, Wallingford, UK, pp. 99–113.

Cannas, A., Tedeschi, L.O., Atzori, A.S. and Fox, D.G. (2007) The Small Ruminant Nutrition System: development and evaluation of a goat submodel. *Italian Journal of Animal Science* 6(Suppl. 1), 609–611.

CSIRO (1990) *Feeding Standards for Australian Livestock. Ruminants*. Commonwealth Scientific and Industrial Research Organization, Melbourne, Australia.

Dunshea, F.R., Bell, A.W. and Trigg, T.E. (1990) Body composition changes in goats during early lactation estimated using a two-pool model of tritiated water kinetics. *British Journal of Nutrition* 64, 121–131.

Fernandes, M.H.M.R., Resende, K.T. Tedeschi, L.O., Fernandes, J.S. Jr, Silva, H.M., Carstens, G.E., Berchielli, T.T., Teixeira, I.A.M.A. and Akinaga, L. (2007) Energy and protein requirements for maintenance and growth of Boer crossbred kids. *Journal of Animal Science* 85, 1014–1023.

Ferreira, A.C.D. (2003) Composição corporal e exigências nutricionais em proteína, energia e macrominerais de caprinos Saanen em crescimento. PhD dissertation, Universidade Estadual Paulista, Jaboticabal, São Paulo, Brazil.

Fox, D.G., Tedeschi, L.O., Tylutki, T.P., Russell, J.B., Van Amburgh, M.E., Chase, L.E., Pell, A.N and Overton, T.R. (2004) The Cornell Net Carbohydrate and Protein System model for evaluating herd nutrition and nutrient excretion. *Animal Feed Science and Technology* 112, 29–68.

INRA (1988) Jarrige, R. (ed.) *Alimentation des bovins ovins et caprins*. Institut National de la Recherche Agronomique, Paris.

INRA (2007) *Alimentation des bovins, ovins et caprins. Besoins animaux. Valeurs des aliments. Tables INRA 2007*. Editions Quae, Versailles, France.

Lachica, M., Aguilera, J.F. and Prieto, C. (1995) A confinement respiration chamber for short gaseous exchange measurements. *Archives of Animal Nutrition* 48, 329–336.

Lachica, M., Prieto, C. and Aguilera, J.F. (1997a) The energy cost of walking on the level and on negative and positive slopes in the Granadina goat (*Capra hircus*). *British Journal of Nutrition* 77, 73–81.

Lachica, M., Aguilera, J.F. and Prieto, C. (1997b) Energy expenditure related to the act of eating in Granadina goats given diets of different physical form. *British Journal of Nutrition* 77, 417–426.

Lachica, M., Barroso, F.G. and Prieto, C. (1997c) Seasonal variation of locomotion and energy expenditure in goats under range grazing conditions. *Journal of Range Management* 50, 234–238.

Lachica, M., Somlo, R., Barroso, F.G., Boza, J. and Prieto, C. (1999) Goats locomotion energy expenditure under range grazing conditions: seasonal variation. *Journal of Range Management* 52, 431–435.

Luo, J., Goetsch, A.L., Nsahlai, I.V., Johnson, Z.B., Sahlu, T., Moore, J.E., Ferrell, C.L., Galyean, M.L. and Owens, F.N. (2004a) Maintenance energy requirements of goats: predictions based on observations of heat and recovered energy. *Small Ruminant Research* 53, 221–230.

Luo, J., Goetsch, A.L., Sahlu, T., Nsahlai, I.V., Johnson, Z.B., Moore, J.E., Galyean, M.L., Owens, F.N. and Ferrell, C.L. (2004b) Prediction of metabolizable energy requirements for maintenance and gain of preweaning, growing and mature goats. *Small Ruminant Research* 53, 231–252.

Luo, J., Goetsch, A.L., Moore, J.E., Johnson, Z.B., Sahlu, T., Ferrell, C.L., Galyean, M.L. and Owens, F.N. (2004c) Prediction of endogenous urinary nitrogen of goats. *Small Ruminant Research* 53, 293–308.

Luo, J., Goetsch, A.L., Nsahlai, I.V., Sahlu, T., Ferrell, C.L., Owens, F.N., Galyean, M.L., Moore, J.E. and Johnson, Z.B. (2004d) Metabolizable protein requirements for maintenance and gain of growing goats. *Small Ruminant Research* 53, 309–326.

Luo, J., Goetsch, A.L., Nsahlai, I.V., Moore, J.E., Galyean, M.L., Johnson, Z.B., Sahlu, T., Ferrell, C.L. and Owens, F.N. (2004e) Voluntary feed intake by lactating, Angora, growing and mature goats. *Small Ruminant Research* 53, 357–378.

McDonald, I., Robinson, J.J., Fraser, C. and Smart, R.I. (1979). Studies on reproduction in prolific ewes. 5. The accretion of nutrients in the foetuses and annexes. *Journal of Agricultural Science (Cambridge)* 92, 591–603.

Medeiros, A.N. (2001) Composição corporal e exigências nutricionais em proteína e energia para caprinos Saanen na fase inicial de crescimento. PhD dissertation, Universidade Estadual Paulista, Jaboticabal, São Paulo, Brazil.

Moore, J.E., Goetsch, A.L., Luo, J., Owens, F.N., Galyean, M.L., Johnson, Z.B., Sahlu, T. and Ferrell, C.L. (2004) Prediction of fecal crude protein excretion of goats. *Small Ruminant Research* 53, 275–292.

Morand-Fehr, P., Sauvant, D. and Brun-Bellut, J. (1987) Recommandations alimentaires pour le caprins. In: *Alimentation des ruminants: revision du systèmes et des tables de l'INRA*. Bulletin Technique de Centre de Recherches Zootechniques et Veterinaires de Theix, No. 70. Institut National de la Recherche Agronomique, Paris, pp. 213–222.

NRC (1981a) *Effect of Environment on Nutrient Requirements of Domestic Animals*. National Academy Press, Washington, DC.

NRC (1981b) *Nutrient Requirements of Goats: Angora, Dairy and Meat Goats in Temperate and Tropical Countries*. National Academy Press, Washington, DC.

NRC (1984) *Nutrient Requirements of Beef Cattle*, 6th edn. National Academy Press, Washington, DC.

NRC (2000) *Nutrient Requirements of Beef Cattle, Update 2000*. National Academy Press, Washington, DC, pp. 16–21.

NRC (2001) *Nutrient Requirements of Dairy Cattle*, 7th revised edn. National Academy Press, Washington, DC, pp. 13–104.

NRC (2007) *Nutrient Requirements of Small Ruminants. Sheep, Goats, Cervids and New World Camelids*. National Academy Press, Washington, DC.

Nsahlai, I.V., Goetsch, A.L., Luo, J., Johnson, Z.B., Moore, J.E., Sahlu, T., Ferrell, C.L., Galyean, M.L. and Owens, F.N. (2004a) Metabolizable energy requirements of lactating goats. *Small Ruminant Research* 53, 253–273.

Nsahlai, I.V., Goetsch, A.L., Luo, J., Johnson, Z.B., Moore, J.E., Sahlu, T., Ferrell, C.L., Galyean, M.L. and Owens, F.N. (2004b) Metabolizable protein requirements of lactating goats. *Small Ruminant Research* 53, 327–337.

Prieto, C., Aguilera, J.F., Lara, L. and Fonollá, J. (1990) Protein and energy requirements for maintenance of indigenous Granadina goats. *British Journal of Nutrition* 63, 155–163.

Prieto, C., Lachica, M., Bermúdez, F. and Aguilera, J.F. (1997) Aplicación del ^{13}C para la estimación del gasto energético en cabras de raza Granadina. In: *Asociación Interprofesional para el Desarrollo Agrario (AIDA), XXIX jornadas de estudio, ITEA – VII jornadas sobre producción animal*, Zaragoza, Spain, pp. 16–18.

Prieto, C., Lachica, M., Nieto, R. and Aguilera, J.F. (2001) The ^{13}C-bicarbonate method: its suitability for estimating the energy expenditure in grazing goats. *Livestock Production Science* 69, 207–215.

Resende, K.T. (1989) Métodos de estimativa da composição corporal e exigências nutricionais de proteína, energia e macroelementos inorgânicos de caprinos em crescimento. PhD dissertation, Universidade Federal de Viçosa, Viçosa, Minas Gerais, Brazil.

Ribeiro, S.D.A. (1995) Composição corporal e exigências em energia, proteína e macrominerais de caprinos mestiços em fase inicial de crescimento. MSc thesis, Universidade Estadual Paulista, Jaboticabal, São Paulo, Brazil.

Sahlu, T., Goetsch, A.L., Luo, J., Nsahlai, I.V., Moore, J.E., Galyean, M.L., Owens, F.N., Ferrell, C.L. and Johnson, Z.B. (2004) Nutrient requirements of goats: developed equations, other considerations and future research to improve them. *Small Ruminant Research* 53, 191–219.

Teixeira, I.A.M.A. (2004) Métodos de estimativa de composição corporal e exigências nutricionais de cabritos F1 Boer×Saanen. PhD dissertation, Universidade Estadual Paulista, Jaboticabal, São Paulo, Brazil.

Tyrrell, H.F. and Reid, J.T (1965) Prediction of the energy value of cow's milk. *Journal of Dairy Science* 48, 1215–1223.

7 Feed Intake

M. Avondo,[1] L. Biondi,[1] R.I. Pagano,[1] A. Bonanno[2] and L. Lutri[3]

[1]*Dipartimento di Scienze Agronomiche, Agrochimiche e delle Produzioni Animali, Università di Catania, Via Valdisavoia 5, 95123 Catania, Italy;* [2]*Dipartimento di Scienze Entomologiche, Fitopatologiche, Microbiologiche e Zootecniche, Università di Palermo, Viale delle Scienze, 90128 Palermo, Italy;* [3]*Istituto Sperimentale Zootecnico per la Sicilia, Via Roccazzo 85, 90136 Palermo, Italy*

Introduction

In the classification of herbivores according to their diet (Hofmann, 1989), the goat species is included in an intermediate class between grass and roughage eaters (cattle, sheep) and concentrate selectors, which do not tolerate high fibre levels and are forced to select less fibrous parts of plants. 'Intermediate feeders' (or 'opportunistic feeders') change their feeding behaviour according to seasonal changes in diet availability (Fedele *et al.*, 1993; Papachristou, 1994) and are much more versatile than the other two categories of animal. 'Intermediate feeders' are characterized by: (i) high saliva secretion and large absorption surface of the rumen epithelium, which protect the animal from the risk of acidosis; and (ii) considerable enlargement of the digestive apparatus when highly fibrous feed is used (Silanikove, 2000). Because of these characteristics, goats can adapt to a broad range of feeding conditions. Moreover, more than any other species, goats are able to choose, among the available feedstuffs, parts of plants with the highest protein content and the highest digestibility, selecting feed on the basis of prehension ease, sensorial characteristics and post-ingestive effects learnt from their own experience (Provenza *et al.*, 2003). It must also be noted that in the presence of certain feeds, such as very young pastures, instead of selecting high-quality parts, goats select more fibrous and dry parts, which can improve rumen functionality. In intensive conditions, goats can tolerate overfeeding. In a free-choice trial Girgentana lactating goats consumed up to 3 kg of dry matter (DM) per day, constantly self-regulating crude protein (CP) and neutral-detergent fibre (NDF) levels at 16.5 and 34%, respectively, with minimal daily variations. In no cases did the goats display noteworthy differences in the main blood parameter levels compared with normal reference values (Avondo *et al.*, personal communication). It has been demonstrated that even in diets with concentrate levels above 60–70%, goats do not greatly alter their productive capacity or their metabolic well-being (Bailoni and Andrighetto, 1995; Economides, 1998; Goetsch *et al.*, 2001; Fedele *et al.*, 2002).

Indeed, once goats perceive the metabolic effects of a diet very rich in concentrates, they change their feeding behaviour to have small and numerous daily meals (Abijaoudé *et al.*, 2000), thereby avoiding the dangerous effects of excessive starch in the rumen. Several experimental trials have been carried out on lactating goats fed in stalls very different diets in terms of NDF (Andrade *et al.*, 1996), CP (Brun-Bellut, 1996), ratio of rumen-degradable protein to rumen-undegradable protein (Mishra and Rai, 1996), concentrate level (Gupta and Atreja, 1999; Goetsch *et al.*, 2001), concentrate type and number of daily administrations (Hadjipanayiotou, 2004), feeding system and, consequently, chemical–nutritive characteristics of the diets (Rapetti *et al.*, 2005) (Table 7.1). In general, these diets had no major effects on total DM intake, sometimes without a clear explanation for this phenomenon as confirmed by the authors. In light of this, predicting goat intake, in relation to diet-related parameters normally used to determine voluntary intake in ruminants, becomes extremely complex. This is further complicated in grazing systems for goats, the most common in the Mediterranean, where 'metabolic homeostasis' is combined with the high selection activity of goats. The latter aims at optimizing the digestive utilization of the pasture, based on 'hedonistic' criteria to choose essences and plant parts, which is hard to predict.

Lactating Goats

Prediction of dry matter intake of goats fed indoors

Experimental data from different studies (46 papers) on stall-fed lactating goats of Alpine and Mediterranean breeds were analysed, in order to develop equations to predict intake in this feeding system. Table 7.2 reports a description of the database used.

Treatment means on intake (n = 133), separated per breed (Alpine versus Mediterranean goats), were subjected to simple regression analysis using animal (live weight, milk production) and feeding factors (feed type, dietary CP and NDF levels) as independent variables (Table 7.3). An initial analysis of the results showed that animal factors were the main causes of variations in intake, especially for Alpine breeds, more specialized for milk production, whose intake level was strongly correlated with milk production or body weight (BW) (Table 7.3, Figs 7.1 and 7.2) and weakly correlated with dietary characteristics. For Mediterranean types, instead, DM intake was strongly correlated with BW, and less correlated with milk production (Table 7.3, Figs 7.3 and 7.4). On the other hand, there was a negative correlation ($R^2 = 0.46$) between DM intake and dietary NDF content in the Mediterranean goats (Table 7.3). The strong correlations between variables, in both data sets, did not allow us to include them in a multiple regression analysis.

These results are in contrast with those reported by Silanikove (2000), who observed that the intake of Black Bedouin breed was less affected by diet fibre

Table 7.1. Dry matter (DM) intake of lactating goats (g/day). Data from various experimental trials carried out in different feeding conditions.

Experimental variable	Group	Breed	Total DM intake (g/day)
Lucerne/sorghum ratio[a]	80:20	Nubian	1470
	70:30		1530
	60:40		1450
	50:50		1550
Type of grain and number of doses/day[b]	maize, 1 dose/day	Damascus	2490
	maize, 3 doses/day		2445
	barley, 1 dose/day		2461
	barley, 3 doses/day		2446
Percentage of concentrate[c]	20	Alpine	1950
	35		2221
	50		2170
	65		2100
RDP/RUP ratio and CP level[d]	72:28; CP = 21.0%	Alpine × Beetal	1140
	55:45; CP = 21.0%		1080
	55:45; CP = 17.5%		1090
Protected fat inclusion, percentage[e]	0	Alpine	1860
	1.5		1800
	3.0		1990
	4.5		2170
	6.0		1960
Protected fat inclusion, percentage[f]	0	Granadina	1395
	9		1349
	12		1274

RDP, rumen-degradable protein; RUP, rumen-undegradable protein; CP, crude protein.
Data source: [a]Andrade *et al.* (1996); [b]Hadjipanaioutou (2004); [c]Goetsch *et al.* (2001); [d]Mishra and Rai (1996); [e]Brown-Crowder *et al.* (2001); [f]Sanz Sampelayo *et al.* (2002).

content than that of Saanen goats. This was attributed to the higher microbial density and dietary degradation rate in the Black Bedouin breed compared with the Saanen. This allowed the Black Bedouin breed to have similar intake to the Saanen, despite the longer rumen retention time. Other authors (Lindberg and Gonda, 1997) also observed better digestive capacity in wild goats (*Nubia ibex*) and in the Bedouin breed compared with Saanen goats.

The results obtained from our analysis could be related to the better dietary conditions of Alpine goats compared with Mediterranean ones.

Prediction of pasture intake of grazing goats

Short grazing is a widely adopted system for small dairy ruminants. This system consists of taking the animals to pasture, on sown swards or natural pastures, for

Table 7.2. Description of the database used to develop regression equations to predict dry matter (DM) intake in stall-fed goats.

Item	BW (kg)	Milk yield (g/day)	DM intake (g/day)	CP (% of DM)	NDF (% of DM)
Alpine breeds					
n	64	64	64	64	41
Mean	53.4	2084	1984	15.5	39.4
SD	9.6	1122	626	2.8	8.1
Min	37.0	469	889	10.0	18.0
Max	70.0	4900	3440	22.3	48.2
Mediterranean breeds					
n	69	69	69	59	43
Mean	49.4	1311	1408	15.9	35.8
SD	12.5	522	474	4.2	7.1
Min	24.0	300	572	10.1	18.0
Max	77.7	2630	3000	25.0	49.4

BW, body weight; CP, crude protein; NDF, neutral-detergent fibre; SD, standard deviation.

Table 7.3. Results of simple linear regression analysis with dry matter intake (g/day) as the dependent variable for Alpine and Mediterranean breeds.

	Alpine breeds			Mediterranean breeds		
Independent variable	*n*	R^2	*P*	*n*	R^2	*P*
Milk (g/day)	64	0.734	<0.001	69	0.396	<0.001
BW (kg)	64	0.754	<0.001	69	0.635	<0.001
Diet CP (% DM)	64	0.204	<0.001	59	0.008	0.504
Diet NDF (% DM)	41	0.001	0.991	43	0.462	<0.001

BW, body weight; CP, crude protein; NDF, neutral-detergent fibre.

4 to 8 h per day between the two milkings. During milkings or during the night, the animals normally receive a feed supplement based on concentrates and/or by-products, and straw or hay.

Estimating pasture intake is desirable for correct feed management, in order to optimize the utilization of feed supplements. In fact, the supplement can cause a substitution effect on herbage intake, sometimes making this practice too costly.

The systems available to estimate feed intake of goats have been developed mainly for stall-fed goats (Luo *et al.*, 2004; Morand-Fehr, 2005). In a detailed study on Mediterranean goat intake at pasture, Fedele *et al.* (1993) proposed some equations to estimate DM intake. In particular, when variables affecting selectivity (chemical and botanical composition of the pasture) were included in the prediction models, intake was predicted more accurately than when only animal variables (e.g. BW and milk production and composition) were used. Since the equations were derived from a limited number of animals grazing in a restricted

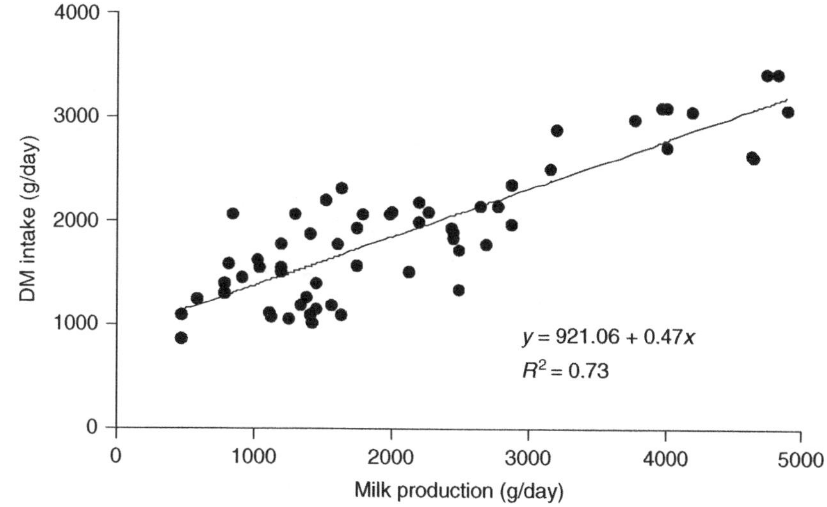

Fig. 7.1 Regression between dry matter (DM) intake and milk production in goats of Alpine breeds ($n = 64$; $P < 0.01$).

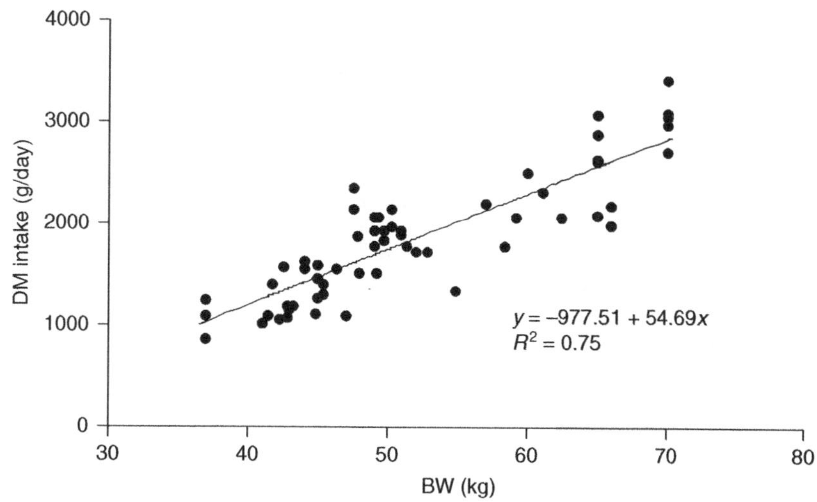

Fig. 7.2 Regression between dry matter (DM) intake and body weight (BW) in goats of Alpine breeds ($n = 64$; $P < 0.01$).

area, characterized by a peculiar botanical composition, they should not be applied to other conditions.

With the aim of formalizing an intake model for lactating goats in short grazing systems, 600 individual pasture intake data were collected over 3 years of trials, together with information on milk production, milk composition, live weights, body condition score and different variables characterizing the pasture, and were statistically analysed.

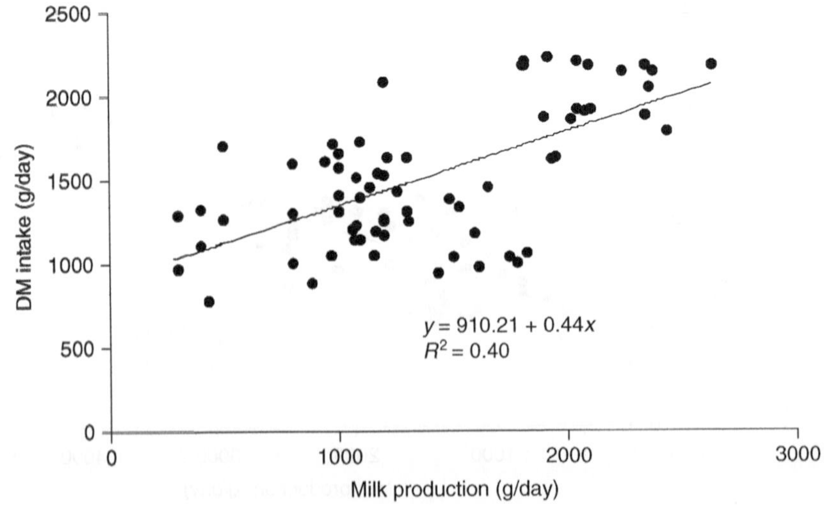

Fig. 7.3. Regression between dry matter (DM) intake and milk production in goats of Mediterranean breeds ($n = 69$; $P < 0.01$).

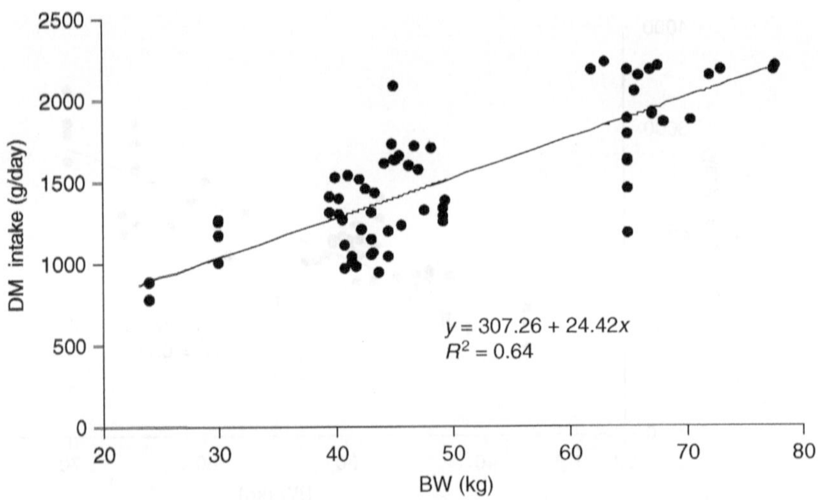

Fig. 7.4. Regression between dry matter (DM) intake and body weight (BW) in goats of Mediterranean breeds ($n = 69$; $P < 0.01$).

The statistical analysis showed that the marked individualistic feeding behaviour of grazing goats did not allow clear identification of a single key factor affecting voluntary intake of DM at pasture. Positive relationships were found between intake and production level or BW, probably owing to the animal's capacity to adjust feed intake to nutritive requirements and rumen volume. Similarly, most reported equations estimating intake for cattle, sheep and goats include BW and/or milk production among the independent variables (Pulina *et al.*, 1996; Peyraud *et al.*, 1998; Pittroff and Kothmann, 2001; Avondo *et al.*, 2002; Luo *et al.*, 2004).

Pasture intake, as expected, was negatively correlated with diet supplements, due to the substitution effect that the latter generally cause on herbage consumption. Such correlation was higher when supplement intake was expressed as grams of CP than when expressed as grams of DM. This had already been observed in sheep and was attributed to their ability to self-regulate intake on the basis of their protein needs (Avondo *et al.*, 2002). Similar results were also found in trials on rams (Avondo *et al.*, 2004a) and goats (Fedele *et al.*, 1993). Recently, the ability of lactating goats to reduce protein intake from pasture when receiving supplements rich in CP was clearly demonstrated (Avondo *et al.*, 2004b). Thus, since goats are highly selective and choose parts of the pasture rich in protein even in very poor feeding conditions, the negative effects of protein supplement on pasture intake should be attributed, with greater likelihood, to the self-regulating protein theory hypothesized for sheep.

In our statistical analyses, all of the variables that were not co-associated, and which were significantly correlated with intake and were biologically meaningful, were included in multiple regression prediction models. The most significant prediction equation ($R^2 = 0.41$; $P < 0.01$) is as follows:

$$\text{Pasture intake (g DM/day)} = 822.11 - 6.188\text{CPS} + 0.138\text{FCM} + 9.131\text{BW}$$

where CPS is the quantity of CP given with supplement (g/day), FCM is the production of 5% fat-corrected milk (g/day) (Pulina *et al.*, 1991) and BW is animal body weight (kg).

Pasture characteristics did not have significant effects on goat intake capacity in our study. Indeed, García *et al.* (1995) observed an increase in goat pasture intake when NDF and lignin content increased. In fact, the poor quality of the pasture is usually not a limiting factor for intake, thanks to the abundant rumen microflora of goats, associated with their intense selective activity. Only in extreme conditions, such as on a pasture containing more than 30% of DM and with a very poor chemical–nutritive composition, was pasture utilization by goats limited. Figure 7.5 shows that DM intake at pasture tends to decrease as DM percentage of herbage increases, particularly over 30%. In these conditions, none of the variables was significantly correlated with intake levels and it was clear that dietary supplements did not have any substitution effect, as normally occurs in good pasture conditions. It is well known that if the herbage is poor in quality, the administration of a supplement does not reduce intake but, in certain circumstances, can actually increase it. Since only a few data on herbage DM content over 30% were available, a specific equation could not be developed. For this reason, it is recommended to use the equation proposed above only for pastures with DM content below 30%.

Growing Goats

Intake in kid goats is linked to the characteristics of the animals and their diet. Lu and Potchoiba (1990) and Luo *et al.* (2004) have developed equations to estimate DM intake for kids, which include BW, weight gain and diet parameters. In particular, Luo *et al.* (2004) used data from about 50 papers on various genotypes,

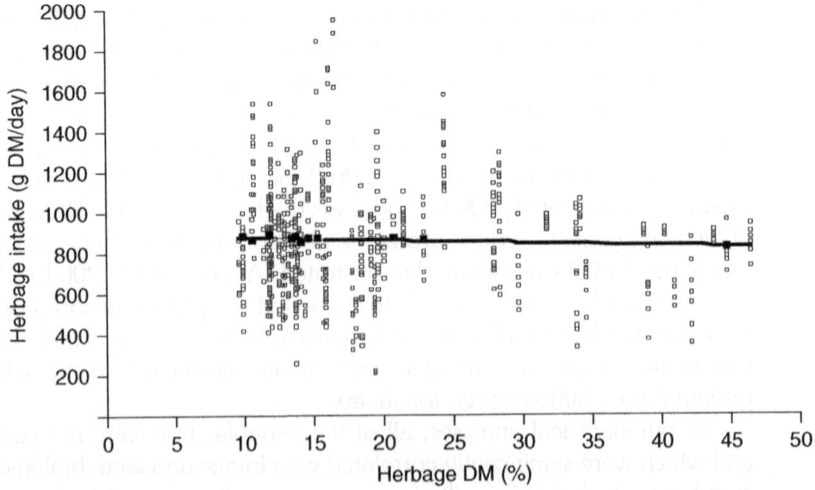

Fig. 7.5. Regression between pasture intake of lactating goats and herbage dry matter (DM) content.

including fibre and dwarf breeds. Sahlu *et al.* (2004), on the basis of the equations proposed by Luo *et al.* (2004), have developed a detailed table reporting the predicted intake values for stall-bred kids belonging to meat breeds, dairy breeds (Saanen, Alpine, Damascus, Norwegian, Swedish Landrace and dairy cross-breeds) and other breeds described as 'indigenous'. Estimated intake values vary with energy concentration of the diet, mean daily weight gain and BW of the animals. Table 7.4 is a simplified version of the table proposed by Sahlu *et al.* (2004), which includes estimated values for DM intake of kids of dairy goats, with different energy concentrations of the diet and a daily weight gain of 100 g. It can be seen that, for equal BW, intake diminishes as the energy concentration of the diet increases. Moreover, even at the lowest energy concentration, the intake estimated by Sahlu *et al.* (2004) for dairy breeds is always lower than that reported by the Institut National de la Recherche Agronomique (INRA) (Morand-Fehr and Sauvant, 1988).

Goetsch *et al.* (2003) found that total DM intake was not affected by feeding management (free-choice availability of concentrate and forage versus mixed diet) in Alpine doelings. However, separate free-choice offering of concentrate and forage increases concentrate and CP daily intake and conversion efficiency.

The equations proposed by Lu and Potchoiba (1990) and by Luo *et al.* (2004) to estimate intake are rather complex as they presuppose the availability of chemical analysis of feed supplied, to calculate the energy concentration of the diet. This complicates their practical application at farm level, particularly in extensive systems in which the characteristics of feed resources vary widely over the year. Moreover, Mediterranean breeds are poorly represented among the genetic types taken into consideration by Luo *et al.* (2004), since out of the 50 trials from which data on growing animals were obtained, only about ten concerned Mediterranean breeds or populations. Therefore, an attempt was made to develop

Table 7.4. Estimated values of dry matter intake (g/day) in kids of dairy goats during growth (100 g/day), in relation to diet energy concentration (calculated by Sahlu *et al.*, 2004). Comparison with data from Institut National de la Recherche Agronomique (INRA, 1988).

	Energy concentration of the diet (MJ/kg)				
Live weight (kg)	7	9	11	13	INRA
10	–	430	380	340	–
15	750	650	580	530	900
20	940	800	720	660	1040
25	1090	930	830	760	1110

Table 7.5. Experiments used for the database created to predict the dry matter intake of growing goats.

Breed	Diet	Reference
Damascus	Hay and concentrate	Hadjipanayiotou (1995)
Nigerian local	Concentrate and by-products	Aregheore (1996)
Greek local	*Carpinus orientalis* or *Fraxinus ornus* leaves	Papachristou (1996)
Spanish	*Phaseolus vulgaris* straw and hay or *Acacia* leaves (*rigidula* or *farnesiana*)	Ramírez and Ledezma-Torres (1997)
Barbari	Straw and whole soy seeds	Mani and Chandra (2003)
Barbari	Hay and concentrate administered as such and as feed block	Samantha *et al.* (2003)
Local Indian	Straw and concentrate	Anbarasu *et al.* (2004)
Anglo-Nubian × Fiji	*Ischaemum aristacum* and *Ipomoea batatas* green forage	Aregheore (2004)
Spanish	Hay and concentrate	Joemat *et al.* (2004)
Spanish	Hay and concentrate	Urge *et al.* (2004)
Girgentana	Hay and concentrate	Avondo (personal communication, 2005)

equations to estimate DM intake using only BW as predictor, based on 87 intake means from different experiments on Mediterranean and African breeds, according to the classification of Flamant and Morand-Fehr (1982). Table 7.5 provides an overview of the diet adopted in each experimental trial of the database.

The correlation between DM intake and BW was significant but very low ($r = 0.21$; $P < 0.05$). In Table 7.6, which reports the mean DM intake per class of live weight of our database, DM intake tends to increase between 10 and 20 kg of live weight; subsequently, however, the intake level diminishes. This trend can be attributed to different feeding conditions of the animals. In this small

Table 7.6. Measured dry matter (DM) intake of kids of Mediterranean goats in relation to live weight and diet characteristics. (Based on the data of the publications reported in Table 7.5.)

Live weight class (kg)	*n*	DM intake (g/day)	DM intake (% of BW)	Mean characteristics of diet: CP (%)	NDF (%)	Concentrate level (% of DM)
10–15	15	616	4.51	15	43	55
15.1–20	21	1203	6.59	11	47	37
20.1–25	43	834	3.79	11	55	36
>25.1	8	1028	3.23	12	41	79

CP, crude protein; NDF, neutral-detergent fibre.

database, the effects of the chemical composition of the diets, within each weight class, could not be tested appropriately. Nevertheless, it could be observed that mean intake in the weight classes above 20 kg was lower than in lighter weight classes, probably due to a particularly high mean dietary NDF concentration (55% between 20.1 and 25 kg of live weight) or, on the contrary, due to an excessive concentrate level in the diet (79% of concentrates for live weights over 25 kg). Moreover, the kids classified as 'Mediterranean' include breeds that differ among themselves and could have different intake and digestive capacity of poor-quality forage. In this regard, Silanikove (1986), Lu and Potchoiba (1990) and Urge *et al.* (2004) demonstrated breed effects on intake.

In conclusion, the prediction of intake of growing goats of Mediterranean breeds using the database created from literature data was unfeasible, because of the wide variability of the breeds and the diets considered in the experimental trials. Therefore, further experimental data or different approaches are required.

Dry and Pregnant Goats

The INRA system considers intake capacity to be constant from drying-off to the fourth month of pregnancy, with a mean reduction of 10% during the fifth month of pregnancy (Morand-Fehr and Sauvant, 1988) (Table 7.7). The reduction in intake as parturition nears is a well-known phenomenon and is associated with the smaller rumen volume caused by fetal growth and the depressive effect of increased secretion of placental oestrogens on appetite (Forbes, 1995). However, Fedele *et al.* (2002) have observed that marked increases in intake can occur at the final stages of gestation in goats fed both hay and concentrate and in goats fed using a free-choice system. Indeed, the animals fed hay and concentrate, in the last period of pregnancy, progressively diminished hay intake in favour of grains, thereby reducing the filling effect responsible for reduced intake. In the same stage of pregnancy, the animals fed using the free-choice method increased their hay intake, showing a preference for better-quality hay. In Table 7.8, reporting results of some trials on Mediterranean non-pregnant dry goats in various feeding conditions, it

Table 7.7. Dry matter (DM) intake of dry and pregnant goats. (Adapted from Morand-Fehr and Sauvant, 1988.)

	DM intake (g/kg BW)	
Live weight (kg)	Dry-off to 4th month of pregnancy	5th month of pregnancy
40	67.3	61.0
50	63.8	58.0
60	61.7	56.1
70	60.7	55.4
80	59.8	54.6

BW, body weight.

Table 7.8. Dry matter (DM) intake and digestibility of non-pregnant dry goats in various feeding conditions.

Breed	Diet	Experimental plan	DM intake ($g/BW^{0.75}$)	DM digestibility (%)	CP (% of DM)	NDF (% of DM)
Blanca Celtiberica[a]	*Atriplex halimus*	[–]	74.7	50.5	13.8	52.4
Sarda[b]	*Pistacia lentiscus* leaves and concentrate	Control	46.6	66.1	12.2	43.2
		25 g PEG/day	50.3	68.3	12.0	43.1
		50 g PEG/day	48.3	68.5	11.8	43.0
Maltese[c]	Hay and concentrates	Traditional	67.9	76.4	12.6	39.9
		Free choice	46.0	59.6	11.4	40.1
Spanish local breed[d]	Pasture with or without supplement	April–May	34.9	–	10.0	40.4
		May–June	61.6	–	12.5	51.1
		June–July	45.5	–	8.7	53.1
Granadina[e]	Shrubs and trees pasture	[–]	35.7	53.1	8.8	51.6

BW, body weight; CP, crude protein; NDF, neutral-detergent fibre; PEG, polyethylene glycol.
Data source: [a]Valderrabano *et al.* (1996); [b]Decandia *et al.* (2000); [c]Fedele *et al.* (2002); [d]García *et al.* (1995); [e]Molina Alcaide *et al.* (1997).

is possible to note some influence of feed characteristics on their intake capacity. In particular, the dietary CP level, generally rather low, was significantly positively correlated with intake ($r = 0.81$). In addition, there was no close correlation between digestibility and NDF in the diet.

References

Abijaoudé, J.A., Morand-Fehr, P., Tessier, J., Schmidely, P. and Sauvant, D. (2000) Contribution of main and secondary meals to the daily intake of stall-housed dairy goats in mid lactation. In: Ledin, I. and Morand-Fehr, P. (eds) Sheep and Goats Nutrition: Intake, Digestion, Quality of Products and Rangeland. *Cahiers Options Méditerranéennes* 52, 33–37.

Anbarasu, C., Dutta, N., Sharma, K. and Rawat, M. (2004) Response of goats to partial replacement of dietary protein by leaf meal mixture containing *Leucaena leucocephala*, *Morus alba* and *Tectona grandis*. *Small Ruminant Research* 51, 47–56.

Andrade, H., Bernal, G. and Llamas, G. (1996) Influence of different alfalfa sorghum ratios in the diet of dairy goats on productivity and rumen turnover. *Small Ruminant Research* 21, 77–82.

Aregheore, M. (1996) Voluntary intake and nutrient digestibility of crop-residue based rations by goats and sheep. *Small Ruminant Research* 22, 7–12.

Aregheore, M. (2004) Nutritive value of sweet potato (*Ipomea batatas* (L) Lam) forage as goat feed: voluntary intake, growth and digestibility of mixed rations of sweet potato and batiki grass (*Ischaemum aristatum* var. *indicum*). *Small Ruminant Research* 51, 235–241.

Avondo, M., Bordonaro, S., Marletta, D., Guastella, A.M. and D'Urso, G. (2002) A simple model to predict the herbage intake of grazing dairy ewes in semi-extensive Mediterranean systems. *Livestock Production Science* 73, 275–283.

Avondo, M., Lutri, L. and Pennisi, P. (2004a) Feeding behaviour of Comisana rams as affected by crude protein level of concentrate. *Small Ruminant Research* 55, 135–140.

Avondo, M., Pagano, R., Di Grigoli, A., Bonanno, A., Formoso, B., Giambalvo, D. and Scarpello, C. (2004b) Comportamento alimentare di capre Girgentane: effetto del tipo di pascolo e del livello proteico del supplemento. In: *Proceedings of the 16th SIPAOC Congress* (Atti del XVI Congresso Nazionale della Società Italiana di Patologia e di Allevamento degli Ovini e dei Caprini), Siena, Italy, p. 232 (Abstract). (Full paper on CD-ROM, 1–4).

Bailoni, L. and Andrighetto, I. (1995) Effetto del rapporto foraggi:concentrati sui parametri digestivo-metabolici e sulle prestazioni produttive di capre primipare in lattazione. *Zootecnia e Nutrizione Animale* 21, 333–344.

Brown-Crowder, I.E., Hart, S.P., Cameron, M., Sahlu, T. and Goetsch, A.L. (2001) Effects of dietary tallow level on performance of Alpine does in early lactation. *Small Ruminant Research* 39, 233–241.

Brun-Bellut, J. (1996) Urea recycling in the rumen of dairy goats: effects of physiological stage and composition of intake. *Small Ruminant Research* 23, 83–90.

Decandia, M., Sitzia, M., Cabiddu, A., Kababya, D. and Molle, G. (2000) The use of polyethylene glycol to reduce the anti-nutritional effects of tannins in goats fed woody species. *Small Ruminant Research* 38, 157–164.

Economides, S. (1998) The nutritive value of sunflower meal and its effect on replacing cereal straw in the diets of lactating ewes and goats. *Livestock Production Science* 55, 89–97.

Fedele, V., Pizzillo, M., Claps, S., Morand-Fehr, P. and Rubino, R. (1993) Grazing behaviour and diet selection of goats on native pasture in Southern Italy. *Small Ruminant Research* 11, 305–322.

Fedele, V., Claps, S., Rubino, R., Calandrelli, M. and Pilla, A.M. (2002) Effect of free-choice and traditional feeding systems on goat feeding behaviour and intake. *Livestock Production Science* 74, 19–31.

Flamant, J.C. and Morand-Fehr, P. (1982) Milk production in sheep and goat. In: Coop, I.E. (ed.) *World Animal Science. C. Production-System Approach. 1. Sheep and Goat Production*. Elsevier, Amsterdam, pp. 275–295.

Forbes, J.M. (1995) *Voluntary Food Intake and Diet Selection in Farm Animals*. CAB International, Wallingford, UK.

García, M.A., Aguilera, J.F. and Molina Alcaide, E. (1995) Voluntary intake and kinetics of degradation and passage of unsupplemented and supplemented pastures from semiarid lands in grazing goats and sheep. *Livestock Production Science* 44, 245–255.

Goetsch, A.L., Detweiler, G., Sahlu, T., Puchala, R. and Dawson, L.J. (2001) Dairy goat performance with different dietary concentrate levels in late lactation. *Small Ruminant Research* 41, 117–125.

Goetsch, A.L., Detweiler, G., Sahlu, T., Hayes, J. and Puchala, R. (2003) Effects of separate offering of forage and concentrate on feed intake and growth of Alpine doelings. *Small Ruminant Research* 48, 209–216.

Gupta, H.K. and Atreja, P.P. (1999) Influence of feeding increasing levels of leucaena leaf meal on the performance of milking goats and metabolism of mimosine and 3-hydroxy-4 (1H) pyridone. *Animal Feed Science and Technology* 78, 159–167.

Hadjipanayiotou, M. (1995) Fractional outflow of soybean meal from the rumen, water intake and ruminal fermentation pattern in sheep and goats at different seasons and age group. *Small Ruminant Research* 17, 137–143.

Hadjipanayiotou, M. (2004) Replacement of barley grain for corn in concentrate diets fed to dairy Damascus goats at different frequencies. *Small Ruminant Research* 51, 229–233.

Hofmann, R.R. (1989) Evolutionary steps of ecophysiological adaptation and diversification of ruminants: a comparative view of their digestive system. *Oecologia* 78, 443–457.

INRA (1988) Jarrige, R. (ed.) *Alimentation des bovines, ovins et caprins*. Institut National de la Recherche Agronomique, Paris.

Joemat, R., Goetsch, A.L., Horn, G.W., Sahlu, T., Puchala, R., Min, B.R., Luo, J. and Smuts, M. (2004) Growth of yearling meat goat doelings with changing plane of nutrition. *Small Ruminant Research* 52, 127–135.

Lindberg, J.E. and Gonda, H.L. (1997) Fibre and protein digestion in goats. In: Morand-Fehr, P. (ed.) Recent Advances in Goat Research. *Cahiers Options Méditerranéennes* 25, 47–58.

Lu, C.D. and Potchoiba, M.J. (1990) Feed intake and weight gain of growing goats fed diets of various energy and protein levels. *Journal of Animal Science* 68, 1751–1759.

Luo, J., Goetsch, A.L., Nsahlai, I.V., Moore, J.E., Galyean, M.L., Johnson, Z.B., Sahlu, T., Ferrell, C.L. and Owens, F.N. (2004) Voluntary feed intake by lactating, Angora, growing and mature goats. *Small Ruminant Research* 53, 357–378.

Mani, V. and Chandra, P. (2003) Effect of feeding irradiated soybean on nutrient intake, digestibility and N-balance in goats. *Small Ruminant Research* 48, 77–81.

Mishra, S. and Rai, S.N. (1996) Effects of different RDP and UDP ratios on voluntary intake, milk production and feed conversion efficiency in lactating goats. *Small Ruminant Research* 20, 31–38.

Molina Alcaide, E., García, M.A. and Aguilera, J.F. (1997) The voluntary intake and rumen digestion by grazing goats and sheep of a low-quality pasture from a semi-arid land. *Livestock Production Science* 52, 39–47.

Morand-Fehr, P. (2005) Recent developments in goat nutrition and application: a review. *Small Ruminant Research* 60, 25–43.

Morand-Fehr, P. and Sauvant, D. (1988) Alimentation des caprins. In: Jarrige, R. (ed.) *Alimentation des bovins, ovins and caprins*. Institut National de la Recherche Agronomique, Paris, pp. 281–304.

Papachristou, T.G. (1994) Foraging behaviour and nutrition of goats grazing on shrublands of Greece. In: Gordon, I.J. and Rubino, R. (eds) Grazing Behaviour of Goats and Sheep. *Cahiers Options Méditerranéennes* 5, 83–90.

Papachristou, T.G. (1996) Intake, digestibility and nutrient utilization of oriental hornbeam and manna ash as browse by goats and sheep. *Small Ruminant Research* 23, 91–98.

Peyraud, J.L., Delaby, L. and Delagarde, R. (1998) Quantitative approach to nutrition of grazing dairy cows: recent developments. In: *Proceedings of the Workshop on 'Pasture Ecology and Animal Intake'*, Dublin. Irish Agriculture and Food Development Authority (TEAGASC), Dunsany, Ireland, pp. 57–75.

Pittroff, W. and Kothmann, M.M. (2001) Quantitative prediction of feed intake in ruminants. III. Comparative example calculations and discussion. *Livestock Production Science* 71, 171–181.

Provenza, F.D., Villalba, J.J., Dziba, L.E., Atwood, S.B. and Banner, R.E. (2003) Linking herbivore experience, varied diets, and plant biochemical diversity. *Small Ruminant Research* 49, 257–274.

Pulina, G., Cannas, A., Serra, A. and Vallebella, R. (1991) Determinazione e stima del valore energetico di latte di capre di razza Sarda. In: *Proceedings of the XLV Congress SISVet*, Altavilla Milicia, Italy. Grafiche Scuderi, Messina, Italy, pp. 1779–1781.

Pulina, G., Bettati, T., Serra, F.A. and Cannas, A. (1996) Razi-O: costruzione e validazione di un software per l'alimentazione degli ovini da latte. In: *Atti 13° Congresso Nazionale SIPAOC*, Varese, Italy. Società Italiana di Patologia e di Allevamento degli Ovini e dei Caprini, Sassari, Italy, pp. 11–14.

Ramírez, R.G. and Ledezma-Torres, R.A. (1997) Forage utilization from native shrubs *Acacia rigidula* and *Acacia farnesiana* by goats and sheep. *Small Ruminant Research* 25, 43–50.

Rapetti, L., Bava, L., Tamburini, A. and Crovetto, G.M. (2005) Feeding behaviour, digestibility, energy balance and productive performance of lactating goats fed forage-based and forage-free diets. *Italian Journal of Animal Science* 4, 71–83.

Sahlu, T., Goetsch, A.L., Luo, J., Nsahlai, I.V., Moore, J.E., Galyean, M.L., Owens, F.N., Ferrell, C.L. and Johnson, Z.B. (2004) Nutrient requirements of goats: developed equations, other considerations and future research to improve them. *Small Ruminant Research* 53, 191–219.

Samantha, A.K., Singh, K.K., Das, M.M., Maity, S.B. and Kundu, S.S. (2003) Effect of complete feed block on nutrient utilisation and rumen fermentation in Barbari goats. *Small Ruminant Research* 48, 95–102.

Sanz Sampelayo, M.R., Perez, L., Martin Alonso, J.J., Gil Extremera, F. and Boza, J. (2002) Effects of concentrates with different contents of protected fat rich in PUFAs on the performance of lactating Granadina goats. 1. Feed intake, nutrient digestibility, N and energy utilization for milk production. *Small Ruminant Research* 43, 133–139.

Silanikove, N. (1986) Interrelationships between feed quality, digestibility, feed consumption, and energy requirements in desert (Bedouin) and temperate (Saanen) goats. *Journal of Dairy Science* 69, 2157–2162.

Silanikove, N. (2000) The physiological basis of adaptation in goats to harsh environments. *Small Ruminant Research* 35, 181–193.

Urge, M., Merkel, R.C., Sahlu, T., Animut, G. and Goetsch, A.L. (2004) Growth performance by Alpine, Angora, Boer and Spanish wether goats consuming 50 or 75% concentrate diets. *Small Ruminant Research* 55, 149–158.

Valderrabano, J., Muñoz, F. and Delgado, I. (1996) Browsing ability and utilization by sheep and goats of *Atriplex halimus* L. shrubs. *Small Ruminant Research* 19, 131–136.

8 Foraging Behaviour and Intake of Goats Browsing on Mediterranean Shrublands

M. DECANDIA,[1] M.D. YIAKOULAKI,[2] G. PINNA,[1] A. CABIDDU[1] AND G. MOLLE[1]

[1]*AGRIS Sardegna - Dipartimento per la Ricerca nelle Produzioni Animali, 07040 Olmedo, Sassari, Italy;* [2]*Department of Range Science, Aristotle University, 54124 Thessaloniki, Greece*

Introduction

The importance of goat production systems in Mediterranean countries is related to the ability of these animals to convert the vegetation of marginal areas into high-quality products. In such environments, shrub species are very often the main available feed resource.

In the Mediterranean basin, the production of goats in extensive systems plays an important socio-economic role and often is one of the few economic activities possible. During the last few years, interest in organic products and their traceability has been increasing. Some woody plants contain volatile compounds that are able to pass from feed to dairy products. These 'markers' may relate the products to the area browsed by the animals (Decandia *et al.*, 2005b).

Among domestic ruminants, goats have a unique ability to adapt to a wide variety of climatic conditions and forage species that very often are characterized by low nutritive value. This is due to the particular behaviour of goats, classified as 'opportunistic feeders', and to their higher digestive efficiency for high-fibre, low-quality forages (Silanikove, 2000) compared with sheep and cattle. The utilization of woody vegetation by goats can also be used as a tool to reduce shrub encroachment and the risk of wildfires.

Since knowledge of the foraging behaviour and intake of goats browsing on Mediterranean shrublands is important for their efficient production, these aspects are discussed in this chapter.

Mediterranean Shrublands

Natural shrublands cover an area of more than 230,000 km^2 (Le Houérou, 1981) and constitute an essential part of traditional production systems in the

Mediterranean basin. The great climatic, edaphic and physiographical heterogeneity of the Mediterranean areas coupled with human intervention have resulted in a large variety of plant species.

There is no commonly accepted classification of Mediterranean shrublands and each country uses its own vernacular names (Papanastasis, 2000). However, from the point of view of range management, shrublands can be subdivided into three categories: (i) maquis or matorral; (ii) garrigues (Tomaselli, 1981; Le Houérou, 1993a); and (iii) phrygana (Papanastasis, 1977).

- Maquis or matorral are xerophilous and schlerophyllous shrublands, growing on relatively deep, mainly siliceous soils (Le Houérou, 1993b; Papanastasis, 1997, 2000). They are tall (1–3 m), dense and impenetrable plant communities dominated by a large variety of species, such as *Arbutus unedo* L., *Arbutus adrachne* L., *Erica arborea* L., *Pistacia lentiscus* L., *Myrtus communis* L., *Quercus ilex* L., *Quercus coccifera* L., *Phillyrea media* L. and *Cistus monspeliensis* L.
- Garrigues are open plant communities of relatively small (0.5–1.5-m tall) schlerophyllous shrubs growing mainly on calcareous soils (Le Houérou, 1993b; Papanastasis, 1997). They are dominated by the evergreen *Q. coccifera* L. and by other evergreen or deciduous woody species, such as *Carpinus orientalis* Miller, *Fraxinus ornus* L., *Thymus capitatus* (L.) Hoffsgg & Link., *Rosmarinus officinalis* L., *Erica multiflora* L. and *Cistus* spp.
- Phrygana are open communities of flammable and dwarf shrubs (shorter than 1 m) growing mainly on rocky soils (Papanastasis, 2000). These are dimorphic species, which replace their large winter leaves with smaller leaves during summer, in order to conserve water and adapt to dry conditions. The most common species are *Sarcopoterium spinosum* (L.) Spach., *Phlomis fruticosa* L., *Cistus* spp., *Coridothymus capitatus* (L.) Reichenb. Fil. and *Salvia officinalis* L.. Phrygana are mainly distributed in the eastern part of the Mediterranean area, including Greece.

Garrigue open shrublands are more available to livestock and thus more productive than maquis. Phrygana plant species are not palatable for livestock, being eaten only in extreme situations. Nevertheless, they are important for the richness of herbaceous species (Papanastasis, 1997).

In the past, since goats were blamed for deforestation and deterioration of rangelands, legislation to limit goat rearing was introduced. Therefore, instead of excluding goats from forests, Papanastasis (1997) suggested an integration policy in which livestock grazing integrates with forestry and other land uses. Nowadays, rearing grazing goats is often considered as an alternative activity which can protect forests and shrublands from wildfires (Yiakoulaki *et al.*, 1999), reduce the cost of animal production and revitalize abandoned rural areas.

Foraging Behaviour of Grazing Goats

Goats are able to thrive on marginal areas, using low-quality natural resources which cannot be efficiently used by other domestic ungulates. Due to their anatomic and physiological characteristics, goats exhibit a very flexible foraging behaviour. Actually, these animals are not considered as exclusive browsers but mainly as 'opportunistic feeders' that have a selective foraging behaviour affected by the interactions among animal, environmental and plant characteristics.

Animal characteristics

Goats are well known for utilizing a wide array of vegetation types, including shrubs, trees and herbaceous species. They are also able to: (i) graze very short herbage, as sheep do (Narjisse, 1991); (ii) feed on plant species with thorns and spines; and (iii) stretch upwards on their hind legs (bipedal stance), selecting foliage from the upper vegetation layer (Fig. 8.1).

This unique foraging behaviour is mainly due to their physical body structure (low body mass, small and sharp-shaped mouth, mobile upper lip and prehensile tongue), high digestive capacity (Hadjipanayiotou and Antoniou 1983; Howe and Barry, 1988) and rumen microflora (Molina Alcaide *et al*., 1997). Furthermore, goats are able to consume much more tannin-rich browse resources and

Fig. 8.1. Goat (local breed) in bipedal stance selecting foliage of kermes oak (*Quercus coccifera* L.) in a Mediterranean shrubland in northern Greece. (Photograph courtesy of M.D. Yiakoulaki.)

to digest them more efficiently than other domestic animals (Dominique *et al.*, 1991; Narjisse, 1991; Silanikove, 2000). This is related to their higher ability to neutralize the negative effects of tannins on palatability and digestibility (Silanikove *et al.*, 1996a). It seems that both salivary (i.e. high secretion of proline-rich proteins) and, above all, ruminal mechanisms are involved in the neutralization of the anti-nutritional effects of tannins in goats. Likewise, ruminal fermentation and adaptation of rumen microbes to terpenes may also enable goats to more efficiently use terpene-rich foods (Landau *et al.*, 2000).

Breed and physiological stage of goats also influence their foraging behaviour. For example, 'rustic' breeds (wild goats, desert Bedouin goats) utilize harsh pastures better than do Swiss Saanen goats (Silanikove *et al.*, 1993). Furthermore, lactating Sarda goats spend more time grazing compared with dry or pregnant ones (Fig. 8.2), due to their higher requirements (Decandia *et al.*, 1997a). By selecting pasture plants or plant parts higher in nutritive value than the average and avoiding consumption of the harmful, goats express a degree of nutritional wisdom (Provenza and Balph, 1990; Provenza, 1995). Forage selection is believed to be genetically transmitted (Cooper *et al.*, 1988) and most information is transferred to the offspring during the weaning period (Thorhallsdottir *et al.*, 1987). An interesting 'maternal effect' on foraging behaviour has been found in kids, since the group reared by ewes on pasture had a higher preference for white clover (*Trifolium repens* L.) than that reared by goats (Orr *et al.*, 1995). However, it seems that preference for a specific feed is not permanent. Food preference is adjusted as sampling continues, depending on the 'internal nutritional value' of different species, such as post-ingestive effects. Animals raised in different environments express different dietary preferences. In fact, Rosenberger and Meuret (1985) found that goats previously adapted to browse oaks had a different dietary selection from those that had never utilized that species before. This indicates that dietary selection is influenced not only by genetics but also by the process of learning (trial and error) and experience. Additionally, animals exhibit different preferences if exposed to forage rich in aromatic compounds, such as phenols and terpenes. Goats discriminate between plants with different levels of monoterpenoids less than sheep (Narjisse, 1991).

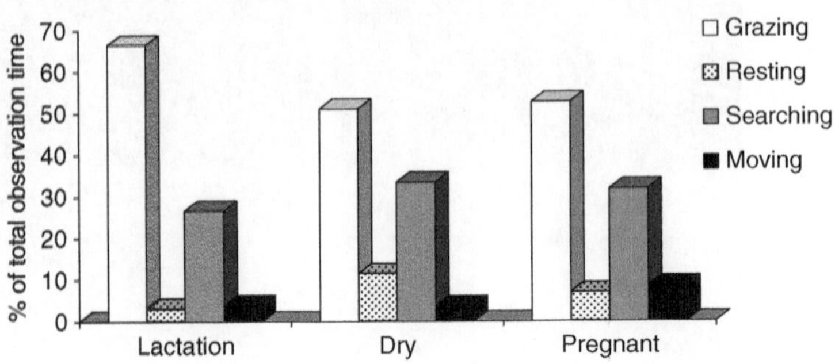

Fig. 8.2. Feeding activities of Sarda goats, expressed as percentage of the total observation time, grazing on Mediterranean shrublands. (Adapted from Decandia *et al.*, 1997a.)

In general, food selection is affected by the interactions between the senses of taste and smell and mechanisms sensing the consequences of food ingestion. Experienced sensations may be, for example, satiety (when animals ingest adequate kinds and amounts of nutritious foods) or malaise (when animals ingest excess of nutrients or toxins or low-nutrient feeds) (Provenza, 1995).

Environmental characteristics

Environmental factors highly affect the foraging behaviour of goats. Although goats adapt to harsh and arid environments better than other ruminants (Shkolnik *et al.*, 1987; Silanikove, 2000), they are sensitive to low temperatures (Narjisse, 1991), high rainfall (Decandia *et al.*, 1997b) and high wind, which reduce their grazing activities. In fact, climatic conditions within each season affect their dietary selection. For example, goats consume more herbaceous species on cold days, whereas they browse more on hot or rainy days, possibly using the woody overstorey as shelter (Landau *et al.*, 2000).

Pasture characteristics

Changes in forage availability throughout the year influence diet selection of grazing goats (Provenza and Malechek, 1983; Rubino *et al.*, 1988; Fedele *et al.*, 1993; Kababya *et al.*, 1998), since they adapt their choice according to what is available in the pasture. When forage availability decreases dramatically (Yiakoulaki *et al.*, 1999), goats shift their diet and may even consume the bark of pine trees (*Pinus brutia* Ten.), which has a very low nutritive value (crude protein (CP) content of 33 g per kilogram of dry matter (DM)). By contrast, high biological diversity within the Mediterranean maquis vegetation increases shrub intake by goats (Rogosic *et al.*, 2006a).

For a long time, goats were considered as obligatory browsers. However, several experimental results in Mediterranean areas have shown that, during the green season, goats tend to select forage resources with higher nutritive value than those selected by obligatory browsers, such as herbaceous species (Narjisse, 1991; Papachristou and Nastis, 1993; Yiakoulaki and Nastis, 1995; Kababya *et al.*, 1998). For example, in Corsica, woody species constituted 75% of goats' diet during winter, but only 15% in spring, during grass regrowth (Leclerc, 1984). In Israel, Mamber goats with a low supplementation level spent 60% of the average eating time browsing the maquis vegetation and 40% grazing grasses in winter, whereas they grazed grasses for 60% of the eating time in spring (Fig. 8.3).

Goats are able to ingest diets with higher nutritive value than what is predicted on the basis of botanical and chemical composition of the pasture. For example, the diet selected by lactating Sarda goats grazing on lentisk-based shrubland (*P. lentiscus* L.) was different from the botanical composition of the pasture (Fig. 8.4) (Decandia *et al.*, 2000a,b). In particular, goats consumed more grasses than other plants during spring, although lentisk was the main component of the pasture. In fact, even if lentisk was the most widespread species in the pasture (60% in summer), it constituted less than 20% of goats' diet,

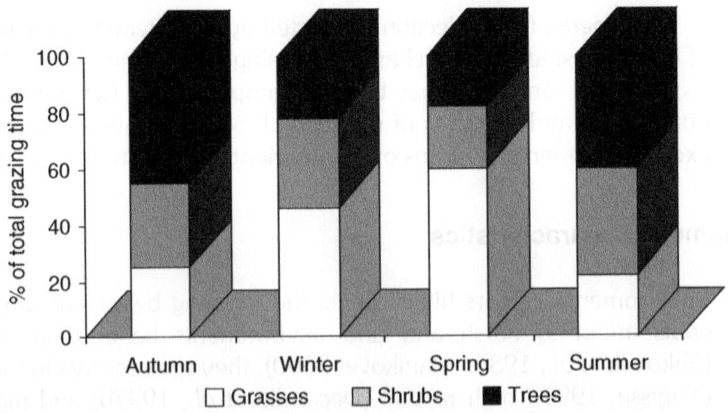

Fig. 8.3. Grazing time of Mamber goats on grasses, shrubs and trees throughout the year. (Adapted from Kababya, 1994.)

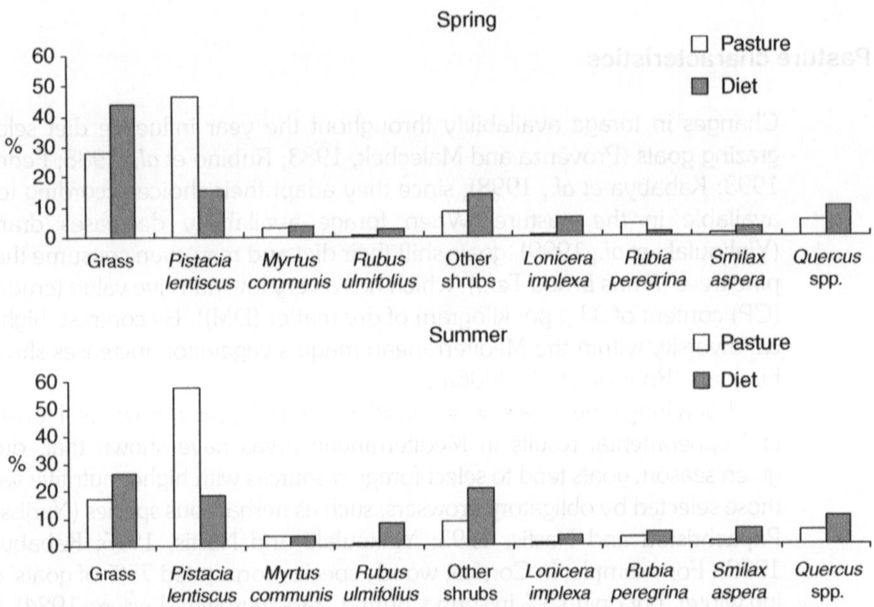

Fig. 8.4. Botanical composition of pasture (contact specific contribution %; Daget and Poissonet, 1969) and diet composition (% of dry matter intake) of lactating Sarda goats browsing on Mediterranean shrubland during spring and summer. (Adapted from Decandia *et al.*, 2000a,b.)

due to its very high tannin level (20%, DM basis). However, during summer goats shifted their diet and ingested fewer grasses, as a result of their lower nutritive value, and more shrubs such as *Rhamnus alaternus* L. (included in the category of other shrubs; Fig. 8.4), which has a relatively high CP content and a low tannin level.

Nutritive Value of Mediterranean Shrublands

The knowledge of the nutritional characteristics of spontaneous plant species in Mediterranean shrublands is fundamental for their proper utilization by grazing animals. The nutritive value of forage species common in this natural resource is variable. Although most shrub species have low to moderate nutritive value (Tsiouvaras and Nastis, 1990; Ben Salem *et al*., 1994, 1999), their advantage over herbaceous species is that they maintain sufficient nutritional levels (CP content and minerals) during critical periods of the year, such as in winter, when herbaceous species are dormant, and in summer, when herbaceous species are dry and depleted (Papanastasis *et al*., 2006).

Chemical composition and digestibility of herbaceous species

Goats grazing on Mediterranean shrublands (Fig. 8.5) prefer woody species. However, as already mentioned, herbaceous species contribute substantially to their diet in some periods of the year. The chemical composition and digestibility of herbaceous species (grasses and legumes) eaten by goats of a local breed in northern Greece is presented in Table 8.1 (Yiakoulaki, 1987; Yiakoulaki and

Fig. 8.5. Goats (local breed) grazing on Mediterranean shrublands (garrigue) in northern Greece. (Photograph courtesy of M.D. Yiakoulaki.)

Table 8.1. Chemical composition (g/kg dry matter) and *in vitro* organic matter digestibility (IVOMD; %) of some herbaceous species commonly found in Mediterranean shrublands. (Adapted from Yiakoulaki, 1987; Yiakoulaki and Nastis, 1993.)

Forage species	Season	CP	NDF	IVOMD	P	Ca
Grasses						
Dactylis glomerata L.	Spring	148	581	67.7	1.7	6.6
	Summer	82	709	56.8	1.3	3.8
	Autumn	92	758	38.9	1.2	3.4
Festuca valesiaca Schleich.	Spring	129	688	57.9	1.9	3.2
	Summer	66	761	41.7	0.9	2.4
	Autumn	39	839	36.6	0.6	2.2
Chrysopogon gryllus (L.) Trin.	Spring	91	710	61.2	1.3	4.6
	Summer	74	749	57.9	1.5	3.9
	Autumn	26	820	40.8	0.6	2.3
Legumes						
Trifolium hirtum All.	Spring	220	550	67.1	1.3	20.0
	Summer	170	576	57.9	1.5	14.6
	Autumn	170	685	39.4	1.4	12.2
Trifolium campestre Shreber.	Spring	173	486	52.9	1.9	14.0
	Summer	163	669	44.3	2.0	10.6
	Autumn	106	722	29.2	1.7	10.0
Trifolium purpureum Loisel.	Spring	177	495	65.1	1.9	12.9
	Summer	158	547	61.6	1.4	12.5
	Autumn	147	705	35.6	1.3	14.2

CP, crude protein; NDF, neutral-detergent fibre.

Nastis, 1993). The nutritive value of grasses decreased rapidly after seed maturation, while that of legumes remained at higher levels (Yiakoulaki, 1987; Yiakoulaki and Nastis, 1993). During the first growing stage, CP of almost all species could satisfy goats' nutritional requirements for lactation. However, after seed maturation CP content of grasses could not satisfy even their maintenance requirements.

Chemical composition of woody species

There are several studies on the chemical composition of woody species in different Mediterranean environments such as Sardinia (Congiu *et al.*, 1978; Cabiddu *et al.*, 2000), Tuscany (Sottini and Geri, 1971), Corsica (Leclerc, 1984), Israel (Kababya *et al.*, 1998; Perevolotsky *et al.*, 1998), Greece (Yiakoulaki, 1987; Yiakoulaki and Nastis, 1993; Papachristou and Papanastasis, 1994), Tunisia (Boubaker *et al.*, 2004; Ammar *et al.*, 2005) and Croatia (Rogosic *et al.*, 2006b). As expected, such studies confirmed that woody species generally have low CP content and high lignin and tannin content. However, such values vary greatly among species and plant phenological stages. For example, CP level is usually higher in young leaves and stems than in other plant parts (Leclerc, 1984), even if

Table 8.2. Dry matter (DM) content (g/kg) and chemical composition (g/kg DM) of woody species commonly found in the Mediterranean shrublands during 1996–2000. (Adapted from Cabiddu *et al.*, 2000.)

Woody species	Season	DM	OM	CP	NDF	ADF	ADL	TP[a]
Cistus spp.	Spring	329	933	113	346	235	114	120
	Summer	412	938	106	331	219	90	155
Pistacia lentiscus L.	Spring	441	942	102	421	287	162	185
	Summer	400	939	95	377	297	163	206
Lonicera implexa Aiton.	Spring	278	923	103	363	230	88	16
	Summer	365	910	99	357	216	94	23
Myrtus communis L.	Spring	401	942	97	347	218	105	118
	Summer	394	951	102	388	255	91	141
Chamaerops humilis L.	Spring	489	928	108	472	284	70	29
	Summer	488	928	87	520	339	76	37
Pyrus amygdaliformis L.	Spring	347	939	143	387	241	100	47
	Summer	447	927	120	429	245	104	28
Quercus ilex L.	Spring	479	957	103	548	351	145	49
	Summer	540	954	93	526	351	136	69
Quercus suber L.	Spring	431	932	98	494	327	142	110
	Summer	459	949	111	505	319	130	106
Rhamnus alaternus L.	Spring	344	917	133	308	189	72	57
	Summer	413	912	110	267	182	88	63
Rubus ulmifolius Schott.	Spring	359	925	132	371	211	77	126
	Summer	405	938	114	374	219	73	121

OM, organic matter; CP, crude protein; NDF, neutral-detergent fibre; ADF, acid-detergent fibre; ADL, acid-detergent lignin; TP, tannic polyphenols.
[a]Determined by the Folin–Ciocalteu method.

it is sometimes associated with higher tannin content (Decandia *et al.*, 2004a). As the growing season progresses and plants mature, the quality of woody species worsens, since CP content decreases and tannin level increases. The chemical composition of some important woody species browsed by Sarda goats during spring and summer is presented in Table 8.2. CP content was usually low with the exception of *Pyrus amygdaliformis* L., *R. alaternus* L. and *Rubus ulmifolius* Schott.

Sunlight strongly influences the nutritive value of shrub species. Leaves of shaded *Q. coccifera* L. shrubs had higher CP, neutral-detergent fibre (NDF) and lignin content than those of sun-exposed shrubs, but tannin level was not significantly different between them (Koukoura, 1988) (Table 8.3). The combined effects of the above changes resulted in lower *in vitro* organic matter digestibility for the leaves of shaded shrubs (Koukoura, 1998). In contrast, Decandia *et al.* (2004a) found that the shaded leaves of *P. lentiscus* L. had higher CP content and a lower tannin level than sun-exposed ones, even if differences were not significant (Table 8.3).

Table 8.3. Chemical composition (g/kg dry matter) and *in vitro* organic matter digestibility (IVOMD; %) of *Quercus coccifera* L. (Koukoura, 1988) and *Pistacia lentiscus* L. (Decandia *et al.*, 2004a) leaves grown in the shade and in the sun.

Species		CP	NDF	ADL	IVOMD	Tannins
Quercus coccifera L.	Sun	85^{b}	434^{b}	115^{b}	40.3^{a}	2.69^{a}
	Shade	101^{a}	539^{a}	158^{a}	35.7^{b}	2.73^{a}
Pistacia lentiscus L.	Sun	89	386	160	–	162
	Shade	103	402	160	–	144

CP, crude protein; NDF, neutral-detergent fibre; ADL, acid-detergent lignin.

[a,b]In each species, mean values in a column with different superscript letters were significantly different ($P \leq 0.05$).

Digestibility of woody species

Digestibility assessment is the most important step for determining forage nutritive value. Measuring digestibility in woody species is often negatively affected by the high presence of ligneous fractions, tannins and other secondary compounds (alkaloids, cumarins). In particular, tannins greatly interfere with *in vitro* digestibility (Makkar *et al.*, 1997). More accurate estimation of the nutritive value of these plants, which are considered as non-conventional forages, has been obtained with *in vivo* digestibility studies. These experiments require, as much as possible, the simulation of real pasture conditions including the daily administration of fresh material, in order to put the animals in a free-choice situation (Meuret, 1989). Due to the low intake of some woody species, the experiments are sometimes carried out using a basic diet of known digestibility. Shrub digestibility is usually underestimated by the *in vitro* compared with the *in vivo* method (Table 8.4), mainly due to the presence of tannins (Sidahmed *et al.*, 1981; Robbins *et al.*, 1987; Nastis and Malechek, 1988; Nastis, 1993). Nevertheless, *in vitro* digestibility values allow one to obtain a high number of estimates of plant nutritive values. Factors that mainly affect the nutritive value of woody plants are different from those of herbaceous species. The relationships between CP content and fibre fractions with the organic matter digestibility (OMD) are stronger in woody species than in herbaceous forages, based on data in sheep and dromedaries (Ben Salem *et al.*, 1994). Species with highest OMD values do not always have higher CP content and lower fibre fractions. A chemical parameter that has an important effect on OMD is the N linked to acid-detergent fibre (ADIN), which is usually very high in shrubs. When OMD of *Q. ilex*, browsed by dry and lactating goats, was predicted on the basis of its chemical composition (CP, acid-detergent fibre, ADIN, acid-detergent lignin and neutral-detergent solubles), the ADIN free N fraction explained about 75% of the OMD variance (Meuret, 1989).

Tannins

Tannins are widely distributed in the plant kingdom. They are complex polyphenolic compounds with relatively high molecular weight (1000–20,000)

Table 8.4. Comparison of intake (g DM/kg BW$^{0.75}$), *in vitro* and *in vivo* DMD values for goats fed woody species.

Breed	Diet composition	Season	Intake (g/kg BW$^{0.75}$)	*In vitro* DMD	*In vivo* DMD
Yearling	100% *Quercus coccifera*	Spring	74	37.0	69.0
Greek[a]	100% *Q. coccifera*	Summer	55	32.0	51.0
Spanish[b]	35% *Quercus dumosa*	–	48	49.0	55.7
	100% *Q. dumosa*	–	19	29.0	46.9
Greek[c]	100% *Q. coccifera*	Spring	79	50.0	70.0
(local breed)	100% *Q. coccifera*	Summer	51	45.0	53.0
	100% *Q. coccifera*	Autumn	59	46.0	55.0
	100% *Q. coccifera*	Winter	58	48.0	56.0
Alpine[d]	100% *Quercus ilex*	–	90	–	47.8
	Q. ilex + soybean meal	–	104	–	51.6
Mamber[e]	100% *Quercus calliprinos*	–	63	35.8	45.4
Sarda[f]	100% *Q. ilex*	Summer	58	–	57.9
	Q. ilex + 200 g supplement[i]	Summer	63	–	61.8
	Q. ilex + 400 g supplement[i]	Summer	57	–	64.6
Sarda[g]	*Pistacia lentiscus* + 200 g supplement[i]	Summer	34	–	67.8
	P. lentiscus + 400 g supplement[i]	Summer	38	–	59.8
Mamber[h]	100% *Q. calliprinos*	–	46	–	25.8
	100% *Ceratonia siliqua*	–	55	–	46.9
	100% *P. lentiscus*	–	32	–	24.5

DM, dry matter; BW, body weight; DMD, DM digestibility.
Data source: [a]Nastis and Malechek (1988); [b]Sidahmed *et al.* (1981); [c]Nastis and Liacos (1982); [d]Meuret (1989); [e]Perevolotsky *et al.* (1993); [f]M. Decandia (unpublished data); [g]Decandia *et al.* (1999); [h]Silanikove *et al.* (1996a). [i]Supplement composition: 66.5% beet pulp and 33.5% soybean meal, on DM basis.

containing sufficient hydroxyl and carboxyl groups to precipitate proteins and to form strong complexes with carbohydrates. According to their structure and reactivity, tannins are classified as hydrolysable and condensed tannins. Hydrolysable tannins are esters of glucose and phenolic acid, with low molecular weight, whereas condensed tannins are polymers of flavonoid units with varying composition and molecular weight. Since the hydrolytic cleavage of condensed tannins yields anthocyanidins, they are now commonly described as proanthocyanidins.

Tannins can protect plants from herbivores (Provenza, 1995) and pathogenic microorganisms by complex and multifactorial mechanisms. Some authors suggest that their toxicity is the basis for plant defence, while others state that deterrence (i.e. rejection based on taste only) is often independent from their toxicity (Silanikove *et al.*, 1996b). The effects of tannins on protein utilization in forage plants is particularly important, since they form strong complexes with proteins and some structural carbohydrates, reducing their availability to the animals and inactivating some enzymes in the digestive tract (Kumar and Vaithiyanathan, 1990). The presence of high tannin concentration in the diet reduces digestibility and feed intake even in goats, known to be more tolerant to tanniferous species than other domestic species (Provenza, 1995; Silanikove *et al.*, 1996a, 1997; Decandia *et al.*, 1999). By contrast, tannins in low concentrations (2–4%) induce beneficial effects, which are associated with suppression of bloat in ruminants (Mueller-Harvey, 2006) and protection of dietary proteins in the rumen.

Tannin effects have been measured by using different compounds which form complexes with hydrolysable and condensed tannins, thus reducing their anti-nutritional activity. Among these compounds, polyethylene glycol (PEG) gave the best results. PEG is a polymer, with molecular weight of 4000–6000, that can bind tannins irreversibly over a wide range of pH (2–8.5), reducing the formation of protein–tannin complexes (Jones and Mangan, 1977). As a result, the availability of certain macronutrients, particularly proteins, increases.

In Sarda goats fed fresh branches of lentisk and supplemented with increasing levels of PEG (0, 25 and 50 g/day), *in vivo* CP digestibility increased significantly (Table 8.5). PEG-supplemented goats spent more time foraging on tanniferous species than on herbaceous (Decandia *et al.*, 2000a,b; Titus *et al.*, 2001) (Fig. 8.6) and had higher intake compared with the goats unsupplemented with PEG (Silanikove *et al.*, 1996a; Decandia *et al.*, 2000a; Titus *et al.*, 2001; Ben Salem *et al.*, 2003). In Mamber goats, supplementation with increasing amounts of PEG (from 0 to 40 g/day; Silanikove *et al.*, 1996a) increased the DM intake

Table 8.5. Effects of different levels of concentrate (concentrate composition: 66.5% beet pulp and 33.5% soybean meal, on dry matter basis) and daily amount (g/day) of polyethylene glycol (PEG) on dry matter intake (DMI) and *in vivo* digestibility of Sarda goats fed lentisk. (Adapted from Decandia *et al.*, 1999.)

	200 g concentrate/day			400 g concentrate/day		
	PEG0	PEG25	PEG50	PEG0	PEG25	PEG50
DMI						
Lentisk (g)	449	491	479	520	604	538
(g/kg $BW^{0.75}$)	34	37	36	38	44	37
Total diet (g)	620	662	634	870	954	887
(g/kg $BW^{0.75}$)	47^{a}	50^{ab}	48^{a}	64^{bc}	71^{c}	63^{bc}
In vivo digestibility						
OM (%)	67.8	70.1	70.2	59.8	65.9	73.1
CP (%)	37.4^{a}	63.3^{ab}	71.5^{b}	30.9^{ac}	61.4^{ab}	73.7^{b}
NDF (%)	45.4	51.5	55.0	31.9	53.1	68.9
ADF (%)	34.5	45.6	47.7	23.4	21.1	59.5

BW, body weight; OM, organic matter; CP, crude protein; NDF, neutral-detergent fibre; ADF, acid-detergent fibre.

^{a,b,c}Mean values in a row with different superscript letters were significantly different ($P \leq 0.05$).

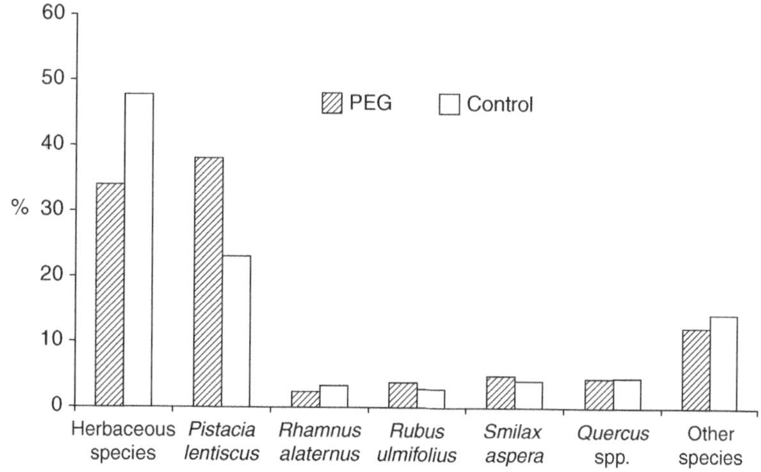

Fig. 8.6. Diet botanical composition of browsing Sarda goats (% of total dry matter intake) either supplemented (PEG) or not (control) with polyethylene glycol 50 g/day. (Adapted from Decandia *et al.*, 2000a.)

(DMI) of leaves of *Quercus calliprinos* (from 660 to 900 g/head per day), of *P. lentiscus* (from 470 to 800 g/head per day) and of *Ceratonia siliqua* (from 800 to 1250 g/head per day). PEG supplementation generally increased DMI and protein and energy digestibility of browsing goats, and the production responses differed among breeds. For example, higher live weight gain of pregnant goats and higher birth weight of their kids occurred in the Mamber breed (Gilboa, 1996), while marked increases of milk yield were observed in Anglo-Nubian (Gilboa *et al.*, 2000) and, to a lower extent, in Sarda goats (Decandia *et al.*, 2000a,b).

Intake

In the heterogeneous Mediterranean environment, grazing animals select their diet from various plant species and plant parts, which differ in digestibility and retention time in the rumen. Thus, the choice of the most appropriate method for intake measurements in such environments should be made carefully. The use of a chromic oxide (Cr_2O_3) marker is an expensive and time-consuming procedure (Yiakoulaki *et al.*, 1997). On the other hand, faecal bags are relatively unsuitable for animals browsing tall and dense maquis vegetation, while the technique of oesophageally fistulated animals can be criticized, since sampling is not continuous and therefore is not representative of the animal's daily foraging behaviour. In addition, the use of the n-alkanes method is questionable in browsing goats, as its accuracy is limited when the number of forage species in the pasture is high. For the above reasons, the direct observation method to evaluate intake of browsing goats is preferred by several authors (Kababya *et al.*, 1998; Perevolotsky *et al.*, 1998). According to this method, intake can be determined as the product of bite frequency (BF; bites/min), bite mass (BM; g DM/bite) and the time spent grazing according to the following equation:

$$\mathrm{DMI} = \sum (\mathrm{BF}_i \times \mathrm{BM}_i \times \mathrm{GT}_i)$$

where DMI is dry matter intake (g), BF_i is bite frequency (number of bites per minute of grazing) of the *i*th species, BM_i is mean bite mass (g DM) of the *i*th species and GT_i is grazing time (min/24 h) of the *i*th species.

Table 8.6 shows some examples of bite frequency and corresponding intake rate (IR, g/min) measured in Sarda goats (M. Decandia, unpublished data), using the method of direct observation of bites (Kababya *et al.*, 1998). The observed mean intake rate of 6.3 g/min was similar to that found in other Mediterranean regions (Perevolotsky *et al.*, 1998).

BF changes according to: (i) animal requirements; (ii) forage availability; and (iii) nutritional and morphological characteristics of the species selected by goats. Mid-lactation goats usually have higher BF and, consequently, higher IR than late-lactation goats, because of their higher nutritive requirements. During summer, due to the higher proportion of browse plants in the diet of goats (Cabiddu *et al.*, 2000), even if BF is lower, IR is almost the same as that in spring. The latter is attributed to the higher BM in woody than herbaceous species. Degree of preference is usually positively correlated with BF. In some cases, low BF of less palatable species, such as *P. lentiscus* L., is due to their high tannin level. The morphological and physical characteristics of the plants also affect BF and IR. For example, the presence of spines in some species (*Rubus* spp.) reduces BF. BM is influenced by the same parameters discussed above.

Forage intake of goats (local breed) browsing on Mediterranean shrublands (*Q. coccifera* L.) in northern Greece was determined by total faecal collection and *in vitro* digestibility (Yiakoulaki and Nastis, 1995). For successful faecal collection, a faecal harness was specially designed for goats (Yiakoulaki and Nastis, 1998), which did not disturb their movements through the dense shrubs, their

Table 8.6. Bite frequency (BF; bites/min) and dry matter intake rate (IR; g/min) observed in Sarda goats in middle and late lactation grazing on Mediterranean shrublands. (Adapted from M. Decandia, unpublished data.)

	Spring		Summer	
Forage species	BF[a]	IR	BF	IR
Herbaceous species	45	7.4	33	4.3
Lonicera implexa Aiton	44	10.3	42	6.0
Myrtus communis L.	38	5.2	33	3.1
Pistacia lentiscus L.	29	9.8	30	7.7
Quercus ilex L.	37	13.0	31	13.6
Ramnus alaternus L.	39	8.5	38	8.8
Rubus spp.	31	4.4	25	5.5

[a]Including chew-bites.

bipedal stance and their feeding behaviour. In another study conducted in the same area, the Cr_2O_3 marker was dosed (Yiakoulaki, 1992). The Cr_2O_3 concentration in goat faeces was determined by a combination of the alkali fusion method (Na_2O_2) with atomic absorption spectrophotometry (Yiakoulaki *et al.*, 1997). Intake values of browsing goats measured in different Mediterranean regions are shown in Table 8.7.

Variables that explain better the variability of DMI in browsing goats are 4% fat-corrected milk (FCM) yield, CP and polyphenolic tannin (PT) level in the diet. Experimental data of lactating Sarda goats browsing a lentisk-based shrubland for 7 h/day were used to calculate the following prediction equations for pasture intake (Decandia *et al.*, 2005a):

$$\text{DMI} = -18.63 + 6.75\ \text{CP} + 0.02\ \text{FCM};\quad n = 38;\quad R^2 = 0.77; P < 0.001 \tag{8.1}$$

$$\text{DMI} = 52.54 + 0.037\ \text{FCM} - 16.44\ \text{PT/CP};\quad n = 40;\quad R^2 = 0.59; P < 0.001 \tag{8.2}$$

where DMI is dry matter intake (g/kg $BW^{0.75}$ per day), FCM is 4% fat-corrected milk yield (g; Sandrucci *et al.*, 1995), CP is crude protein in the diet (%), and PT/CP is the percentage ratio between polyphenolic tannins and crude protein.

Table 8.7. Dry matter intake values of browsing goats in different Mediterranean regions.

Breed	Country	Period	Intake (g DM/kg $BW^{0.75}$)	Main browsed species
Rove[a]	France		44–76	*Quercus coccifera, Rosa officinalis, Juniperus oxycedrus*
Mamber[b]	Israel	Dec–Aug	64–88	*Quercus calliprinos, Stirax officinalis, Pistacia palaestina*
Native breed × Alpine[c]	Tunisia		102	*Acacia cyanofilla, Atriplex nummularia, Ceratonia siliqua, Pistacia lentiscus*
Sarda[d]	Italy	Mar–Jun	71–104	*P. lentiscus, Quercus* spp., *Myrtus communis*
Sarda[e]	Italy	Mar–Jul	48–123	*P. lentiscus, Quercus* spp., *M. communis*
Sarda[f]	Italy	Jun–Jul	55–91	*P. lentiscus, Quercus* spp., *M. communis*
Native breed[g]	Greece	Nov–Feb	42–66	*Q. coccifera, Carpinus orientalis, Fraxinus ornus, Cistus* spp.

DM, dry matter; BW, body weight.
Data source: [a]Dumont *et al.* (1995); [b]Kababya *et al.* (1998); [c]Ben Salem *et al.* (2000a); [d]Decandia *et al.* (1997a,b); [e]Decandia *et al.* (2000a,b); [f]Decandia *et al.* (2004a); [g]Yiakoulaki and Nastis (1995).

In those experimental conditions, CP and PT/CP in the diet were related to the percentage of grass in the pasture, determined by the method of Daget and Poissonet (1969), and milk urea (MU), as follows:

$$CP = 5.10 + 0.20\ \text{Grass} + 0.119\ \text{MU}; \quad n = 31;\ R^2 = 0.82;\ P < 0.001 \tag{8.3}$$

$$PT/CP = 1.21 + 0.019\ \text{MU} - 0.012\ \text{Grass}; \quad n = 29;\ R^2 = 0.49;\ P < 0.001 \tag{8.4}$$

where CP is crude protein in the diet (%), PT/CP is the percentage ratio between polyphenolic tannins and crude protein, Grass is the percentage of herbaceous species in the pasture and MU is milk urea level (mg/100 ml).

The regression equations (8.1) and (8.3) have higher R^2 than equations (8.2) and (8.4) and seem more suitable for application at farm scale. Using these relationships, a prediction model of DMI of browsing goats on shrublands rich in high-tannin species is proposed (Fig. 8.7). By estimating the percentage of herbaceous species in the pasture, and measuring the urea level in the milk, CP (%) in the diet can be predicted using Eqn (8.3) (Fig. 8.7a). If the dietary CP level is known, the measurement of FCM yield allows the prediction of DMI on pasture (Eqn (8.1); Fig. 8.7b). For instance, if the grass component in the pasture is 20% and milk urea is 25 mg/100 ml, the predicted CP (%) should be 12 (Fig. 8.7a). If the FCM is on average 1200 (g), the DMI should be 80 (g/kg $BW^{0.75}$ per day; Fig. 8.7b). Even if this model is not general, it can be applied in conditions similar to those in which the experiments used to fit the equations were conducted, that is: (i) goat breed of low–medium production level; (ii) limited access time to the pasture (5–7 h/day); (iii) shrubland with high-tannin species (e.g. based on *P. lentiscus* L.); (iv) low contribution of grass in the botanical composition of shrubland (10–30%); and (v) low–moderate supplementation level (200–400 g DM/head per day).

Supplementation

It is clear that intake of forage from shrublands is not able to satisfy energy and protein requirements of grazing goats all year around, energy being less limited than protein (Fig. 8.8). As a consequence, supplementation is needed to fill in the nutritional gaps (Liakos *et al.*, 1980). Higher differences between requirements and supply occur in autumn and winter, corresponding to the last 2 months of pregnancy until early lactation. In this period, the use of forages and/or concentrates with a medium to high level of CP is important (Hadjipanayiotou, 1986), particularly if the percentage of herbaceous species in the shrubland is very low.

The effect of supplementation depends on the breed and its browsing ability. In Israel, supplementation with different energy levels and the same protein amount did not influence kid growth and milk yield of local goats, which normally produce medium–low amounts of milk. By contrast, high supplementation level increased milk yield but reduced milk quality of highly productive goat breeds, e.g. Anglo-Nubian × Damascus (Fig. 8.9).

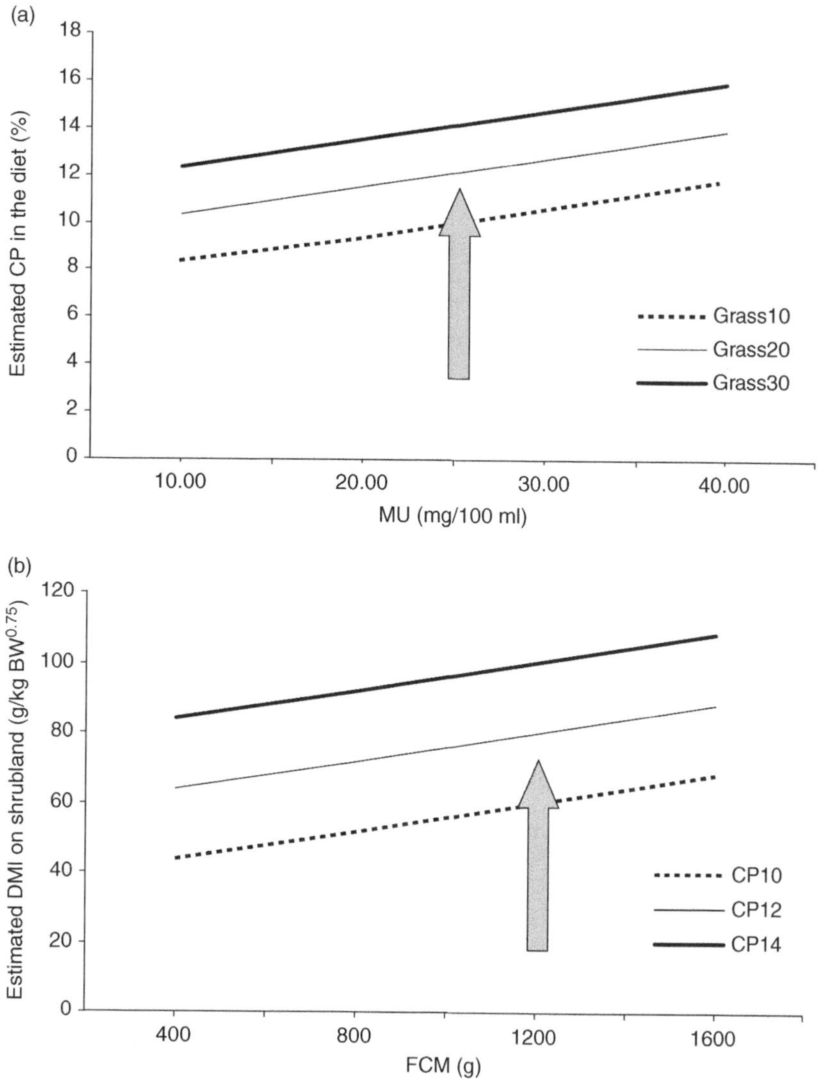

Fig. 8.7. Prediction model of goat intake browsing on shrublands with high-tannin species. (a) Estimated crude protein (CP) in the diet as a function of percentage of herbaceous species in the pasture (Grass) and milk urea (MU) level: CP = 5.10 + 0.20 Grass + 0.119 MU; $n = 31$; $R^2 = 0.82$; $P < 0.001$. (b) Estimated dry matter intake (DMI) as a function of CP in the diet and 4% fat-corrected milk (FCM): DMI = –18.63 + 6.75 CP + 0.02 FCM; $n = 38$; $R^2 = 0.77$; $P < 0.001$ (see text). BW, body weight.

In general, a high level of energy supplement (cereal-based concentrates) allows the recovery of body condition but usually reduces the grazing activity of goats. In fact, in behaviour studies, goats fed low concentrate amounts spent more time grazing than those fed high concentrate amounts (75 and 59% of total observation time, respectively) (Landau *et al.*, 1993).

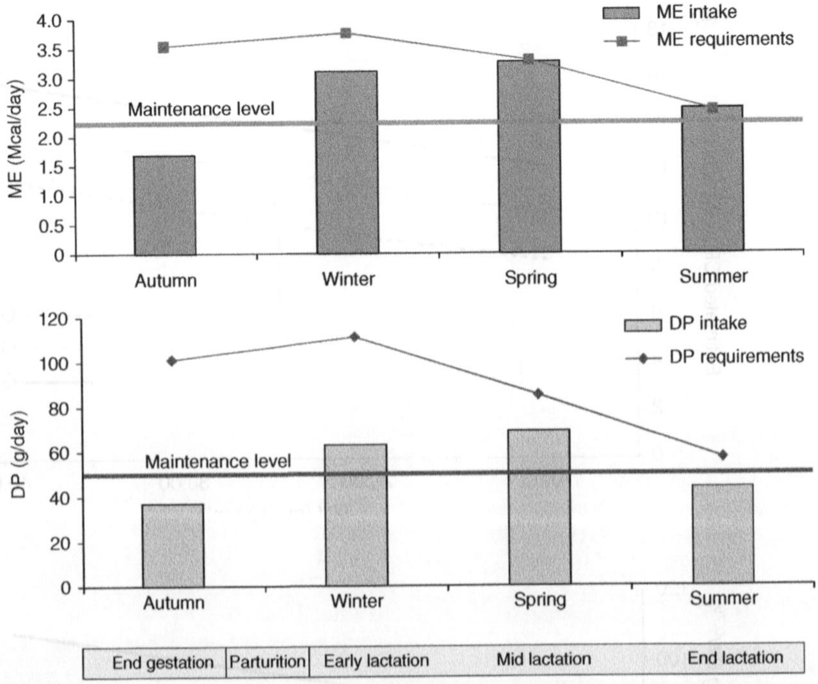

Fig. 8.8. Estimated metabolizable energy (ME) and digestible protein (DP) requirements and supply (calculated according to NRC, 1981) of Sarda goats grazing on Mediterranean shrubland throughout the year. (Adapted from Decandia *et al.*, 2004b.)

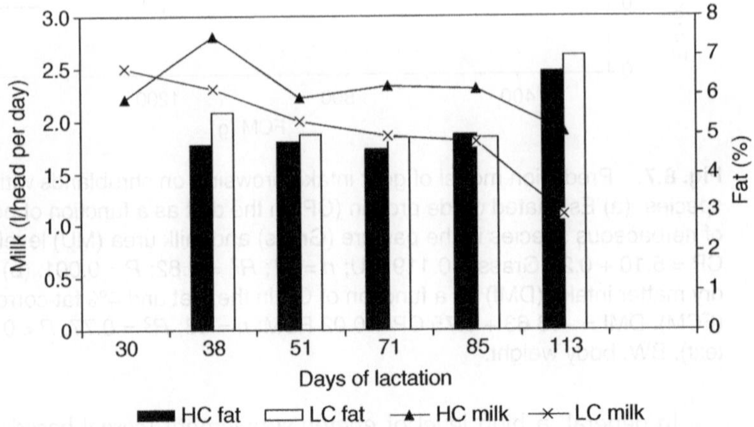

Fig. 8.9. Milk yield (l/head per day) and milk fat content (%) of Damascus goats supplemented with high (HC, 1500 g/head per day) and low (LC, 750 g/head per day) levels of commercial concentrate (metabolizable energy = 2.6 Mcal, crude protein = 175 g/kg). (Adapted from Landau *et al.*, 1993.)

Protein level of goats' diet is affected by pasture botanical composition and, particularly, by the herbage/shrubs ratio. In Greece, local goats browsing on three shrublands (A, B and C) with varying shrubby and herbaceous vegetation cover (A: 53 versus 32%; B: 62 versus 20%; C: 66 versus 12%, respectively) selected a diet with higher CP content from the pasture with the highest proportion of herbaceous species (Yiakoulaki and Nastis, 1995). In general, the reduction of shrub cover from 66 to 53% improved the nutritive value and increased intake of goats' diet.

Although MU level of goats browsing high-tannin species is an index of CP intake, this index is not as good as in sheep (Cannas *et al.*, 1998; Molle *et al.*, 2000). In fact, MU level in Sarda goats browsing a lentisk-based shrubland was positively correlated with CP intake only in animals that received PEG as supplement (Fig. 8.10).

In the Mediterranean region spring is very short and in 3 months the nutritive value of grass decreases rapidly. As a consequence, goats spend more time searching for plants and plant parts with higher nutritive value, thus reducing their total energy intake. In this situation, provision of moderate amounts of concentrates with highly degradable fibre and ruminally degradable protein, i.e. formulated on the basis of the specific digestive kinetics of browsing animals, increased milk yield by 10% (Meuret, 1989). Sarda goats increased their intake of lentisk (from 449 to 520 g/head per day) when the supplementation level (mixture of sugarbeet pulp, 66.5% of DM, and soybean meal, 33.5% of DM) doubled (from 200 to 400 g/day) (Decandia *et al.*, 1999). In general, supplemental energy and protein enable animals to consume more foods with plant toxins such as terpenes, tannins and saponins (Provenza *et al.*, 2003).

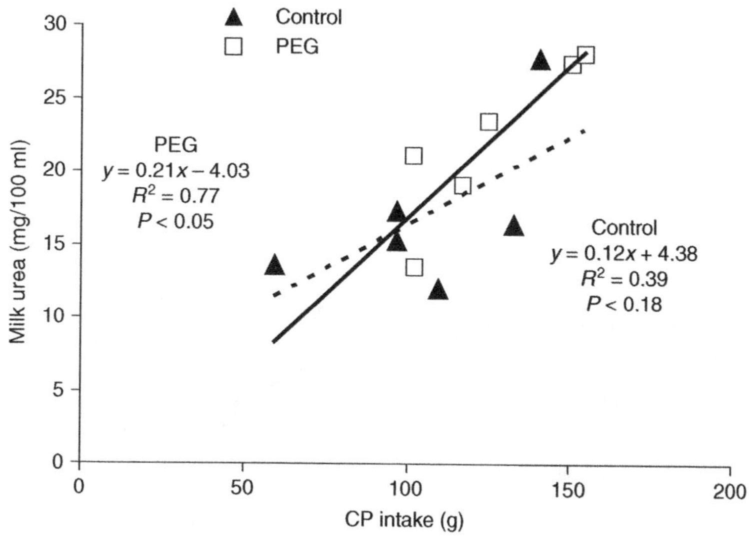

Fig. 8.10. Relationships between crude protein (CP) intake (g) and milk urea (mg/100 ml) in Sarda goats either supplemented (PEG, continuous line) or not (control, dashed line) with polyethylene glycol 50 g/day. (Adapted from Decandia *et al.*, 2000a.)

Grazing Management

A sustainable grazing management should both supply adequate forage nutrients, to meet animal requirements, and preserve natural resources. However, a good knowledge of how animals interact with the vegetation is necessary to achieve this goal.

Given that the Mediterranean shrublands are heterogeneous (patchy) environments, their spatial and temporal heterogeneity greatly influences distribution, dietary selectivity and intake of grazing animals. An animal's choice to graze an area (patch choice) depends on the balance between satiation and motivation to eat, both affected by the animal's nutritional status (Fig. 8.11). Patch choice and, therefore, intake rate are also based on patch characteristics, especially on spatial distribution and potential intake rate (Fig. 8.11). The latter is associated with abundance, ease of prehension and instantaneous 'palatability' of the patchy vegetation. An efficient grazing management should consider the above relationships, to take advantage of the patchy and varied environmental characteristics which can positively affect intake.

Within a grazing day, animals may come across extremely diverse vegetation, which can influence their feeding behaviour. For example, animal motivation to eat increases if the diversity of plants on offer is high and the palatability is medium or low during grazing (Meuret and Bruchou, 1994). When a shepherd drives the flock on a grazing circuit, his strategy of resource utilization interacts with the flock's foraging strategy and, consequently, affects animal decisions. For this reason, effective shepherding can increase dietary motivation on heterogeneous pastures (Meuret, 1996).

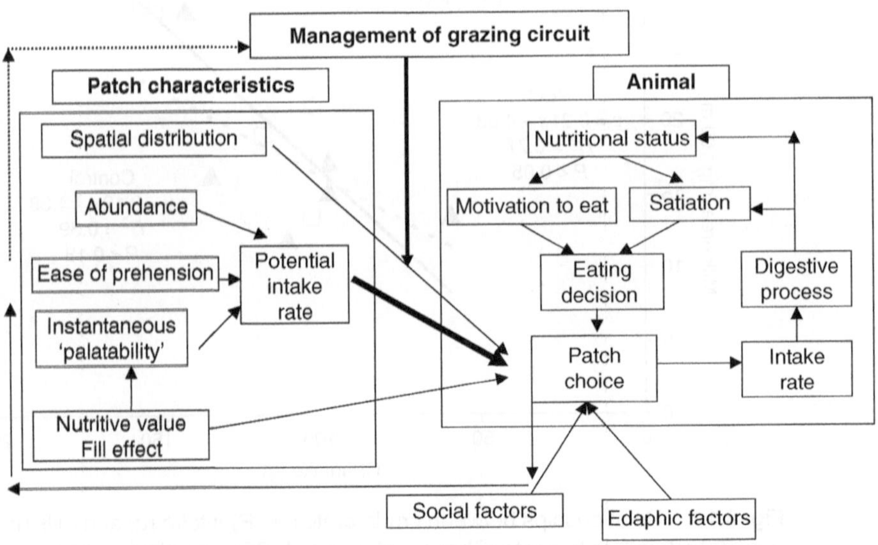

Fig. 8.11. Framework of a mechanistic model of patch choice and intake behaviour within a feeding site. (Adapted from Baumont *et al.*, 2000.)

The parameters that affect the interaction between grazing animals and pasture characteristics are summarized in Fig. 8.11 (Baumont *et al.*, 2000). The success of this model is largely dependent on the shepherd's understanding of forage heterogeneity and its influence on animal foraging strategies. A model of a shepherded circuit viewed as a sequence of patches characterized by different degrees of forage palatability and abundance, named the 'menu model', is described in Fig. 8.12. The aim of this model is to improve animal intake, allowing a high use of abundant but not highly palatable plants and also protecting the highly palatable but rare resources. Such a circuit has a moderation patch (M), with plants abundant but not highly palatable, at the beginning (Fig. 8.12), in order to reduce the appetite of the flock. Successively, an appetite promotion patch (A), with highly palatable but less abundant plants, is offered, to stimulate a low appetite. After that, a target patch with medium plant abundance and palatability is used as a main course (MC) for the bulk of the meal.

When goats start to lose interest in this patch, a booster patch (B) can be offered to increase their appetite. A booster patch may have either very low palatability and medium abundance or high palatability and low abundance. After that, animals can be moved to another target patch, with slightly better instantaneous palatability and abundance than the MC, considered as the second course (SC)

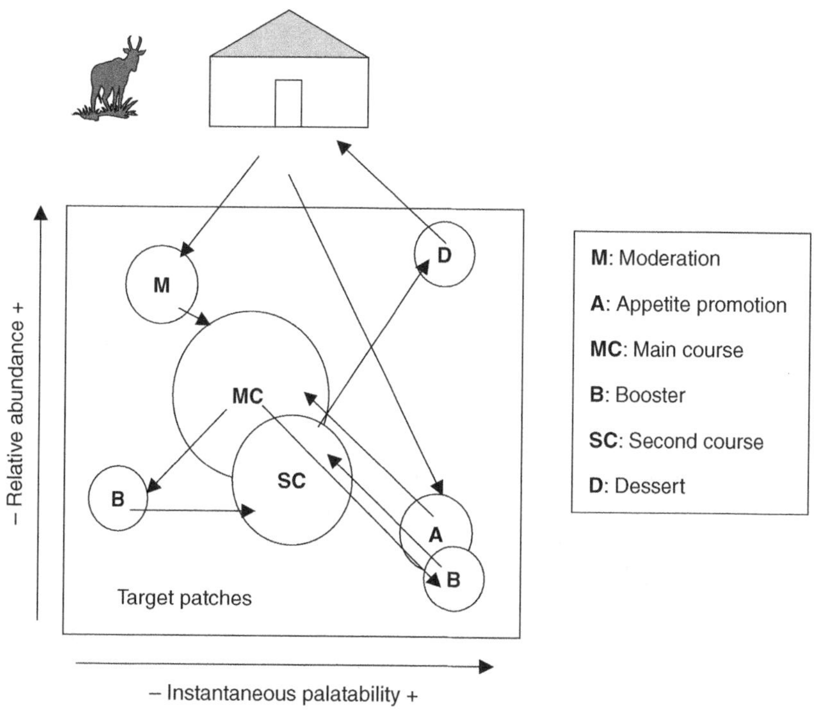

Fig. 8.12. The menu model: a grazing circuit with an ordered sequence of offered patches during one main meal, whose objective is to constantly increase the animal's eating motivation on resources of medium or low quality. (Adapted from Meuret *et al.*, 1995.)

of the diet. At the end of the circuit, when animals are already full, a dessert patch (D) with high forage availability and palatability can be used to reduce the risk of over-grazing.

The application of this model requires frequent pasture modifications by the shepherd and adaptation to local conditions. In order to reduce human work, several strategies that stimulate goat circulation in the shrubland could be used. For example, the placement of fences, gates, shadow points, watering points and mineral licks in strategic areas of the pasture could force the animals to use certain areas. Moreover, the introduction of some grass sward, even in a small area, or the increase of natural herbaceous plants in some areas of the shrubland may affect animal distribution and dietary selection. In fact, use of grasses at the vegetative stage increases shrub intake by goats (Leclerc, 1984), and a high level of degradable protein in the herbage can counteract the negative effects of tannins on ruminal environment (Landau *et al.*, 2000). Grazing management should also take into consideration grazing time, especially during warm seasons, when high temperatures reduce grazing activity of goats during the day. In these periods, goats should be allowed to pasture preferably in early morning or late evening, when they usually eat more woody and shrubby species (Leclerc, 1984).

If goats are supplemented, the level and the composition of the feed additives modify their feeding behaviour. The use of 'alternative' supplements, rich in highly digestible fibres (e.g. sugarbeet pulp and soybean hulls) and N (urea), administered to goats in small amounts twice a day before access to the shrubland, can increase pasture intake (Meuret *et al.*, 1994; Ben Salem *et al.*, 2000b).

Grazing management of goats in Mediterranean shrublands could also deal with the use of biological control of shrubby vegetation as an alternative to mechanical and chemical methods, since shrub encroachment is not a rare phenomenon in these areas. In fact, in Mediterranean grasslands of northern Greece, shrub encroachment was a consequence of reduced grazing by goats and abandonment of traditional management practices, such as shrub use for fuelwood and charcoal (Zarovali *et al.*, 2007).

Due to their diet diversity, high resistance to plant toxins and anti-nutritive factors and their morpho-physiological characteristics, goats can eat plants that are less palatable and/or not attainable for other animals (Hart, 2001). Most of these plants (e.g. *Rubus* spp., *Rosa* spp., *Euphorbia* spp., *Smilax* spp. and *Rubia* spp.), if not managed effectively, tend to spread over the pasture and form some 'non-grazeable patches', thus increasing the risk of wildfires. Goats preferentially eat flowers and seeds of immature weeds, such as *Silybum* and *Cardus* spp. (Yiakoulaki and Papanastasis, 2003), reducing their spread and perpetuation (Allan and Holst, 1996).

The use of goats associated or not with other methods for shrub and/or weed control is a management practice that has started to gain attention in the USA and Australia. A balanced utilization of shrub and/or weed-invaded areas by mixed grazing ruminants (goats, sheep, cattle, horses) associated with additional practices, such as shrub cutting for fuelwood or prescribed burning, may be particularly helpful. However, to increase the effectiveness of weed control by grazing, further research is needed on stocking rate, season and duration of grazing, effects on animal performance and cost of application of this practice.

References

Allan, C.J. and Holst, P.J. (1996) The ecological role of the goat in maintaining pasture and range. In: *Proceedings of the VI International Conference on Goats*. International Academic Publishers, Beijing, China, pp. 427–435.

Ammar, H., López, S. and González, J.S. (2005) Assessment of the digestibility of some Mediterranean shrubs by *in vitro* techniques. *Animal Feed Science and Technology* 119, 323–331.

Baumont, R., Prache, S., Meuret, M. and Morand-Fehr, P. (2000) How forage characteristics influence behaviour and intake in small ruminants: a review. *Livestock Production Science* 64, 15–28.

Ben Salem, H., Nefzaoui, A. and Abdouli, H. (1994) Palatability of shrubs and fodder trees measured on sheep and dromedaries: 1. Methodological approach. *Animal Feed Science and Technology* 46, 143–153.

Ben Salem, H., Nefzaoui, A. and Ben Salem, L. (1999) Sheep and goat preferences for fodder shrubs. Relationship with the nutritive characteristics. *Options Méditerranéennes* 52, 155–159.

Ben Salem, H., Nefzaoui, A., Ben Salem, L. and Tisserand, J.L. (2000a) Deactivation of condensed tannins in *Acacia cyanofilla* Lindl foliage by polyethylene glycol in feed blocks. Effect on feed intake, diet digestibility, nitrogen balance, microbial synthesis and growth by sheep. *Livestock Production Science* 64, 51–60.

Ben Salem, H., Nefzaoui, A. and Ben Salem, L. (2000b) Supplementing range goats in central Tunisia with feed blocks or a mixture of *Opuntia ficus indica* var. *inermis* and *Atriplex nummularia*. Effects on behavioural activities and growth. In: *Proceedings of the VII International Conference on Goats*, Tours, France. Institut de l'Elevage and Institut National de la Recherche Agronomique, Paris, pp. 988–989.

Ben Salem, H., Ben Salem, L., Nefzaoui, A. and Ben Saïd, M.S. (2003) Effect of PEG and olive cake feed blocks supply on feed intake, digestion, and health of goats given kermes oak (*Quercus coccifera* L.) foliage. *Animal Feed Science and Technology* 110, 45–59.

Boubaker, A., Kayouli, C., Buldgen, A. and Boukary, A. (2004) Chemical and biological characterisation of some woody species browsed by goats in the North-West of Tunisia. *Options Méditerranéennes* 59, 147–151.

Cabiddu, A., Decandia, M., Sitzia, M. and Molle, G. (2000) A note on the chemical composition and tannin content of some Mediterranean shrubs browsed by Sarda goats. *Options Méditerranéennes* 52, 175–178.

Cannas, A., Pes, A., Mancuso, R., Vodret, B. and Nudda, A. (1998) Effect of dietary energy and protein concentration on the concentration of milk urea nitrogen in dairy ewes. *Journal of Dairy Science* 81, 499–508.

Congiu, F., Dattilo, M. and Serra, A. (1978) Indagine preliminare sulla composizione chimica e sul valore nutritivo delle principali essenze arbustive presenti in Sardegna. *Zootecnia e Nutrizione Animale* 4, 433–441.

Cooper, S.M., Owen-Smith, N. and Bryant, J.P. (1988) Foliage acceptability to browsing ruminants in relation to seasonal changes in the leaf chemistry of woody plants in a South African savanna. *Oecologia* 75, 336–342.

Daget, P. and Poissonet, J. (1969) *Analyse Phytologique des Prairies. Application Agronomique.* Document No. 48. Centre National de la Recherche Scientifique– Centre d'Etudes Phytosociologiques et Ecologiques Louis Emberger, Montpellier, France.

Decandia, M., Molle, G., Sitzia, M., Ruiu, P.A., Pampiro, F., Pintus, A. and Kababya, D. (1997a) Feeding behaviour of goats browsing on a Mediterranean shrubland. In: *Book of Abstracts of the 48th Annual Meeting of the European Association for Animal Production*, Wien, Austria. Abstract S3.13, p. 293.

Decandia, M., Sitzia, M., Molle, G., Fois, N. and Fois, G. (1997b) Effect of weather conditions on browsing behaviour of goats in Mediterranean shrubland. In: *Proceedings of the International Workshop on 'Ecological Basis of Livestock Grazing in Mediterranean Ecosystems'*, Thessaloniki, Greece. European Commission of Science, Research and Development, Brussels, pp. 227–230.

Decandia, M., Molle, G., Ligios, S. and Scanu, G. (1999) Effect of different levels of polyethylene glycol on the nutritional value of *Pistacia lentiscus* L. fed to goats. In: *Proceedings of the XIII ASPA Congress*, Piacenza, Italy. Franco Angeli s.r.l., Milan, Italy, pp. 384–386.

Decandia, M., Sitzia, M., Cabiddu, A., Kababya, D. and Molle, G. (2000a) The use of polyethylene glycol to reduce the anti-nutritional effects of tannins in goats fed woody species. *Small Ruminant Research* 38, 157–164.

Decandia, M., Molle, G., Sitzia, M., Cabiddu, A., Ruiu, P.A., Pampiro, F. and Pintus, A. (2000b) Responses to an antitannic supplementation by browsing goats. In: *Proceedings of the VII International Conference on Goats*, Tours, France. Institut de l'Elevage and Institut National de la Recherche Agronomique, Paris, pp. 71–73.

Decandia, M., Molle, G., Sitzia, M., Cabiddu, A., Pampiro, F. and Ruiu, P.A. (2004a) Feeding behaviour of dairy goats supplemented with polyethylene glycol browsing on bushland with different herbage cover. *Options Méditerranéennes* 59, 29–33.

Decandia, M., Cabiddu, A., Sitzia, M. and Molle, G. (2004b) Gestione delle risorse alimentari nell'allevamento estensivo di caprini nelle regioni del Mediterraneo. *Informatore Zootecnico* 18, 176–196.

Decandia, M., Pinna, G., Cabiddu, A. and Molle, G. (2005a) Intake by lactating goats browsing on Mediterranean shrubland. In: *Proceedings of the XX International Grassland Congress*, Dublin. Wageningen Academic Publishers, Wageningen, The Netherlands, p. 540.

Decandia, M., Cabiddu, A., Molle, G., Branca, A., Epifani, G., Pintus, S., Tavera, F., Piredda, G., Pinna, G. and Addis, M. (2005b) Effect of different feeding systems on fatty acid composition and volatile compound content in goat milk. 11th Seminar of the FAO–CIHEAM Sub-Network on Sheep and Goat Nutrition, Catania, Italy. *Options Méditerranéennes*, in press.

Dominique, M.F., Dellow, D.W. and Barry, T.N. (1991) Voluntary intake and rumen digestion of a low-quality roughage by goats and sheep. *Journal of Agricultural Science* 117, 117–120.

Dumont, B., Meuret, M. and Prud'hon, M. (1995) Direct observation of biting for studying grazing behavior of goats and llamas on garrigue rangelands. *Small Ruminant Research* 16, 27–35.

Fedele, V., Pizzillo, M., Claps, S., Morand-Fehr, P. and Rubino, R. (1993) Grazing behaviour and diet selection of goats on native pasture in Southern Italy. *Small Ruminant Research* 11, 305–322.

Gilboa, N. (1996) The negative effects of tannins and its neutralization in livestock. PhD thesis, The Hebrew University of Jerusalem, Jerusalem.

Gilboa, N., Perevolotsky, A., Landau, S., Nitsan, Z. and Silanikove, N. (2000) Increasing productivity in goats grazing Mediterranean woodland and scrubland by supplementation of polyethylene glycol. *Small Ruminant Research* 38, 183–190.

Hadjipanayiotou, M. (1986) Studies on the response of lactating Damascus goats to dietary protein. *Journal of Animal Physiology and Animal Nutrition* 57, 41–52.

Hadjipanayiotou, M. and Antoniou, T. (1983) A comparison of rumen fermentation patterns in sheep and goats given a variety of diets. *Journal of the Science of Food and Agriculture* 34, 1319–1322.

Hart, S.P. (2001) Recent perspectives in using goats for vegetation management in the USA. *Journal of Dairy Science* 84(E Suppl.), E170–E176.

Howe, J.C. and Barry, T.N. (1988) Voluntary intake and digestion of gorse (*Ulex europeus*) by goats and sheep. *Journal of Agricultural Science* 111, 107–114.

Jones, W.T. and Mangan, J.L. (1977) Complexes of the condensed tannins of sainfoin (*Onobrychis viciaefolia* Scop) with fraction 1 leaf protein and submaxillary mucoprotein and their reversal by PEG and pH. *Journal of the Science of Food and Agriculture* 28, 126–136.

Kababya, D. (1994) Grazing behavior and nutrition of goats in Mediterranean woodland. Masters thesis, Faculty of Agriculture of the Hebrew University of Jerusalem, Jerusalem.

Kababya, D., Perevolotsky, A., Bruckental, I. and Landau, S. (1998) Selection of diets by dual-purpose Mamber goats in Mediterranean woodland. *Journal of Agricultural Science* 131, 221–228.

Koukoura, Z. (1988) Composition of kermes oak browse as affected by shade and stage of maturity. *Animal Feed Science and Technology* 21, 1–9.

Kumar, S. and Vaithiyanathan, S. (1990) Occurrence, nutritional significance and effect on animal productivity of tannins in tree leaves. *Animal Feed Science and Technology* 30, 21–38.

Landau, S., Vecht, J. and Perevolotsky, A. (1993) Effects of two levels of concentrate supplementation on milk production of dairy goats browsing Mediterranean scrubland. *Small Ruminant Research* 11, 227–237.

Landau, S., Provenza, F. and Silanikove, N. (2000) Feeding behaviour and utilization of vegetation by goats under extensive systems. In: *Proceedings of the VII International Conference on Goats*, Tours, France. Institut de l'Elevage and Institut National de la Recherche Agronomique, Paris, pp. 47–52.

Leclerc, B. (1984) Utilisation du maquis corse par des caprins et des ovins. *Acta Ecologica, Ecologia Applicata* 5, 383–406.

Le Houérou, H.N. (1981) Impact of man and his animals on Mediterranean vegetation. In: Di Castri, F., Goodall, D.W. and Specht, R.L. (eds) *Mediterranean-type Shrublands. Ecosystems of the World*, Vol. 11. Elsevier, Amsterdam, pp. 479–521.

Le Houérou, H.N. (1993a) Environmental aspects of fodder trees and shrubs plantation in the Mediterranean basin. In: Papanastasis, V. (ed.) *Fodder Trees and Shrubs in the Mediterranean Production Systems: Objectives and Expected Results of the EC Research Contract*. Agriculture, Agrimed Research Programme EUR 14459 EN. Commission of the European Communities, Brussels, pp. 11–34.

Le Houérou, H.N. (1993b) Grazing lands of the Mediterranean basin. In: Coupland, R.T. (ed.) *Natural Grassland. Ecosystems of the World*, Vol. 8. Elsevier Science, Amsterdam, pp. 171–196.

Liakos, L.G., Papanastasis, V.P. and Tsiouvaras, C.N. (1980) Contribution to the conversion of kermes oak bushlands to grasslands and comparison of their production with improved bushlands in Greece. *Dasiki Erevna* 1, 97–141.

Makkar, H.P.S., Blummel, M. and Becker, K. (1997) *In vitro* rumen apparent and true digestibilities of tannin-rich forages. *Animal Feed Science and Technology* 67, 245–251.

Meuret, M. (1989) Feuillages fromages et flux ingérés. Thesis in Agronomic Sciences, Gembloux, Belgium. Institut National de la Recherche Agronomique–Sciences pour l'Action et le Développement, Avignon, France, p. 229.

Meuret, M. (1996) Organizing a grazing route to motivate intake on coarse resources. *Annales de Zootechnie* 45(Suppl.), 87–88.

Meuret, M. and Bruchou, C. (1994) Modélisation de l'ingestion selon la diversité des choix alimentaires réalisés par la chèvre au paturage sur parcours. *Rencontres Recherches Ruminants* 1, 225–228.

Meuret, M., Lasseur, J., Molénat, G. and de Simiane, M. (1994) Un complément pour stimuler l'appétit. *Reussir-Patre* 416, 23–26.

Meuret, M., Bellon, S., Guerin, G. and Hanus, G. (1995) Faire paturer sur parcours. *Rencontres Recherches Ruminants* 2, 27–36.

Molina Alcaide, E., Garcia, M.A. and Aguilera, J.F. (1997) The voluntary intake and rumen digestion by grazing goats and sheep of a low-quality pasture from a semi-arid land. *Livestock Production Science* 52, 39–47.

Molle, G., Sitzia, M., Decandia, M., Fois, N., Scanu, G. and Ligios, S. (2000) Intake and performance of dairy ewes grazing Mediterranean forages either as pure or mixed swards. *Options Méditerranéennes* 52, 187–192.

Mueller-Harvey, I. (2006) Unravelling the conundrum of tannins in animal nutrition and health. *Journal of the Science of Food and Agriculture* 86, 2010–2037.

Narjisse, H. (1991) Feeding behaviour of goats on rangelands. In: Morand-Fehr, P. (ed.) *Goat Nutrition*. Pudoc, Wageningen, The Netherlands, pp. 13–24.

Nastis, A.S. (1993) Nutritive value of fodder shrubs. In: Papanastasis, V.P. (ed.) *Fodder Trees and Shrubs in the Mediterranean Production Systems: Objectives and Expected Results of the EC Research Contract.* Agriculture, Agrimed Research Programme EUR 14459 EN. Commission of the European Communities, Brussels, p. 207.

Nastis, A.S. and Liacos, L.G. (1982) Consumption, digestion, and utilization by yearling goats of oak (*Quercus coccifera*) foliage at three phonological stages. In: *Proceedings of the Symposium on Dynamics and Management of Mediterranean-type Ecosystems.* General Technical Report PSW-58. US Department of Agriculture, Forest Service, Berkeley Southwest Forest and Range Station, San Diego, California, p. 622.

Nastis, A.S. and Malechek, J.C. (1988) Estimating digestibility of oak browse diets for goats by *in vitro* techniques. *Journal of Range Management* 41, 255–258.

NRC (1981) *Nutrient Requirements of Goats: Angora, Dairy, and Meat Goats in Temperate and Tropical Countries*. National Academy Press, Washington, DC, pp. 2–9.

Orr, R.J., Penning, P.D., Parsons, A.J., Harvey, A. and Newman, J.A. (1995) The role of learning and experience in the development of dietary choice by sheep and goats. *Annales de Zootechnie* 44(Suppl.), 111.

Papachristou, T.G. and Nastis, A.S. (1993) Nutritive value of diet selected by goats grazing on kermes oak shrublands with different shrub and herbage cover in Northern Greece. *Small Ruminant Research* 12, 35–44.

Papachristou, T.G. and Papanastasis, V.P. (1994) Forage value of Mediterranean deciduous woody fodder species and its implication to management of silvopastoral systems for goats. *Agroforestry Systems* 27, 269–282.

Papanastasis, V.P. (1977) Fire ecology and management of phrygana communities in Greece. In: Mooney, H.A. and Conrad, C.E. (coords) *Proceedings of the Symposium on Environmental Consequences of Fire and Fuel Management in Mediterranean Ecosystems*. General Technical Report WO-3. US Department of Agriculture, Forest Service, Washington, DC, pp. 476–482.

Papanastasis, V.P. (1997) Effect of weather condition on browsing behaviour of goats in Mediterranean shrubland. In: *Proceedings of the International Workshop on 'Ecological Basis of Livestock Grazing in Mediterranean Ecosystems'*, Thessaloniki, Greece. European Commission of Science, Research and Development, Brussels, pp. 227–230.

Papanastasis, V.P. (2000) Shrubland management and shrub plantations in southern Europe. In: Gintzburger, G., Bounejmate, M. and Nefzaon, A. (eds) *Proceedings of the Workshop on Native and Exotic Fodder Shrubs in Arid and Semi-arid Zones*, Hammamet, Tunisia, 27 October–2 November 1996. International Centre for Agricultural Research in the Dry Areas, Aleppo, Syria, Vol. I, pp. 54–66.

Papanastasis, V.P., Yiakoulaki, M.D., Decandia, M. and Dini-Papanastasi, O. (2006) Potential of fodder trees and shrubs as animal feeds in the Mediterranean areas of Europe. In: Lloveras, J., Gonzalez-Rodriquez, A., Vazquez-Yanea, O., Pineiro, J.,

Santamaria, O., Olea, L. and Poblaciones, M.J. (eds) *Proceedings of the 21st General Meeting of the European Grassland Federation*, Badajoz, Spain. Sociedad Española para el Estudio de los Pastos, Madrid, pp. 428–438.

Perevolotsky, A., Brosh, A., Ehrlich, O., Gutman, M., Henkin, Z. and Holzer, Z. (1993) Nutritional value of common oak (*Quercus calliprinos*) browse as fodder for goats: experimental results in ecological perspective. *Small Ruminant Research* 11, 95–106.

Perevolotsky, A., Landau, S., Kababia, D. and Ungar, E.D. (1998) Diet selection in dairy goats grazing woody Mediterranean rangeland. *Applied Animal Behaviour Science* 57, 117–131.

Provenza, F.D. (1995) Postingestive feedback as an elementary determinant of food selection and intake in ruminants. *Journal of Range Management* 48, 2–17.

Provenza, F.D. and Balph, D.F. (1990) Applicability of five diet-selection models to various foraging challenges ruminant encounter. In: Hughes, R.N. (ed.) *Behavioural Mechanisms of Food Selection*. NATO ASI Series G: Ecological Sciences, Vol. 20. Springer–Verlag, Berlin, pp. 423–459.

Provenza, F.D. and Malechek, J.C. (1983) Tannin allocation in blackbrush (*Coleogyne ramosissima*). *Biochemical Systematics and Ecology* 11, 233–238.

Provenza, F.D., Villalba, J.J., Dziba, L.E., Atwood, S.B. and Banner, R.E. (2003) Linking herbivore experience, varied diets, and plant biochemical diversity. *Small Ruminant Research* 49, 257–274.

Robbins, C.T., Mole, S., Hafgerman, A.E. and Hanley, T.A. (1987) Role of tannins in defending plants against ruminants: reduction in dry matter digestion? *Ecology* 68, 1606–1615.

Rogosic, J., Estell, R.E., Skobic, D., Martinovic, A. and Maric, S. (2006a) Role of species diversity and secondary compound complementarity on diet selection of Mediterranean shrubs by goats. *Journal of Chemical Ecology* 32, 1279–1287.

Rogosic, J., Pfister, J.A., Provenza, F.D. and Grbesa, D. (2006b) Sheep and goat preference for and nutritional value of Mediterranean shrubs. *Small Ruminant Research* 64, 169–179.

Rosenberger, S. and Meuret, M. (1985) Une histoire de coups de dents. *La chèvre* 151, 21–27.

Rubino, R., Fedele, V. and Pizzillo, M. (1988) La capra al pascolo: attitudine al pascolamento ed efficienza di utilizzazione. *Agricoltura Ricerca* 81, 73–86.

Sandrucci, A., Rapetti, L. and Galli, A. (1995) Stima del valore energetico del latte di capra. In: *Proceedings of the XLIX National SISVet Meeting*, Salsomaggiore Terme (Parma), Italy. Italian Society for Veterinary Science, Italy, p. 257.

Shkolnik, A., Brosh, A. and Choshniak, I. (1987) Goats and the desert ecosystem. In: *Proceedings of the 4th International Conference on Goats*, Brasilia, Brazil. International Goat Association, Little Rock, Arkansas, Vol. 1, pp. 115–129.

Sidahmed, A.E., Morris, J.G., Koong, L.J. and Radosevitch, S.R. (1981) Contribution of three chaparral shrubs to the protein and energy requirements of Spanish goats. *Journal of Animal Science* 53, 1391–1400.

Silanikove, N. (2000) The physiological basis of adaptation in goats to harsh environments. *Small Ruminant Research* 35, 181–193.

Silanikove, N., Tagari, H. and Shkolnik, A. (1993) Comparison of rate passage, fermentation rate and efficiency of digestion of high fiber diet in desert black Bedouin goats as compared to Swiss Saanen goats. *Small Ruminant Research* 12, 45–60.

Silanikove, N., Gilboa, N., Nir, I., Perevolotsky, A. and Nitsan, Z. (1996a) Effect of daily supplementation of polyethylene glycol on intake and digestion of tannin-containing leaves (*Quercus calliprinos*, *Pistacia lentiscus* and *Ceratonia siliqua*) by goats. *Journal of Agriculture and Food Chemistry* 44, 199–205.

Silanikove, N., Gilboa, N., Nitsan, Z. and Perevolotsky, A. (1996b) Goats fed tannin-containing leaves do not exhibit toxic syndrome. *Small Ruminant Research* 25, 195–201.

Silanikove, N., Gilboa, A., Nitsan, Z. and Perevolotsky, A. (1997) Effect of foliage-tannins on feeding in goats. *Options Méditerranéennes* 34, 43–45.

Sottini, E. and Geri, G. (1971) Composizione chimica, valore nutritivo ed utilizzazione pascoliva dei germogli e delle foglie di alcune essenze tipiche della macchia mediterranea. *Informatore Agrario* 3, 27–38.

Thorhallsdottir, A.G., Provenza, F.D. and Balph, D.F. (1987) Food aversion learning in lambs with or without a mother: discrimination, novel and persistence. *Applied Animal Behaviour Science* 18, 324–340.

Titus, C.H., Provenza, F.D., Perevolotsky, A., Silanikove, N. and Rogosic, J. (2001) Supplemental polyethylene glycol influences preferences of goats browsing blackbrush. *Journal of Range Management* 54, 161–165.

Tomaselli, R. (1981) Main physiognomic types and geographic distribution of shrub systems related to Mediterranean climates. In: Di Castri, F., Goodall, D.W. and Specht, R.L. (eds) *Mediterranean-type Shrublands. Ecosystems of the World*, Vol. 11. Elsevier, Amsterdam, pp. 95–104.

Tsiouvaras, C.N. and Nastis, A.S. (1990) Browse production and nutritive value of some fodder shrubs and trees in a semi-arid environment in Greece. In: *Proceedings of the 6th Meeting of the FAO European Sub-Network on Mediterranean Pastures and Fodder Crops*, Bari, Italy. Food and Agriculture Organization of the United Nations, Rome, pp. 169–172.

Yiakoulaki, M.D. (1987) Nutritive value of range species as determined with laboratory procedures and their contribution to meet animal demands. MS thesis, Aristotle University of Thessaloniki, Greece.

Yiakoulaki, M.D. (1992) Effect of different proportions of shrubby and herbaceous vegetation for goats. PhD dissertation, Aristotle University of Thessaloniki, Greece.

Yiakoulaki, M.D. and Nastis, A.S. (1993) Mineral content of selected forage species common in the Mediterranean shrublands. In: *Proceedings of the 7th Meeting of the FAO Sub-Network on Mediterranean Pastures and Fodder Crops*, Chania, Crete. REUR Technical Series No. 28. Food and Agriculture Organization of the United Nations, Rome, pp. 137–140.

Yiakoulaki, M.D. and Nastis, A.S. (1995) Intake by goats grazing kermes oak shrubland with varying cover in Northern Greece. *Small Ruminant Research* 17, 223–228.

Yiakoulaki, M.D. and Nastis, A.S. (1998) A modified faecal harness for grazing goats on Mediterranean shrublands. *Journal of Range Management* 51, 545–546.

Yiakoulaki, M.D. and Papanastasis, V.P. (2003) Diet selection of sheep and goats grazing on cereal stubble in northern Greece. In: 'Sustainable grazing, nutritional utilization and quality of sheep and goat products'. First Joint Seminar of the Sub-Networks FAO–CIHEAM on Sheep and Goat Nutrition and on Mountain and Mediterranean Pastures, Granada, Spain, 2–4 October 2003. *Options Méditerranéennes* 67, 245–250.

Yiakoulaki, M.D., Papadoyannis, I.N. and Nastis, A.S. (1997) Determination of marker chromic oxide in feces of grazing goats on Mediterranean shrublands by AAS. *Animal Feed Science and Technology* 67, 163–168.

Yiakoulaki, M.D., Kodona, M.S. and Nastis, A.S. (1999) Effect of prescribed burning and management systems on diet selection and intake of goats grazing in *Pinus brutia* forest. In: Waterhouse, A. and McEwan, E. (eds) *Proceedings of the EQULFA Project Meeting on 'Landscapes Livestock and Livelihoods in European Less Favoured Areas'*, Thessaloniki, Greece. Scottish Agricultural College, Auchincruive, UK, pp. 85–89.

Zarovali, M.P., Yiakoulaki, M.D. and Papanastasis, V.P. (2007) Effects of shrub encroachment on herbage production and nutritive value in semi-arid Mediterranean grasslands. *Journal of Grass and Forage Science*, 62, 355–363.

9 Grazing Management of Dairy Goats on Mediterranean Herbaceous Pastures

A. BONANNO,[1] V. FEDELE[2] AND A. DI GRIGOLI[1]

[1]*Dipartimento S.EN.FI.MI.ZO., Sezione di Produzioni Animali, Università degli Studi di Palermo, Viale delle Scienze, 90128 Palermo, Italy;*
[2]*CRA – Istituto Sperimentale per la Zootecnia, Via Appia, Bella Scalo, 85054 Muro Lucano, Potenza, Italy*

Introduction

For a long time, goats have typically been reared in areas not suitable for intensive animal production systems. In fact, goats are more capable of utilizing natural pastures, often dominated by trees and shrubs, than other productive animal species. In reality, depending on the agro-climatic and social–economic conditions, different systems for rearing goats are used, ranging from extensive ones, based on spontaneous forage resources, to intensive types carried out either outdoors, often based on grazing cultivated pastures, or indoors. Whereas in extensive systems more rustic breeds and populations are used, in intensive ones selected breeds (e.g. Saanen, Alpine) are required. Between these two extremes, the technical–economic efficiency of a wide range of mixed systems for goats is strictly linked to the agro-climatic environment and the productive potential of the breeds.

Indeed, in many cases of productive goat breeds reared in favourable environments, feeding is based on natural pastures integrated with forage crops, such as single- or multi-species meadows and swards, and conserved forage and concentrate during periods of scarce grass availability (Fedele *et al.*, 1997).

In all grazing goat systems, as with other species, feeding strategies should be adequate for the productive potential of rustic or selected breeds. For this reason, recovery and improvement of natural pastures, correct grazing management of natural or cultivated pastures, and feed integration are all important methods of avoiding nutrition deficit in goats.

Among the options for pasture renovation techniques (stones removal, control of weeds and shrubs, fertilization, over-sowing), only very few are practicable (e.g. fertilization and mowing of ageing vegetation), while many of them are too expensive or do not increase yield (Fedele *et al.*, 1988). Pasture improvement should be done differently for goats than for other grazing species. Since goats consume preferably grasses (e.g. ryegrass, rescue grass and cocksfoot) and the so-called 'weeds' (e.g. chicory, geranium and wild carrot), and less frequently leguminous species (e.g. clovers and medics), pasture improvement should mainly aim to thicken the vegetation and improve soil fertility. During summer, goats are able to consume species often refused by other ruminants (*Rumex* spp., *Polygonum aviculare, Daucus carota, Convolvulus arvensis*), which satisfy part of their nutritional requirements, and thus need less feed integration.

A rational approach to grazing management should allow goats, within the limits of the seasonal fluctuations in grass availability, to express fully and for as long as possible their selective behaviour and maximum intake capacity, to satisfy their nutritional requirements for each physiological stage, and to optimize their production. Therefore, correct grazing management requires knowledge of: (i) goats' behaviour at pasture; (ii) the best forage species to be sown in cultivated pastures; (iii) proper grazing techniques (continuous or rotational); (iv) beginning and length of the grazing period, and frequency of grazing turns at pasture, often based on grass height but greatly depending on seasonal grass growth rate; (v) the intensity of utilization, based on daily grazing duration and stocking rate; and (vi) the timing, amount and quality of feed integration. Finally, a rational and balanced management is fundamental for maximizing the efficiency of a grazing productive system, and protecting the local environment and landscape.

Feeding Behaviour on Natural Pasture

The natural vegetation of pastures (i.e. herbaceous, shrubby or woody) modifies the feeding behaviour of ruminants. A common saying of goat farmers is 'grasses are for cattle, shrubs are for goats'. Even though goats are good biological transformers of woody-type forages, such feeds are not always their preferred ones (Table 9.1).

Co-grazing different animal species on natural mixed vegetation (grass, trees and shrubs) enables better evaluation of the behaviour of an animal species and its differences with other species. In one study of this type, goats made better use of trees and shrubs than sheep or cattle, but when grass was abundant (in spring) they preferred grass, as much as did the other two species (Fedele, 2001a). The observed behaviours were related to the different levels of preference of these animal species for various plants. Among several shrubs (e.g. hawthorn, rest-harrow, lentisk, European dewberry and dog rose) and trees (e.g. hedge maple, European beech and holly oak) common to the Mediterranean woodlands, goats consumed all of them, while sheep and cattle made a much more specific choice (Table 9.2).

Due to anatomic and morphological adaptations, goats are able to exploit less accessible resources. In fact, it is not rare to see an animal up on its hind legs or jumping up to rip the leaves from branches of trees, climbing on to tree trunks to browse the ivy, penetrating the tangles of shrubby vegetation to nibble at tufts of grass or new shoots, picking up with its tongue seeds that have fallen to the ground and pulling out roots that stick up out of the ground (Claps *et al.*, 2001).

In natural pastures composed exclusively of herbaceous species, the selective feeding behaviour of goats is influenced by several, and often interacting,

Table 9.1. Contribution of different vegetation types to the diet of grazing ruminants (% of class out of the total number of species present in the diet). (Adapted from Fedele, 2001a.)

Vegetation type/ruminant	Spring	Summer	Autumn	Winter
Woody species				
Goats	5	76	45	38
Sheep	3	18	15	4
Cattle	2	10	5	3
Shrubby species				
Goats	7	11	27	28
Sheep	2	11	10	4
Cattle	2	4	4	3
Grasses				
Goats	88	13	28	34
Sheep	95	71	75	92
Cattle	96	86	91	94

factors such as: (i) dry matter (DM) content; (ii) grass availability; (iii) floristic composition; and (iv) plant phenological stage. A DM content ranging from 18 to 26% favours intake, whereas the presence of latex, whose concentration is particularly high in the first stages of development of certain plants (e.g. *Cichorium* spp. and *Sochus*), strongly restricts it. The presence of thorns, harshness or excessive pubescence in flowers, leaves and stems (e.g. *Dipsacum*, *Carduus* and *Picris*) limits the intake of these epigeal parts by goats (Fedele *et al.*, 1996). Even though these factors influence the selective behaviour of goats, the availability of grass and the floristic composition of the pasture seem to be even more decisive. A decrease in pasture DM availability, from 2000–3000 kg/ha during spring to 700–1500 kg/ha during autumn, reduced DM intake from 0.73–0.92 kg/day per head to 0.49–0.59 kg/day per head (Fedele *et al.*, 1996).

A detailed analysis of the composition of the oesophageal bolus (samples of diet collected from goats with an oesophageal fistula) showed that, on average, goats chose a diet composed of 70% grasses, 25% forbs (from various botanical families) and only 5% legumes. It seems that this choice is related to the chemical composition of the grass and to the fact that goats tend to increase their level of intake by selecting whole plants or the parts richest in digestible organic matter and crude protein (CP) (Morand-Fehr *et al.*, 1993; Fedele *et al.*, 1996). This does not result in an increase of legume intake for the following reasons: (i) legumes have a negative effect on digestibility and palatability because of their secondary metabolites (Barry and Duncan, 1984; Reed, 1995); (ii) when legumes have a high CP content (winter), their DM and fibre contents are lower in comparison to that of other species, especially grasses. In this phenological stage, pasture vegetation is very rich in fermentable compounds and very poor in fibre.

Table 9.2. Dietary preference of grazing ruminants in the Cilento region (Campania, Italy) expressed as time spent grazing each species during 3 h[a]. (Adapted from Claps *et al.*, 2001.)

Plant	Goats	Sheep	Cattle
Acer campestre (hedge maple)	very high	high	low
Quercus ceris (turkey oak)	high	low	low
Fagus sylvatica (European beech)	very high	medium	medium
Fraxinus excelsior (ash)	high	low	low
Quercus ilex (holly oak)	low	null	null
Ulmus campestris (mountain elm)	very high	high	high
Alnus cordata (Italian alder)	high	low	low
Robinia pseudoacacia (black locust)	very high	high	null
Crataegus oxyacantha (hawthorn)	high	low	null
Cytisus scoparius (Scotch broom)	very high	high	ND
Colutea arborescens (bladder senna)	null	null	null
Coronilla emerus (scorpion senna)	medium	medium	high
Phillyrea latifolia	medium	ND	medium
Pistacia lentiscus (lentisk)	high	null	null
Ligustrum vulgare (European privet)	medium	low	medium
Pyrus pyraster (wild pear)	high	ND	low
Prunus spinosa (blackthorn)	high	low	low
Cuscus aculeatus (butcher's broom)	low	null	null
Rosa canina (dog rose)	medium	low	null
Rubus fruticosus (blackberry)	medium	low	null
Graminaeous	very high	very high	very high
Lolium perenne (perennial ryegrass)	very high	very high	very high
Avena barbata (slender oat)	very high	very high	very high
Dactylis glomerata (cocksfoot)	high	very high	high
Bromus spp. (brooms)	high	high	high
Legumes	low	very high	very high
Vicia spp. (vetches)	high	very high	very high
Medicago spp. (lucernes)	medium	very high	very high
Triflolium spp. (clovers)	low	high	very high
Weeds	high	low	null
Daucus carota (wild carrot)	high	low	null
Cichorium spp. (chicories)	high	null	null
Ranunculus bulbosus (bulbous buttercup)	low	null	null
Borago officinalis (borage)	very high	medium	low

ND, not determined.

[a]Null, ≤ 5 min; low, from 6 to 15 min; medium, from 16 to 44 min; high, from 45 to 90 min; very high, from 91 to 180 min.

As a consequence, goats tend to refuse these plants, to avoid digestive disorders (Fedele *et al.*, 1996); and (iii) selection of leaf apexes, young shoots, flowers and other parts of grasses, and other species enables goats to raise the degradable protein level of the diet and to create a better balance with the non-structural carbohydrates (NSC) (Fedele *et al.*, 1996). Some plant species and their parts browsed by goats are shown in Figs 9.1, 9.2 and 9.3, while the indices of preference of goats for certain plant categories (grasses, forbs, legumes) of natural pastures in different seasons are shown in Table 9.3.

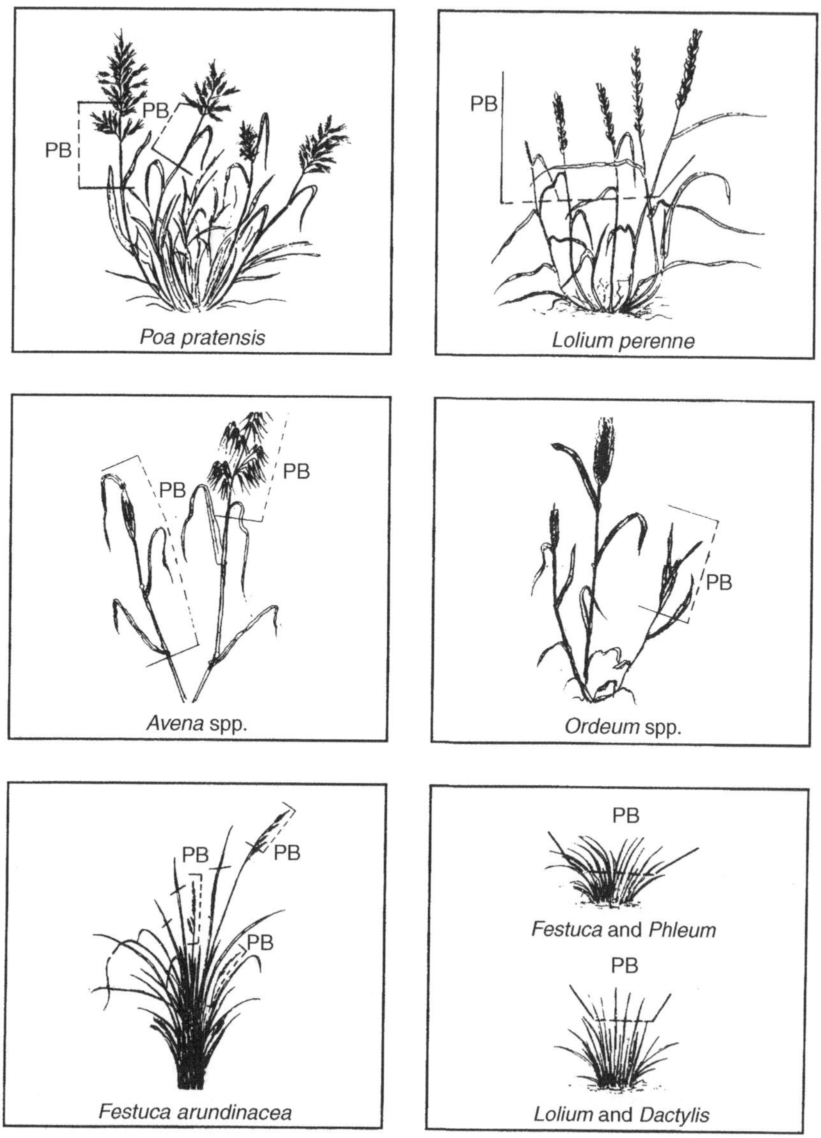

Fig. 9.1. Plant parts browsed (PB) by goats (drawn by Lucia Sepe). (Adapted from Fedele, 1996.)

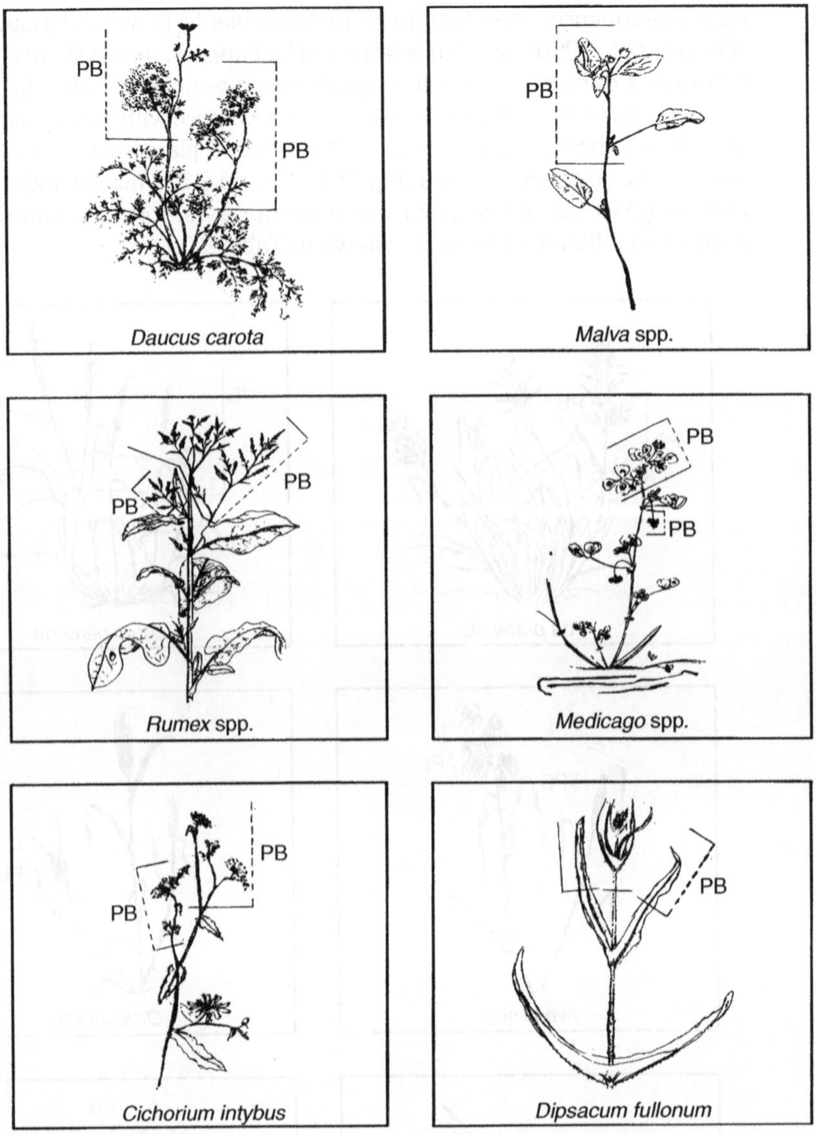

Fig. 9.2. Plant parts browsed (PB) by goats (drawn by Lucia Sepe). (Adapted from Fedele, 1996.)

The maximum presence of forb species, the so-called 'weeds', in the diet occurs in summer, when the other plant categories dry up and partly disappear. Apart from being a strategic resource during the summer, forbs play an important role in the aromatic quality of milk and cheese (see Fedele, Chapter 5, this volume). Even though pasture improvement has often aimed to restrict the spread of forbs and increase that of legumes, these changes might not be always beneficial for grazing goats.

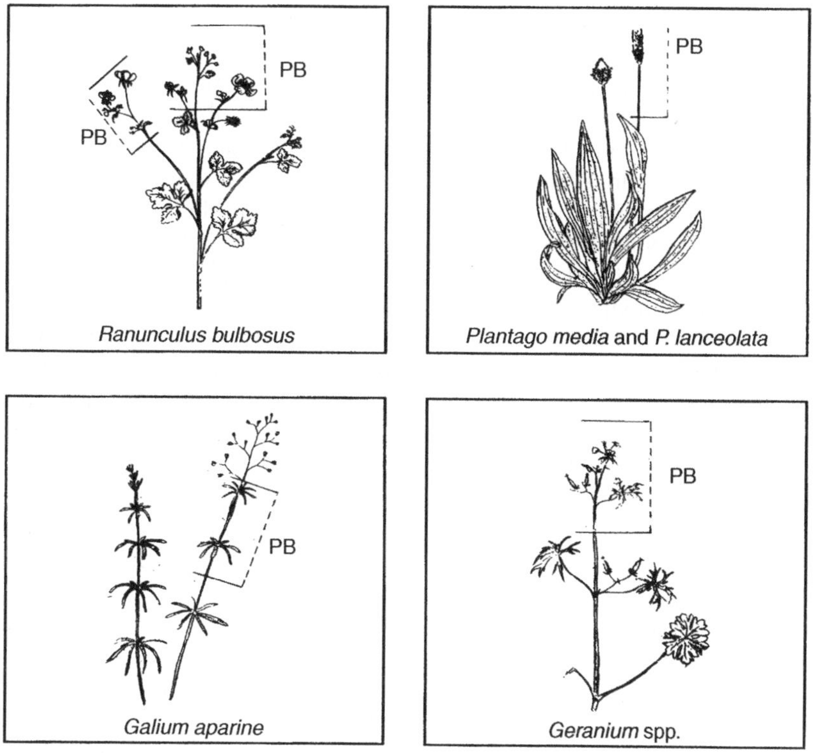

Fig. 9.3. Plant parts browsed (PB) by goats (drawn by Lucia Sepe). (Adapted from Fedele, 1996.)

Table 9.3. Indices of preference of goats for different plant categories. (Adapted from Fedele *et al.*, 1993.)

	Plants on offer (*n*)				Refused plants (*n*)				Mean index for browsed plants			
Category	SP	SU	AU	WI	SP	SU	AU	WI	SP	SU	AU	WI
Grasses	17	12	10	7	0	3	3	0	3.6	2.4	3.9	4.4
Forbs	30	24	19	6	14	7	9	4	3.3	3.6	3.2	1.1
Legumes	9	9	7	3	5	2	5	3	1.9	1.4	1.0	0.4

SP, spring; SU, summer; AU, autumn; WI, winter.

Index of preference is the ratio between the percentage of one species in the extrusa collected from oesophageally fistulated goats and the percentage of the same species available in the pasture. Botanical composition was determined by cutting and manually separating the species from 2 m × 2 m areas randomly distributed in several points of the pasture. The index of preference was thus classified in five classes: 5 = ratio > 1, plants grazed by goats: 81–100% of available plants in the pasture; 4 = ratio 0.81–1, plants grazed: 61–80%; 3 = ratio 0.61–0.8, plants grazed: 41–60%; 2 = ratio 0.41–0.6, plants grazed: 21–40%; 1 = ratio < 0.4, plants grazed: 1–20%.

Feeding Behaviour on Cultivated Pasture

With the aim of selecting their favourite plant species and better-quality plant parts, at pasture goats move much more and, thus, spend more energy than sheep and cattle (Harrington, 1982; Animut *et al.*, 2005a).

When dry sheep and goats grazed ryegrass (*Lolium perenne*) and white clover (*Trifolium repens*) during the whole day (Table 9.4) (Penning *et al.*, 1997), goats spent less time eating than did ewes. This is due to their lower overall DM intake associated with their lower body weight (BW), even though the voluntary intake of goats per kilogram of BW is analogous to that of sheep. This has been attributed to faster rumen feed passage rate and higher ruminal feed degradability of goats, compared with sheep, as a consequence of their more efficient processes of particle size reduction by chewing and rumination (García *et al.*, 1995). Goats spend less time ruminating, and consequently spend more time being inactive, than sheep (Penning *et al.*, 1997). Moreover, goats have lower intake rate and smaller bite size than sheep, due to their greater selectivity, whereas both species show a faster intake rate and greater quantity of legumes consumed per bite in comparison with grasses (Penning *et al.*, 1997).

The time goats spend grazing depends on: (i) number of hours on pasture; (ii) herbage availability; (iii) grass botanical and nutritional composition; and (iv) feed supplementation.

Throughout a whole day, grazing activity is generally concentrated in two periods, at dawn and dusk, corresponding to the two main meals of maximum intake, whereas it is greatly reduced during the night (Penning *et al.*, 1997; Abbadessa *et al.*, 2000; Baumont *et al.*, 2000). 'Active' grazing, which associates herbage intake with search for vegetation, was the main activity (77.3% of the time spent at pasture) when goats grazed on an improved herbaceous pasture for 7 h/day (Di Grigoli *et al.*, 2003). In fact, goats compensate for an access to the pasture restricted to few hours a day by increasing time spent grazing and intake rate (Romney *et al.*, 1996).

Table 9.4. Time spent in different activities during 24 h and intake during daylight of dry goats and ewes grazing ryegrass and white clover. (Adapted from Penning *et al.*, 1997.)

Variable	Goats	Sheep
BW (kg)	46.0	71.2
Eating (min/24 h)	520	663
Ruminating (min/24 h)	208	234
Idling (min/24 h)	712	543
Intake rate (g DM/min)		
Ryegrass	3.5	4.3
White clover	4.0	5.2
Intake (kg DM/day)	1.5	2.4
Level of intake (g DM/kg BW)	32.7	33.4

BW, body weight; DM, dry matter.

A reduction in plant biomass causes an increase in time spent grazing and the relative movements (Bordi *et al.*, 1994; Romney *et al.*, 1996), an increase in the frequency and number of bites, and a decrease in bite size (Milne *et al.*, 1981). An increase in goat stocking rate, which leads to a reduction of biomass availability, increases grazing time (Animut *et al.*, 2005a), reduces selectivity and increases bite frequency (Abbadessa *et al.*, 2000). Keeping access to browsed species constant, as grass height decreases, grazing time declines and browsing time increases (Orihouela and Solano, 1999). Furthermore, goats respond to a decrease in grass height or an increase in stocking rate by reducing their eating time (Animut *et al.*, 2005a) and herbage intake (Penning *et al.*, 1997) more rapidly than sheep.

In grazing systems based on grass and legume monocultures, goats, contrary to sheep, graze mainly on the first rather than the second. The different selective behaviour of these two animal species has been attributed to the fact that goats graze higher and more apical strata of vegetation than sheep (Del Pozo *et al.*, 1997). Since grasses predominate in these strata, their different selectivity might only reflect a different distribution of plant species within the grazing strata. Other explanations are based on goats' preference for forage plants with high NSC content and digestibility, such as grass species (Fedele *et al.*, 1993, 1996), and their refusal of forages with secondary metabolites (tannin, saponins and terpenes), which render them less digestible and less palatable, such as legumes (Provenza, 1995). When wet pellets of straw with essence extracts of grasses and legumes were offered, goats also showed high preference for the first and no interest for the second essence, suggesting that feed flavour plays a specific role in the feeding preference of goats (De Rosa *et al.*, 2002).

Goats prefer multi-species pastures to mono-specific ones. In fact, the presence of several botanical species, with complementary characteristics, allows more selective feeding behaviour depending on palatability, and nutritional and productive requirements (Baumont *et al.*, 2000). The utilization of an improved herbaceous pasture, compared with a ryegrass sward, allowed the goats to select a more heterogeneous diet, rich in grasses and species of other botanical families, which were preferred to legumes. This favoured a higher intake level of herbage (811 versus 618 g DM/day) with more protein (23.5 versus 20.2% of DM) and lower fibre content (neutral-detergent fibre (NDF) 30.9 versus 40.9% of DM) in the improved pasture than in the ryegrass sward (Avondo *et al.*, 2004). In another study, the higher availability and better quality of biomass of a mixed sward of ryegrass and berseem clover, compared with the daily separate and sequential utilization of both monocultures, allowed grazing goats to select more ryegrass (83 versus 69% of DM) and have higher herbage intake (900 versus 742 g DM/day) in the mixed sward (Bonanno *et al.*, 2004).

Based on the selective behaviour of grazing goats, the use of multi-species pasture is preferable, because it allows a better nutritional balance of the selected diet. Greater species variability can be obtained by: (i) rational fertilization in natural pastures; (ii) over-sowing the most preferred species in improved pastures; or (iii) seeding a mixture of two or more species in cultivated pastures.

The time spent grazing is inversely correlated with the protein concentration of the diet (Harrington, 1982). A longer grazing time is due to a higher seeking of plant parts with higher protein content, to satisfy nutritional requirements. However, the behaviour of goats is less altered by a lower protein level of the available vegetation than that of sheep, which indicates that goats have lower protein requirements (Kronberg and Malecheck, 1997). This is due to their greater N recycling, and thus greater efficiency in N utilization (Tisserand, 2003).

Usually, feeding supplementation reduces grazing time (Bordi *et al.*, 1994), since unsupplemented goats spend more time consuming greater quantities of forage or selecting a balanced diet (Provenza and Malecheck, 1986).

Climatic conditions also affect goats' grazing behaviour. When it rains, goats stop grazing and look for shelter, thus reducing grazing time, while sheep seem to be less bothered (Penning *et al.*, 1997).

Another factor influencing grazing behaviour, and therefore productivity, is animal ranking within the permanent hierarchical scale of a flock of goats, where the presence of horns, greater age and larger size favour dominance. Competitions between dominant and subordinate animals occur especially when resources are scarce. In fact, competition is less important on pasture, whereas it is more relevant in the stable, where goats compete during the distribution of feeds or for a place to rest (Di Grigoli *et al.*, 2003). Competition during grazing arises in conditions of high forage availability; by contrast, forage shortage determines less competition in selecting the most preferred species, because selective activity would lead to a greater energy expenditure, which could not be compensated by the low quality of the selected diet (Barroso *et al.*, 2000).

Grazing Techniques

In many conditions, grazing-based productive systems are the most convenient, since the herbage used directly by the animals is the cheapest feed source. Therefore, in order to maximize the efficiency of a goat farm, a prime objective would be to optimize the utilization of this natural resource.

The advantages of grazing are linked mainly to the possibility of: (i) exploiting resources in areas that cannot be mechanized, thus reducing the costs of materials and manpower; (ii) ensuring better animal welfare; and (iii) maintaining soil fertility, especially through the deposition of animal excreta. Nevertheless, the irregular forage availability throughout the year is an important limiting factor for grazing, which influences farm productivity. For this reason, it is important to apply grazing techniques that enable pasture to maintain adequate and constant herbage amount and quality for as long as possible, in order to prolong and maximize the intake levels of animals, and thus reduce the costs of feeding supplementation.

Description of grazing techniques

The efficiency of different grazing approaches depends on the achievement of the best interaction between forage resources and the grazing animal, and depends on specific practical conditions. Briefly, the various grazing techniques available derive from two principal methods (continuous and rotational) (Cavallero and Ciotti, 1991; Molle *et al.*, 2004).

In the continuous grazing system, the animals graze on pasture grass continuously for the whole season; therefore the vegetation does not have periods of undisturbed growth. *Continuous stocking* (or *set stocking*) is when either the grazing area or the stocking rate remains unchanged during the season, thus favouring feed selection by the animals. By contrast, *intensive continuous stocking* (or *continuous variable stocking*) is when the surface or the stocking rate is adjusted during the season, as a function of grass availability (often in terms of plant height) or animal requirements. This modality should be preferred to the first one, because of the better balance between animal needs and grass availability.

In the rotational grazing system, periods of grazing are alternated with periods of vegetation rest. The animals are kept in a fenced sector of pasture with an instantaneous stocking rate such that their intake capacity exceeds the herbage growth rate; once utilization is completed, they are moved to another sector. If, after the first grazing turn, grass regrowth allows a successive period of utilization, this is referred to as *rotational grazing*. When the sector used is sufficient to satisfy the animals' needs for periods of a few days, this is referred to as *strip grazing* or *daily rotational grazing*, which can be set with mobile electrified fencing. This system avoids excessive or low utilization of vegetation and, because of the frequent changes of paddock, minimizes the risks of spreading parasitic infestations.

In all grazing systems, the choice of the beginning of the grazing season, which depends mainly on the amount of forage on offer per head, as affected by agro-climatic conditions, is determinant for the herd productivity. An earlier utilization at grass height of 6–8 cm increased milk yield of goats compared with a later utilization at grass height of 10–15 cm (Le Frileux *et al.*, 2000). However, it is important to consider that, when animals graze grass in early stages of growth, they tend to select herbage with high protein concentration and low fibre, and thus the risks of digestive disturbances increase.

Regardless of the technique adopted, grazing can have a daily duration restricted to a few hours or lasting the whole day. When grazing lasts less than 12 h/day, it is referred to as *short grazing*; this is the most common condition on herbaceous pasture in Mediterranean environments, where grazing normally occurs between the two daily milkings, for a time varying from 4 to 8 h/day. Generally, an increase in daily grazing duration, especially when no concentrate is supplied, increases grass intake and milk production (Bonanno *et al.*, 2007c). However, with adequate grass availability, a short grazing period does not always result in a reduction in intake that might limit the satisfaction of animals' needs; in fact it is compensated by faster rates and higher levels of herbage intake, in comparison to grazing for the whole day (Romney *et al.*, 1996).

The use of nocturnal grazing, to integrate daily grazing, is one of the most important management practices to improve animal performance (Bayer *et al.*, 1987). The benefits of such integration are related to: (i) better body condition of the animals; (ii) higher tolerance to heat stress; (iii) higher grass intake; (iv) better quality and digestibility of the selected diet; and (v) higher milk production (Iason *et al.*, 1999, Bonanno *et al.*, 2007c). This type of grazing is certainly advisable under arid or semi-arid conditions, when day-time high temperatures greatly restrict the duration of grazing activity. However, the advantages of this system are counter-balanced by some disadvantages, such as: (i) greater need for manpower; (ii) higher risk of animals being stolen; and (iii) more damage to vegetation due to trampling by animals. Therefore, in certain conditions, day-and-night grazing is unfeasible.

The diffusion of mixed farming systems, in which goats coexist with cattle and/or sheep, has led to the development of mixed grazing systems. In these systems, groups of animals of different species or at different physiological stages utilize the same pasture, either concomitantly or successively. The benefits are greater when differences in nutritional requirements and feeding behaviour of the groups are higher. The diversity and complementarity in herbivores' feeding and selective behaviour, especially on heterogeneous pastures, can greatly modify the botanical and structural composition of the pasture. Because of the differences in feeding preferences between goats and sheep or cattle, mixed grazing systems including goats are usually more productive than systems with a single animal species. Such advantages are due to: (i) greater presence of legumes refused by goats, resulting in a more intense N fixation; (ii) better pasture development; (iii) limited risks of pasture over-grazing; and, consequently, (iv) higher animal performance (Harrington, 1982).

Sequential grazing consists of allowing the most needy or selective animal species to graze first, and then allowing the less demanding animal species or category to utilize the pasture. The introduction of sheep after goats was shown to be an effective management strategy for ryegrass and white clover-based pasture, because it allowed ewes to have higher intake levels and better performance, in terms of growth of suckling lambs, compared with a sheep–sheep grazing sequence. This can be explained by the reduced consumption of white clover by goats, subsequently utilized by sheep (Del Pozo *et al.*, 1996, 1997; Osoro and Del Pozo, 1996).

Choice of grazing techniques for goats

The choice of grazing technique for goats is not a simple issue. Experiments carried out on various animal species have shown that the environmental and practical conditions are not always ideal for clearly demonstrating the validity of one technique rather than another. Having to make a choice between the two principal grazing techniques described (i.e. continuous and rotational), there is no doubt that the second is capable of expressing a greater productive potential of grassland, especially if the intervals of regrowth do not bring about an excessive ageing of the leaf tissue. Generally, the production of biomass is greater with rotational than with continuous grazing, with average increases of 15–20% and peaks of up to 50%. Nevertheless, there are numerous comparisons where this higher pasture productivity is not converted into the expected increase and improvement in animal performance (Cavallero and Ciotti, 1991). Continuous grazing makes available to animals a much higher percentage of leaves than stems, as compared with rotational grazing, and this is converted into higher digestibility and protein content of the ingested grass (Reyneri, 1989). Moreover, rotational grazing is less practical, because it requires a more complex and onerous management than continuous grazing (e.g. installation of fixed or mobile fencing, greater need for manpower). Experiments on goats grazing shrublands in the Mediterranean area have shown that using the rotational system, rather than continuous grazing, increased vegetation availability and weight gain of goats, probably due to greater vegetation regrowth, better plant nutritional quality and lower energy expenditure by the animals to move around (Tsiouvaras *et al.*, 1999). On the other hand, in the few experiments carried out on goats grazing herbaceous pastures in Mediterranean conditions, the use of rotational grazing had limited success in terms of grass intake, milk production (Masson *et al.*, 1991) and BW gain (Gutman and Seligman, 1979).

On natural pasture, grass intake and milk production did not differ greatly between the continuous and rotational systems, with a slightly higher yield and fat and protein content of milk for the continuous system (Rubino *et al.*, 1988) (Table 9.5). Analogous results were obtained on grass meadow utilized by goats under continuous or rotational grazing (Table 9.5). Although the latter expressed slightly higher milk production (+6%) than the continuous, no relevant differences emerged in grass height and availability, milk quality, and goat behaviour, weight and body condition (Van Quackebeke *et al.*, 1997).

It needs to be underlined that the adoption of one or the other grazing technique, even on the same pasture, does not always give the same results over the years. For example, a similar mixed sward of Italian ryegrass and berseem clover, set on adjacent land plots over 2 successive years, was utilized in both years, starting from April, by groups of goats at the same lactation stage and receiving equal concentrate integration. In the first year, the sward was subdivided into sub-plots rotationally grazed, whereas in the following year continuous grazing was activated over the whole surface (Bonanno *et al.*, 2004, 2007d) (Table 9.5).

Table 9.5. Comparison between continuous and rotational grazing of lactating goats on herbaceous pastures.

Pasture type	Grazing technique	Stocking rate (goats/ha)	Duration (days)	Biomass availability (DM)	Intake[a] (g DM/day per head)	Milk yield	Reference
Natural pasture	Rotational	6	110		914	231 kg/ lactation	Rubino *et al.* (1988)
	Continuous	6	110		980	246 kg/ lactation	Rubino *et al.* (1988)
Grass meadow	Rotational		112	2.8 kg/day		+6%	Van Quackebeke *et al.* (1997)
	Continuous		112	2.8 kg/day			
Mixed sward of ryegrass and berseem clover	Rotational	32	50	3.6 t/ha	1095	1027 g/day per head	Bonanno *et al.* (2004)
	Continuous	36	87	4.3 t/ha	1240	1332 g/day per head	Bonanno *et al.* (2007d)

[a]Intake of forage and concentrate.

Continuous grazing, due to the more favourable temperature and precipitation conditions in the second year, led to higher growth rates, more lasting growth of the grass and a markedly higher animal productivity, compared with rotational grazing, in spite of a slightly higher stocking rate (36 versus 32 goats/ha for continuous and rotational grazing, respectively).

In theory, rotational grazing, as compared with the continuous system, should ensure higher productivity and availability of grassland, better plant distribution over time, higher intake levels, and more efficient pasture utilization, by limiting the goats' selection activity. In practice, the few experiments focusing on rotational grazing on herbaceous pasture have not demonstrated effective improvements in terms of grass intake and milk yield, at least in Mediterranean environments.

Therefore, continuous grazing remains an adequate choice for goats, also because of its fewer organizational and management difficulties. When continuous grazing does not allow adequate levels of utilization, as in pastures with high grass availability, it is advisable to reduce the area grazed by the animals and utilize the non-grazed area for forage storage. This system ensures higher levels of pasture utilization (20% more grass and 22% more energy) in comparison to continuous systems, and the use of hay or silage during the periods of grazed forage shortage (Rubino *et al.*, 1988).

Stocking Rate

Animal products from grazing systems are the result of a complex interaction between the amount and the composition of available forage and the efficiency with which nutrients consumed by animals are transformed into products (Hart *et al.*, 1988). Grazing intensity is one of the main management factors influencing this interaction, being the major determinant of forage and animal productivity. Thus, an important aspect of grazing management is to choose the stocking rate, i.e. the 'number of animals to graze per unit area of land for a specific amount of time' (Malecheck, 1982). Both overstocking and understocking have adverse effects on the botanical composition, digestibility and nutritive value of the biomass and the diet selected by the animals at pasture. An excessive stocking rate induces animals to consume most of the leafy structure of plants, which is essential for reconstituting the root reserves necessary for plant regrowth, and may also lead to the reduction, and even the extinction, of the more palatable botanical species. Thus, overgrazing can reduce forage yield, herbage intake and animal performance, and increase soil degradation and erosion. By contrast, undergrazing worsens forage quality and can cause consistent forage loss, as a result of tissue ageing. In fact, as plants develop, the amount of stems and fibre increases, whereas protein content and digestibility decrease. Moreover, due to their higher fibre concentration, older plants become harder and less palatable. Therefore, animals grazing on grass at advanced developmental stages select a diet of poor nutritive value. A correct adjustment of the stocking rate allows animals to consume forage that has optimal palatability and composition, and limits the loss of refused vegetation.

Studies on the effect of stocking rate have focused mainly on goats grazing on woody and shrubby lands (Njwe *et al.*, 1995; Tsiouvaras *et al.*, 1999; Abbadessa *et al.*, 2000; Animut *et al.*, 2005a,b). In goats browsing Mediterranean shrubland, the increase in grazing pressure lowered the incidence of the more palatable herbaceous or shrubby species, and reduced forage yield and voluntary intake (Tsiouvaras *et al.*, 1999). In a pasture composed of various grasses and forbs, the increased stocking rate reduced goat growth performance, and this effect seemed to be linked more to the limited ability of goats to select forage of adequate nutritive value than to the reduction of DM intake as a consequence of decreasing forage availability (Animut *et al.*, 2005b).

Even though there are only a few studies on goats grazing on pastures composed exclusively of herbaceous species, some indications can be given based on an experiment on milking Girgentana goats grazing a cultivated pasture (Bonanno *et al.*, 2005). From 9 April, the goats were allowed to continuously graze a mixed sward of Italian ryegrass (*Lolium multiflorum* Lam. subsp. *westerwoldicum*, var. Elunaria) and berseem clover (*Trifolium alexandrinum* L., var. Lilibeo) for 7 h/day, at different stocking rates (72, 48 and 36 goats/ha), and supplemented daily with 500 g of barley meal. Commonly, in an on-farm forage system, cultivated pastures integrate the natural vegetal resources, as either grazed or stored forage and, providing high forage yield, their exploitation with grazing is done using higher instantaneous stocking rates than natural pastures. In this experiment, the stocking rate influenced the length of the grazing period. In fact, grazing under the highest stocking rate was suspended 25 days earlier (62 versus 87 days) in comparison with the lower stocking rates (Fig. 9.4), due to a faster deterioration and reduction of the available vegetation. During the period in which the three different stocking rates were used simultaneously (first 62 days of trial), the highest stocking rate reduced herbage height and amount of available

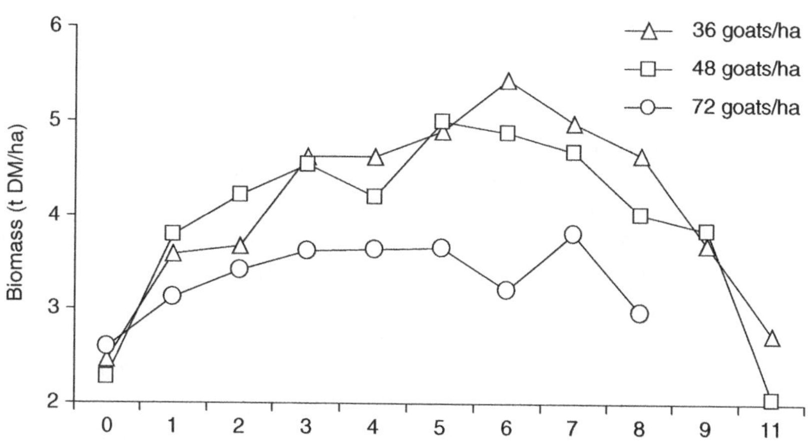

Fig. 9.4. Evolution of biomass (tonnes of dry matter (DM) per hectare) in a mixed sward of Italian ryegrass and berseem clover grazed under different stocking rates. (Adapted from Bonanno *et al.*, 2007d.)

biomass, and affected sward botanical composition. Higher contents of DM and NDF were observed under the lowest and highest stocking rates, whereas the protein content decreased slightly with the lowest stocking rate (Table 9.6). As stocking rate increased, forage yield tended to decrease (6.4, 6.1 and 5.8 t DM/ha for 36, 48 and 72 goats/ha, respectively), due to slower herbage growth (63.2, 61.6 and 51.5 kg DM/day per ha). Thus, the highest stocking rate reduced daily herbage allowance per goat (2.8, 2.0 and 1.3 kg DM/day per goat) and increased animal utilization (27, 34 and 49% of available biomass). On the whole, the utilization level was rather low, even at the highest stocking rate, but consistent with the values observed in goats grazing natural pastures, which ranged from 34 to 63% of available biomass (Fedele *et al.*, 1988).

As stocking rate increased, intake of nutrients from the selected herbage (i.e. DM, CP and NDF) and diet digestibility decreased (Table 9.7) (Bonanno *et al.*, 2007d). The NDF percentage in the selected herbage was not influenced by stocking rate, whereas protein content was slightly lower at the lowest stocking rate. However, the NDF content was lower, and protein content higher, in comparison with available forage, independent of stocking rate. Goats grazing herbaceous pastures tend to select the most tender and proteic and the least fibrous parts of the plant, even when the quality of available vegetation is poor (Pizzillo *et al.*, 1988; Rubino *et al.*, 1988). When biomass quantity or quality does not allow this selective behaviour, goats prefer to reduce their herbage intake (Animut *et al.*, 2005b).

Table 9.6. Average height, availability, chemical and botanical composition of biomass in a mixed sward of Italian ryegrass and berseem clover grazed during the spring by goats at different stocking rates. (Adapted from Bonanno *et al.*, 2007d.)

	Stocking rate (goats/ha)		
	36	48	72
Herbage height (cm)	21.3^{a}	18.5^{b}	12.4^{c}
Biomass (t DM/ha)	4.3^{a}	4.2^{a}	3.3^{b}
Chemical composition (% of DM)			
DM	20.4^{a}	18.4^{b}	19.8^{a}
CP	16.0^{b}	17.0^{a}	17.0^{a}
NDF	49.8^{a}	47.7^{b}	50.3^{a}
Botanical composition (% of DM)			
Ryegrass	24.9^{a}	23.9^{a}	15.1^{b}
Berseem clover	46.0	45.3	51.1
Other species	6.7^{b}	11.4^{a}	8.3^{ab}
Dead matter	22.3^{b}	19.4^{c}	25.4^{a}

DM, dry matter; CP, crude protein; NDF, neutral-detergent fibre.
^{a,b,c}Mean values in a row with different superscript letters were significantly different ($P \leq 0.05$).

Stocking rate also influenced the botanical composition of the selected herbage (Table 9.7). At high stocking rates, goats were obliged to reduce their intake of ryegrass, which, being more palatable, gradually started to disappear from the pasture, and to increase their intake of legumes, which, being less preferred, became more abundant.

The lower herbage intake at the highest stocking rate corresponded to a reduction in milk yield (Table 9.7), without variations in milk composition.

The relationship between stocking rate and 3.5% fat-corrected milk (FCM) produced over the entire grazing period (87 days) is shown in Fig. 9.5. The latter was calculated according to the equation: FCM = milk (g) × (0.634 + 0.1046 × fat(%)) (Pulina *et al.*, 1991). As grazing pressure increased, daily milk yield per hectare increased. However, total milk yield per goat declined drastically at the highest stocking rate, whereas the amount of milk per hectare was highest at a medium stocking rate (Bonanno *et al.*, 2007d). From an economic point of view (Hart *et al.*, 1988), maximum profit from a pasture system is achieved at the point where milk production per hectare is at a maximum and individual milk yield begins to decline, a condition that generally corresponds to a medium stocking rate, as in this specific case. Thus, the best compromise between maximum production per animal and maximum production per hectare is achieved at a moderate stocking rate.

Table 9.7. Average intake, chemical and botanical composition of selected herbage, diet digestibility and milk yield of goats grazing a mixed sward of Italian ryegrass and berseem clover at different stocking rates. (Adapted from Bonanno *et al.*, 2007d.)

	Stocking rate (goats/ha)		
	36	48	72
Herbage intake (g/day per head)			
DM	806^{a}	716^{b}	635^{c}
CP	206^{a}	190^{b}	170^{b}
NDF	269^{a}	232^{b}	202^{c}
Herbage chemical composition (% of DM)			
DM	18.7^{a}	19.1^{a}	16.5^{b}
CP	25.6^{b}	26.4^{a}	26.8^{a}
NDF	33.0	32.7	31.7
Diet DM digestibility (%)	70.1^{a}	69.8^{a}	67.1^{b}
Herbage botanical composition (% of DM)			
Ryegrass	54.7^{a}	45.8^{b}	36.8^{c}
Berseem clover	35.7^{b}	34.9^{b}	51.2^{a}
Weeds	9.6^{b}	19.3^{a}	11.9^{b}
Milk yield (g/day per head)	1332^{a}	1328^{a}	1004^{b}

DM, dry matter; CP, crude protein; NDF, neutral-detergent fibre.
^{a,b,c}Mean values in a row with unlike superscript letters were significantly different ($P \leq 0.05$).

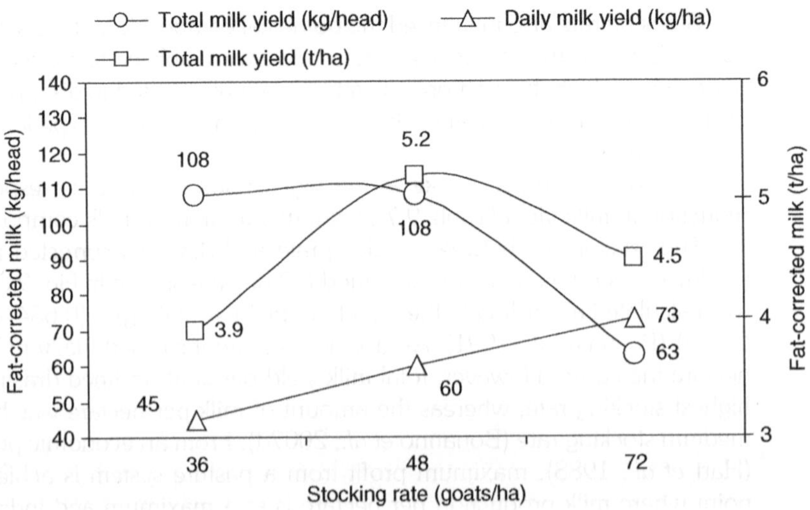

Fig. 9.5. Effect of grazing a mixed sward of Italian ryegrass and berseem clover under different stocking rates on 3.5% fat-corrected milk production of goats during the entire utilization period (87 days). (Adapted from Bonanno *et al.*, 2007d.)

In practice, the adoption of a moderate stocking rate allows better forage exploitation and animal performance. Moderate grazing intensity ensures high forage availability and quality for the animals and, consequently, the opportunity to fully express goats' selective behaviour and optimize the level of herbage intake and production. In this way, such a stocking rate improves the overall productivity of a pasture system.

Obviously, the level of stocking rate which is considered as 'moderate' varies mainly with pasture type and climatic conditions. For example, in goats grazing on a 1-year sulla (*Hedysarum coronarium*) meadow, as stocking rates increased (30, 50 and 70 goats/ha), herbage intake (1353, 1185 and 1069 g DM/day per goat) and milk yield (1763, 1516 and 1361 g/day per goat) decreased, without relevant changes in milk composition (Bonanno *et al.*, 2007b). However, in this case, at the lowest grazing pressure, the highest total milk yield per goat corresponded to the lowest daily and total milk production per hectare. This might have resulted from the rapid deterioration of sulla vegetation even under less intense utilization, due to the warm climate, with consequent insufficient amount of pasture to satisfy the energy requirements of goats for milk production.

Therefore, in order to guarantee a level of grass intake adequate for goats' production requirements, a sustainable animal stocking rate should be established over a given period of time. However, there are no methods of estimating stocking rate that could be applied to all grazing systems. The available methods are based on the knowledge of grass mass at pasture or grass growth rate, and take into account the level of pasture utilization by the animals. Grass availability can be estimated as a function of grass height, measured by a rising plate meter, using linear regression equations found in the literature for various forage species or pasture types (Molle *et al.*, 2004). Alternatively, such an estimation can be more empirical, by drying and weighing grass samples cut from areas of known size.

On the other hand, it is more difficult to determine grass growth rate, since it varies greatly, especially with temperature and precipitation conditions, pasture botanical composition and grazing technique. The coefficient of utilization of available forage depends on several factors, such as: (i) slope and exposure of the pasture land; (ii) floristic composition; (iii) nutritive value; (iv) grazing technique; (v) stocking rate; and (vi) selectivity of the grazing animal species. Since the value of this coefficient was 0.65 for sheep (Molle *et al.*, 2004), then a value no higher than 0.55 could be proposed for goats grazing herbaceous pasture, considering their high selectivity and physical activities and high refusal rate of vegetation.

Pasture Integration with Concentrates

The exclusive exploitation of pasture does not always assure adequate production levels of goats, especially because of marked seasonal variations in availability and quality of grass and climatic conditions, not always favourable to grazing. In this case, the use of supplements is necessary. However, in grazing systems it is not easy to decide when and how to provide animals with integration, because the level and quality of herbage intake are not always known. In spite of these limits, the information available today allows one to find solutions capable of improving the overall efficiency of a goat farm.

Choice of concentrate supplements

During the grazing season, the chemical and nutritional characteristics of pasture grass change substantially, thus influencing the quality of intake, especially of very selective animals like goats. Therefore, improving the productive efficiency of the 'grazing system' not only means satisfying the animals' energy requirements, but also optimizing the digestive process by synchronizing the ruminal fermentation of nutrients of the grass with those supplied by feed integration. As a consequence, simply increasing the amount of concentrate supplied has not always been successful in goats. In fact, in experiments on natural (Rubino *et al.*, 1995) (Table 9.8) or cultivated pasture (Min *et al.*, 2005) (Table 9.9), the effects of a high supply of concentrates on milk production were absent or almost negligible, while in the experiments of Claps *et al.* (1994), the highest milk production was reached with the highest concentrate intake (1050 g/day) (Table 9.8). In these last experiments the possibility of a free choice among barley, chickpeas, broad beans and beet pulp allowed the goats to optimize the nutrient content of their diet, as also observed by Fedele *et al.* (2002) in indoor conditions.

Generally, herbage intake decreased, as concentrate supplementation increased. On natural pasture, goats receiving an integration of 150 g concentrate/day with 18% of CP consumed 51% more herbage than those supplied with 550 g of the same concentrate daily (Fedele *et al.*, 1993). On the other hand, herbage intake decreased from 53.1 g/kg of metabolic weight (MW) to 14.8 g/kg MW going from zero to 1050 g of concentrate supplementation per day (Claps *et al.*, 1994) (Table 9.8). Goats grazing on cultivated pastures (Table 9.9), such

Table 9.8 Effects of the supplementation of concentrate to goats grazing herbaceous natural pastures on grass intake, milk yield and composition.

Concentrate		Intake (g/kg MW per day)		Milk			
Type and CP (% of DM)	Amount (g/day per head)	Grass DM	Grass CP	Yield (kg/lactation or g/day*)	Fat (%)	Protein (%)	Reference
Mix, 12%	550	45.4^{bc}	7.4^{b}	333	4.0^{a}	2.9	Fedele *et al.* (1993)
Mix, 18%	550	37.3^{c}	6.4^{b}	328	4.0^{a}	2.9	
Mix, 18%	150	56.4^{a}	9.4^{a}	313	3.9^{ab}	2.9	Rubino *et al.* (1995)
Mix, 38%	150	48.9^{ab}	7.6^{b}	282	3.7^{b}	3.0	
Barley, 11.5%	450	31.4	3.7	160^{d}	4.3^{b}	3.3^{a}	Claps *et al.* (1994)
Chickpeas, 22%	*ad libitum* (mean = 700)	34.1	4.7	238^{b}	4.6^{a}	3.2^{b}	
Mixed grains offered free choice, 13%	*ad libitum* (mean = 1050)	14.8	1.9	328^{a}	4.2^{b}	3.3^{a}	
	0	53.1	7.5	226^{c}	4.1^{b}	3.3^{a}	
Barley and chickpeas, 14%	600			1480^{a*}	2.9	3.3	Fedele *et al.* (2000)
Maize and broad beans, 15%	600			1350^{a*}	3.1	3.4	
	0			1070^{b*}	3.1	3.3	

MW, metabolic weight; CP, crude protein; DM, dry matter.
[a,b,c,d]Within each study, mean values in a column with different superscript letters were significantly different ($P \leq 0.05$).
*, g/day

Table 9.9. Effects of the supplementation of concentrate to goats grazing cultivated pastures on grass intake, milk yield and composition.

	Concentrate		Intake (g/day per head)			Milk			
Pasture type	Type and CP (% of DM)	Amount (g/day per head or g/kg milk over 1.5 kg/day*)	Diet OM	Grass DM	Grass CP	Yield (g/day per head)	Fat (%)	Protein (%)	Reference
Improved pasture	Barley and tickbean, 15%	500		811^{a}	189^{a}	1088	4.3^{a}	3.8^{a}	Avondo *et al.* (2004)
	Barley and tickbean, 20%	500		725^{ab}	160^{b}	1100	4.2^{a}	3.8^{a}	
Ryegrass sward	Barley and tickbean, 15%	500		618^{b}	129^{c}	1110	4.2^{a}	3.9^{a}	
	Barley and tickbean, 20%	500		657^{b}	138^{bc}	1133	3.6^{b}	3.5^{b}	
Mixed sward of ryegrass and berseem clover	Barley and tickbean, 15%	500		661^{b}	126	1027^{a}	3.8^{b}	3.6^{b}	Bonanno *et al.* (2004)
		0		900^{a}	179	899^{bc}	4.0^{b}	3.4^{c}	
Monocultures of ryegrass and berseem clover	Barley and tickbean, 15%	500		519^{c}	86	953^{ab}	3.9^{b}	3.5^{bc}	
		0		742^{b}	124	846^{c}	4.7^{a}	3.7^{a}	

(*Continued*)

Table 9.9. *Continued.*

	Concentrate		Intake (g/day per head)			Milk			
Pasture type	Type and CP (% of DM)	Amount (g/day per head or g/kg milk over 1.5 kg/day*)	Diet OM	Grass DM	Grass CP	Yield (g/day per head)	Fat (%)	Protein (%)	Reference
Low-quality mixed vegetative forages (year 2000)	Mixed, 11%	660*	1800[a]			2950[a]	3.1	3.07[a]	Min *et al.* (2005)
		330*	1400[b]			2480[b]	3.1	3.02[b]	
		0	1200[bc]			2090[b]	3.0	2.80[b]	
High-quality mixed vegetative forages (year 2001)	Mixed, 11%	660*	2100			4120	3.0	3.06[b]	Min *et al.* (2005)
		330*	2100			4270	3.0	3.10[a]	
		0	2500			3770	2.9	3.00[b]	

CP, crude protein; DM, dry matter; OM, organic matter.
[a,b,c]Within each study, mean values in a column with different superscript letters were significantly different ($P \leq 0.05$).

as a mixed sward of Italian ryegrass and berseem clover or monospecific swards of the same species, also showed a reduction in herbage intake due to concentrate supplementation (Bonanno *et al.*, 2004), with substitution rates of herbage with concentrate intake equal to 0.54 and 0.58, respectively. The effects of increasing amounts of concentrate supplementation on herbage intake vary with pasture forage quality. For example, in goats grazing mixed vegetation, when concentrate supply increased from 0.33 to 0.66 g/day, diet total intake (forage and concentrate) increased for the low-quality forage, while it did not change for the high-quality forage (Min *et al.*, 2005).

These results were attributed not only to a lower grass intake, but also to a lack of synchronization between nutrients in ruminal fermentations. During the winter rest, the grass is rich in soluble compounds (especially sugars and N) and poor in fibre. As spring begins, a positive nutrient balance is achieved, which lasts until the first hot days of late spring–summer. From this moment on, there is a progressive fall in soluble compounds and increase in fibre, particularly lignin (Fedele, 2001a). In fact, the highest nutrient imbalance is observed in periods of low grass availability (winter and summer), and can be corrected by feeding supplements.

Goats' responses to concentrates

The supply of concentrates to grazing goats affects their feeding behaviour. When goats grazing on natural pasture were fed concentrate 600 g/day, they selected more grasses and plants of other families and fewer legumes, and reduced the total number of species selected (Claps *et al.*, 2000). This effect is less evident when goats graze on cultivated pasture, which has a more uniform botanical composition. In fact, when goats grazed a mixed sward of Italian ryegrass and berseem clover, the daily supplementation with 500 g of a mixture of barley and tickbean did not change their selective behaviour. By contrast, when monocultures of the two species were utilized in sequence during the day, restricting the animals' selective freedom, the concentrate reduced clover intake to the benefit of ryegrass (Bonanno *et al.*, 2004).

Changes in selective behaviour influence not only the plants selected, but also the nutritional composition of the parts selected. Generally, when pasture conditions are adequate, goats are able to choose a diet as balanced as possible in nutrients, due to their high selective behaviour. Depending on the amount or protein content of the concentrate supplied, goats select grass containing different nutritive elements. For example, an increase from 12 to 18% of CP content of concentrates (used in a dose of 550 g/day) decreased the intake of protein from grass by more than 1 g/kg MW (Table 9.8) and of fibre by almost 2 g/kg MW (Fedele *et al.*, 1993). Similarly, in the study of Avondo *et al.* (2004) on Girgentana goats, an increase from 15 to 20% in CP content of the concentrate caused a 15% lower protein intake of an improved pasture (189 versus 160 g/day) (Table 9.9).

On the other hand, such compensations in nutrient intake at pasture, for the effects of concentrate or its protein content, are not equally evident with cultivated resources, due to their greater homogeneity. Concentrate supply did not modify the CP percentage of the selected forage of Italian ryegrass and berseem clover, as either mixed sward (19.1 and 19.9% of CP with or without concentrate) or monoculture (16.6 and 16.7% of CP with or without concentrate), presumably because the goats' selectivity was limited by the high NDF content (more than 50% of DM) of the forage, which induced a great filling effect (Bonanno *et al.*, 2004). In the same way, a concentrate rich in protein did not reduce the protein intake of goats grazing a pure sward of Italian ryegrass, probably due to selective difficulties connected with the botanical and nutritional homogeneity of the pasture (Avondo *et al.*, 2004) (Table 9.9).

By contrast, increasing the quantity of concentrate with 18% of CP from 150 to 550 g/day decreased protein intake from grass by about 30% (9.4 versus 6.4 g/kg MW) and fibre intake by 26% (11.0 versus 8.1 g/kg MW) (Fedele *et al.*, 1993). Increasing concentrate from 0 to 1050 g/day, the protein intake from pasture decreased by four times (7.5 versus 1.9 g/kg MW) and the fibre by 3.5 times (17.6 versus 4.8 g/kg MW) (Claps *et al.*, 1994) (Table 9.8).

The selective behaviour of goats cannot always compensate for a nutritional imbalance, especially when they are fed simple concentrates, such as legume or cereal grains. For example, supplying chickpeas or barley alone as pasture integration leads to diets excessively rich (22% of CP) or poor (11.5% of CP) in protein, respectively (Claps *et al.*, 1994). A free choice among concentrate grains (barley, chickpeas, broad beans) and by-products rich in ruminable fibre (beet pulps) enabled the goats to choose a well-balanced diet for CP content (13.2%) and NDF (41.8%). As a consequence, the amount of milk produced per head per lactation was 160 kg, 238 kg and 328 kg for goats supplemented with barley (protein deficient), chickpeas and mixed concentrates (barley, chickpeas, broad beans and beet pulp), respectively, compared with the 226 kg of milk per head per lactation of unsupplemented grazing goats (Claps *et al.*, 1994) (Table 9.8).

For goats grazing either natural or cultivated pastures, generally the supply of concentrate increases energy intake and milk production (Morand-Fehr *et al.*, 2007). Nevertheless, since concentrate reduces grass intake, it can have considerable effects on milk quality. In particular, when the animals cannot balance nutrients at pasture, the consequent nutritional imbalance worsens milk composition and cheese yield. For example, a lower intake of structural carbohydrates from a cultivated pasture induced by the supply of concentrates decreased milk fat concentration (Bonanno *et al.*, 2004) (Table 9.9), probably due to a lower rumen production of acetic acid, the precursor of short- and long-chain fatty acids in milk. By contrast, when goats have the opportunity to balance nutrients at pasture, as occurred when grazing natural pasture (Fedele *et al.*, 2000) (Table 9.8) or mixed vegetation (Min *et al.*, 2005) (Table 9.9), the concentrate improved milk yield and did not modify its fat content.

Furthermore, when the energy provided with concentrate is not balanced with dietary degradable protein, the consequent limited synthesis of microbial protein in the rumen reduces milk casein (Bonanno *et al.*, 2004). Therefore, when the energy intake from concentrate exceeds that of dietary N, it is probably used for adipose tissue formation, thus inducing animal fattening and negatively affecting its productivity. By contrast, a protein excess from concentrates, especially under high protein supply from pasture, leads to an imbalance between N and energy in the diet, which increases the level of ammonia in the rumen, and consequently the risks of alkalosis, and the levels of urea and non-protein N in the milk.

Practical Indications for Feeding Supplementation

In order to properly supplement the diet of grazing goats, it is important to know their intake and nutritive requirements, such as those reported in Tables 9.10 and 9.11, re-elaborated using experimental data collected over the years by the CRA (Istituto Sperimentale per la Zootecnia, Potenza, Italy) (V. Fedele, unpublished data), in Table 9.12 (re-elaborated from Fedele, 2001b) and in Table 9.13 (Claps *et al.*, 1994; Fedele, 1994). Apart from seasonal grass availability, grass intake depends on animal physiological stage (dry or lactating), BW and milk yield (Table 9.10). In summer and winter, for reasons explained in the previous paragraphs, even Mediterranean goats, such as Maltese and Siriana breeds, having high production requirements (with milk yield higher than 1 kg/day) cannot consume on pasture, on average, more than 500 g DM/day. During these two seasons, the low quality of grass causes or worsens the nutritional deficit. In summer, the marked decrease in fermentable compounds, especially sugars, and the increase in lignin greatly reduce the nutritive value of the grass, which changes

Table 9.10. Dry matter (DM) intake of goats grazing herbaceous natural pasture. (Adapted from V. Fedele, unpublished data.)

Physiological stage	Body weight (kg) or milk yield (kg)	DM intake (g/day per head) Spring	Summer	Winter
Dry	< 40	540–630	290–370	310–390
	> 40	610–750	310–470	330–440
Lactating	< 1	660–810	390–470	410–530
	> 1	830–1060	420–490	440–590

Table 9.11. Nutritional composition of grass selected by goats. (Adapted from V. Fedele, unpublished data.)

Season	Digestible protein (% of DM)	NDF (% of DM)	UFL[a]/kg DM
Spring	12–16	44–51	0.64–0.70
Summer	7–11	65–71	0.56–0.61
Winter	14–18	34–41	0.69–0.78

DM, dry matter; NDF, neutral-detergent fibre.
[a]UFL = milk forage unit, energy unit of the INRA system corresponding to 1.70 Mcal of net energy for lactation estimated at maintenance feeding level.

Table 9.12. Dry matter (DM) intake and nutritive value of woody vegetation (leaves and shoots) and woody and herbaceous vegetation selected by goats grazing natural pasture in Mediterranean environment. (Adapted from Fedele, 2001b.)

Season		DM intake (g/day per head)	Digestible protein (g/day per head)	UFL[a]/kg DM
Spring	Trees and shrubs	430–950	42–68	0.36–0.66
	Trees, shrubs and grasses	550–1250	48–91	0.44–0.72
Summer	Trees and shrubs	550–1250	37–62	0.31–0.62
	Trees, shrubs and grasses	420–910	44–86	0.38–0.67

[a]UFL = milk forage unit, energy unit of the INRA system corresponding to 1.70 Mcal of net energy for lactation estimated at maintenance feeding level.

Table 9.13. Nutritional requirements[a] of grazing goats. (Adapted from Claps *et al.*, 1994; Fedele, 1994.)

Dry matter (% of BW)	
For maintenance	3.3
Per kilogram of milk	1.1
Crude protein (%)	13–14
NDF (%)	39–41
UFL[b]	0.91–0.95
Net energy (kcal/kg MW)	
For maintenance	100
Per kilogram of milk	730
Digestible protein (g/kg MW)	
For maintenance	4.2
Per kilogram of milk	40

BW, body weight; NDF, neutral-detergent fibre; MW, metabolic weight
[a]The requirements were estimated on milking goats grazing natural pasture and subjected to a free choice of the type and the amount of concentrates offered.
[b]UFL = milk forage unit, energy unit of the INRA system corresponding to 1.70 Mcal of net energy for lactation estimated at maintenance feeding level.

from 0.64–0.70 milk forage units (UFL)/kg DM in spring to 0.56–0.61 UFL/kg DM in summer (Table 9.11). In winter, even though the nutritive value of the grass is high (0.69–0.78 UFL/kg DM), its excess of fermentable substances, i.e. sugars reach values of 18–22% of DM (Fedele, 2001a), and low fibre content often cause disturbance in rumen fermentation.

When areas covered exclusively by woody vegetation are alternated with others dominated by herbaceous vegetation, the goats' intake level changes from spring to summer. In spring, the presence of herbaceous vegetation causes an increase of goats' intake from 430–950 g/day (only woody vegetation) to 550–1250 g/day in woody and herbaceous vegetation; in summer the effect is opposite (from 550–1250 to 420–910 g/day), because a large part of the herbs are dry and less grazed by animals (Table 9.12).

The nutritive value of the diet consumed by goats can go from an average value of about 0.67 UFL, when grasses are dominant, to 0.48 UFL, when woody vegetation is dominant. This great variability is due to the quality of the woody species present in the pasture. For example, the predominance of species such as holly oak, hawthorn, turkey oak, juniper and heather reduces considerably animal intake and the nutritive value of the vegetation consumed. By contrast, the marked presence of hedge maple, purple coronilla, flowering ash, ash, heather and black locust, all with a good nutritive value (0.55–0.75 UFL), leads to high intake levels.

The intake of nutrients can be estimated on the basis of the information described above. Then, on the basis of animal requirements, it is possible to establish how much concentrate should be supplied. Indications for energy and protein requirements of goats are discussed by Cannas *et al.* (Chapter 6, this volume). Values measured specifically in Mediterranean conditions for milking goats fed on natural pastures were reported (Table 9.13) by Claps *et al.* (1994) and Fedele (1994). These requirements are on average 20% higher than those estimated for animals at the same physiological stage fed indoors with the same method, i.e. with animals being free to choose between simple concentrates offered in unlimited quantities (Fedele *et al.*, 2002).

In addition to the aspects discussed above, other factors should be taken into consideration when choosing the quantity and type of concentrate to feed grazing goats. Even though goat's milk can be consumed fresh, it is mostly transformed into cheese. Therefore, the farmer's goal should be not only to increase the animals' productive efficiency, but also to find the best combination of milk production, cheese yield and product quality.

In optimal grazing conditions, i.e. under intense selective activity and good quality of diet, integration with concentrates only allows a substitution effect, which worsens milk composition and therefore may not be justified. On the other hand, when pasture is nutritionally poor, the use of concentrate is certainly convenient, since balancing the nutrients of this diet increases milk yield and improves its composition.

Furthermore, the use of a degradable energetic concentrate is useful when animals graze a highly proteic grass, especially at an early stage when it is rich in soluble N. A diet with an excess of N not balanced with available energy causes a surplus of N for microbial growth in the rumen, which enhances the urea level in plasma and milk (Harmeyer and Martens, 1980). This imbalance, which results in high urea level, increases the risk of mastitis and lameness, and worsens the reproductive performance of animals (Mellado *et al.*, 2004).

For goats grazing cultivated pasture, a strong relationship between the diet CP and the level of milk urea N (MUN) was found: CP(% of DM) $= 6.91 \pm 1.42 + 0.61 \pm 0.06 \times$ MUN (mg/dl); $R^2 = 0.79$ (Bonanno *et al.*, 2007a). Therefore, similar to sheep and cattle, milk urea measurements can be an easy and useful tool for detecting an excess of protein concentration in the diet selected at pasture, and thus balancing the ration of grazing goats through adequate supplementation.

In many areas, pasture is the principal component of the diet for goats. It has two important roles: (i) to supply fresh herbage at low cost, in comparison to the other forage resources; and (ii) to guarantee the quality of the typical and traditional productions. Considering that the grazed herbage is often not able to meet the requirements of lactating goats, the supplementation with concentrates is necessary. In many cases, the supplementation of no more than 600 g of concentrates per day per head is sufficient to cover the requirements of grazing lactating Mediterranean or indigenous goats without worsening the quality of their products (see Fedele, Chapter 5, this volume).

References

Abbadessa, V., Trabalza Marinucci, M. and Cavallucci, C. (2000) Behavioural patterns of free-ranging Angora goats under different stocking rates. *Options Méditerranéennes* 52, 151–154.

Animut, G., Goetsch, A.L., Aiken, G.E., Puchala, R., Detweiler, G., Krehbiel, C.R., Merkel, R.C., Sahlu, T., Dawson, L.J., Johnson, Z.B. and Gipson, T.A. (2005a) Grazing behaviour and energy expenditure by sheep and goats co-grazing grass/forb pastures at three stocking rates. *Small Ruminant Research* 59, 191–201.

Animut, G., Goetsch, A.L., Aiken, G.E., Puchala, R., Detweiler, G., Krehbiel, C.R., Merkel, R.C., Sahlu, T., Dawson, L.J., Johnson, Z.B. and Gipson, T.A. (2005b) Performance and forage selectivity of sheep and goats co-grazing grass/forb pastures at three stocking rates. *Small Ruminant Research* 59, 203–215.

Avondo, M., Pagano, R., Di Grigoli, A., Bonanno, A., Formoso, B., Giambalvo, D. and Scarpello, C. (2004) Comportamento alimentare di capre Girgentane: effetto del tipo di pascolo e del livello proteico del supplemento. In: *Proceeding of the 16th SIPAOC Congress* (Atti del XVI Congresso Nazionale della Società Italiana di Patologia e di Allevamento degli ovini e dei caprini), Siena, Italy, p. 232 (Abstract). (Full paper on CD-ROM, 1–4).

Barroso, F.G., Alados, C.L. and Boza, J. (2000) Social hierarchy in the domestic goat: effect on food habitus and production. *Applied Animal Behaviour Science* 69, 35–53.

Barry, T.N. and Duncan, S.J. (1984) The role of condensed tannins in the nutritional value of *Lotus pedunculatus* 1. Voluntary intake. *British Journal of Nutrition* 51, 485–491.

Baumont, R., Parche, S., Meuret, M. and Morand-Fehr, P. (2000) How forage characteristics influence behaviour and intake in small ruminants: a review. *Livestock Production Science* 64, 15–28.

Bayer, W., Suleiman, H., Kaufmann, R. and Water-Bayer, A. (1987) Resource use and strategies for development of pastoral systems in sub-humid West Africa. The case of Nigeria. *Quarterly Journal of International Agriculture* 26, 58–71.

Bonanno, A., Di Grigoli, A., Alicata, M.L., Avondo, M., Pagano, R., De Vita, G., Stringi, L. and Giambalvo, D. (2004) Ingestione al pascolo e produzione di latte di capre Girgentane in funzione della combinazione spaziale delle specie foraggere e dell'integrazione con concentrato. In: *Proceeding of the 16th SIPAOC Congress* (Atti del XVI Congresso Nazionale della Società Italiana di Patologia e di Allevamento degli ovini e dei caprini), Siena, Italy, p. 241 (Abstract). (Full paper on CD-ROM, 1- 4.)

Bonanno, A., Di Grigoli, A., Alicata, M.L., Tornambè, G., Avondo, M., Pagano, R., Giambalvo, D., Stringi, L. and Di Miceli, G. (2007d) Effect of stocking rate on selective behaviour and milk production of Girgentana goats grazing a ryegrass and berseem clover mixture. *Options Méditerranéennes* Série A: Séminaires Méditerranéennes, 74, 351–357.

Bonanno, A., Todaro, M., Di Grigoli, A., Tornambè, G. and Alicata, M.L. (2007a) Dietary factors associated with the concentration of milk urea nitrogen in grazing goats. In: *Proceedings of the 5th International Symposium FIL–IDF on 'The Challenge to Sheep and Goats Milk Sectors'*, Alghero, Italy, p. 58 (Abstract).

Bonanno, A., Di Grigoli, A., Stringi, L., Di Miceli, G., Giambalvo, D., Tornambè, D., Vargetto, D. and Alicata, M.L. (2007b) Intake and milk production of goats grazing sulla forage under different stocking rates. *Italian Journal of Animal Science* 6(Suppl. 1), 605–607.

Bonanno, A., Di Grigoli, A., Vargetto, D., Tornambè, D., Di Miceli, G. and Giambalvo, D. (2007c) Grazing sulla and/or ryegrass forage for 8 or 24 hours daily. 1. Effects on ewes feeding behaviour. In: *Proceedings of the 14th Symposium of the European Grassland Federation (EGF)*, Ghent, Belgium. *Grassland Science in Europe* 12, 208–211.

Bordi, A., De Rosa, G., Napoletano, F., Vesce, G. and Rubino, R. (1994) Influence of feeding supplementation on goats grazing behaviour. *Options Méditerranéennes* 50, 57–68.

Cavallero, A. and Ciotti, A. (1991) Aspetti agronomici dell'utilizzazione dei prati e dei pascoli. *Rivista di Agronomia* 25, 81–126.

Claps, S., Rubino, R., Cogliandro, E. and Fedele, V. (1994) Fabbisogni alimentari e nutritivi, comportamento alimentare e produzione di latte di capre allevate al pascolo con differenti integrazioni alimentari. In: Rubino, R. (ed.) *Atti Giornata Scientifica Miglioramento dell'efficienza produttiva degli ovini e dei caprini*. Ars Grafica, Villa D'Agri, Italy, pp. 1–16.

Claps, S., Fedele, V., Sepe, L. and Di Trana, A. (2000) Effect of concentrate grain source and herbage intake on diet degradability and milk production in grazing goats. In: *7èmes Rencontres Recherches Ruminants*. Institut National de la Recherche Agronomique, Paris, pp. 319–322.

Claps, S., Donato, R., Fedele, V., Morone, G. and Rosati, A. (2001) *L'allevamento caprino nel Parco Nazionale del Cilento e Vallo di Diano*. Ars Grafica, Villa D'Agri, Italy, p. 62.

Del Pozo, M., Wright, A., Whyte, T.K. and Colgrove, P.M. (1996) Effects of grazing by sheep or goats on sward composition in ryegrass/white clover pasture and on subsequent performance of weaned lambs. *Grass and Forage Science* 51, 142–154.

Del Pozo, M., Wright, A. and Whyte, T.K. (1997) Diet selection by sheep and goats and sward composition changes in a ryegrass/white clover sward previously grazed by cattle, sheep or goats. *Grass and Forage Science* 52, 278–290.

De Rosa, G., Moio, L., Napoletano, F., Grasso, F., Cubitosi, L. and Bordi, A. (2002) Influence of flavor on goat feeding preferences. *Journal of Chemical Ecology* 28, 269–281.

Di Grigoli, A., Bonanno, A., Alabiso, M., Brecchia, G., Russo, G. and Leto, G. (2003) Effects of housing system on welfare and milk yield and quality of Girgentana goats. *Italian Journal of Animal Science* 2(Suppl. 1), 542–544.

Fedele V. (1994) L'alimentazione dei caprini. In: Roberto, R. (ed.) *Atti Giornata Divulgativa Miglioramento dell'efficienza produttiva degli ovini e dei caprini*. Ars Grafica, Villa D'Agri, Italy, pp. 1–21.

Fedele, V. (1996) L'alimentazione. In: Rubino, R. (ed.) *L'Allevamento Caprino*. Associazione Nazionale della Pastorizia, Rome, pp. 51–101.

Fedele, V. (2001a) Alimentazione, tra pascolo e integrazione. *Caseus* 3, 36–43.

Fedele, V. (2001b) Un'alimentazione secondo natura. In: *Il Caciocavallo Podolico e la Manteca: un grande formaggio del Sud*. Associazione Nazionale per la Tutela Valorizzazione dei Formaggi Sotto il Cielo, Potenza, Italy, pp. 35–44.

Fedele, V., Zarriello, G. and Di Trana, A. (1988) Osservazioni preliminari sul pascolamento di capre di ceppo locale. *Agricoltura Ricerca* 81, 63–68.

Fedele, V., Pizzillo, M., Claps, S., Morand-Fehr, P. and Rubino, R. (1993) Grazing behaviour of goats on native pasture in Southern Italy. *Small Ruminant Research* 11, 305–322.

Fedele, V., Rubino, R., Claps, S. and Sepe, L. (1996) Effect of plant characteristic, herbage availability and quality on grazing goat intake. In: *Proceedings of the 6th International Conference on Goats*. International Academic Publishers, Beijing, Vol. 2, pp. 608–611.

Fedele, V., Rubino, R. and Claps, S. (1997) Produzione foraggera, utilizzazione dell'erba e 'performances' produttive di capre allevate con differenti sistemi alimentari. *Rivista di Agronomia* 31, 333–337.

Fedele, V., Signorelli, F., Brancaleoni, E., Ciccioli, P. and Claps, S. (2000) Effect of concentrate grain source and herbage intake on physical chemical features and milk aroma in grazing goats. In: *Proceedings of the 7th International Conference on Goats*, Tours, France. Institut de l'Elevage and Institut National de la Recherche Agronomique, Paris, Vol. 1, pp. 152–154.

Fedele, V., Claps, S., Rubino, R., Calandrelli, M. and Pilla, A.M. (2002) Effect of free-choice and traditional feeding systems on goat feeding behaviour and intake. *Livestock Production Science* 74, 19–31.

García, M.A., Aguilera, J.F. and Molina Alcaide, E. (1995) Voluntary intake and kinetic of degradation and passage of unsupplemented and supplemented pastures from semiarid lands in grazing goats and sheep. *Livestock Production Science* 44, 245–255.

Gutman, M. and Seligman, N.G. (1979) Grazing management of herbaceous Mediterranean foothill range in the Upper Jorgan Valley. *Journal of Range Management* 32, 86–92.

Harmeyer, J. and Martens, H. (1980) Aspects of urea metabolism in ruminants with reference to the goat. *Journal of Dairy Science* 63, 1707–1728.

Harrington, G.N. (1982) Grazing behaviour of the goat. In: *Proceedings of the 3rd International Conference of Goat Production and Disease*, Tucson, Arizona. Dairy Goat Journal Publishing, Scottsdale, Arizona, pp. 398–403.

Hart, R.H., Samuel, M.J., Test, P.S. and Smith, M.A. (1988) Cattle, vegetation, and economic responses to grazing systems and grazing pressure. *Journal of Range Management* 41, 282–286.

Iason, G.R., Mantecon, A.R., Sim, D.A., Gonzalez, J., Foreman, E., Bermudez, F.F. and Elston, D.A. (1999) Can grazing sheep compensate for a daily foraging time constraint? *Journal of Animal Ecology* 68, 87–93.

Le Frileux, Y., Le Mens, P. and Guérin, B. (2000) Influence de la mise l'herbe et de la nature azotée du concentré sur la transformation fromagére. In: *Proceedings of the National Conference of Cheese Makers*, Digne, France, pp. 1–8.

Kronberg, S.L. and Malecheck, J.C. (1997) Relationships between nutrition and foraging behaviour of free-ranging sheep and goats. *Journal of Animal Science* 75, 1756–1763.

Malecheck, J.C. (1982) Grazing management of goats in extensive rangeland production systems. In: *Proceedings of the 3rd International Conference of Goat Production and Disease*, Tucson, Arizona. Dairy Goat Journal Publishing, Scottsdale, Arizona, pp. 404–407.

Masson, C., Rubino, R. and Fedele, V. (1991) Forage utilization in goats. In: Morand-Fehr P. (ed.) *Goats Nutrition*. Pudoc, Wageningen, The Netherlands, pp. 145–159.

Mellado, M., Valdez, R., Lara, L.M. and García, J.E. (2004) Risk factors involved in conception, abortion, and kidding rates of goats under extensive conditions. *Small Ruminant Research* 55, 191–198.

Milne, J.A., Maxwell, T.J. and Souter, W.G. (1981) The effect of supplementary feeding and herbage mass on the intake and performance of grazing ewes in early lactation. *Animal Production* 32, 185–195.

Min, B.R., Hart, S.P., Sahlu, T. and Satter, L.D. (2005) The effects of diets on milk production and composition and on lactation curves in pastured dairy goats. *Journal of Dairy Science* 88, 2604–2615.

Molle, G., Decandia, M., Ligios, S., Fois, N., Treacher, T.T. and Sitzia, M. (2004) Grazing management and stocking rate with particular reference to the Mediterranean environment. In: Pulina, G. (ed.) *Dairy Sheep Nutrition*. CAB International, Wallingford, UK, pp. 191–211.

Morand-Fehr, P., Fedele, V., Schmidely, P. and Rubino, R. (1993) Relation between chemical composition of pasture plants and dietary preferences of goats. In: *Workshop Report on 'Sheep and Goat Nutrition'*, Aristotle University, Thessaloniki, Greece. Commission of the European Communities, Luxemburg, pp. 1–5.

Morand-Fehr, P., Fedele, V., Decandia, M. and Le Frileux, Y. (2007) Influence of farming and feeding systems on composition and quality of goat and sheep milk. *Small Ruminant Research* 68, 20–34.

Njwe, R.M., Ikwuegbu, O.A., Taratali, G. and Little, D.A. (1995) Effect of stocking rate on the botanical composition and nutritive value of diets selected by West African dwarf goats maintained on researcher-managed and farmer-managed stylo fodder banks during the cropping season in Central Nigeria. *Animal Feed Science and Technology* 51, 317–328.

Orihouela, A. and Solano, J.J. (1999) Grazing and browsing times of goats with three levels of herbage allowance. *Applied Animal Behaviour Science* 61, 335–339.

Osoro, K. and Del Pozo, M. (1996) Effect of different grazing strategies on performance of sheep and goats in perennial ryegrass-white clover pasture. In: *Proceedings of the 6th International Conference on Goats*. International Academic Publishers, Beijing, Vol. 2, pp. 601–603.

Penning, P.D., Newman, J.A., Parsons, A.J., Harvey, A. and Orr, R.J. (1997) Diet preferences of adult sheep and goats grazing ryegrass and white clover. *Small Ruminant Research* 24, 175–184.

Pizzillo, A., Di Trana, A. and Fedele, V. (1988) Comportamento alimentare al pascolo di capre Maltesi e Locali. *Agricoltura Ricerca* 81, 69–78.

Provenza, D.F. (1995) Postingestive feedback as an elementary determinant of food preference and intake in ruminants. *Journal of Range Management* 48, 2–17.

Provenza, D.F. and Malecheck, J.C. (1986) A comparison of food selection and foraging behaviour in juvenile and adult goats. *Applied Animal Behaviour Science* 16, 49–61.

Pulina, G., Cannas, A., Serra, A. and Vallebella, R. (1991) Determinazione e stima del valore energetico di latte di capre di razza Sarda. In: *Proceeding of the 45th SISVet Congress* (Atti della Società Italiana delle Scienze Veterinarie, Vol. 45), Altavilla Milicia, Palermo, Italy, pp. 1779–1781.

Reed, J.D. (1995) Nutritional toxicity of tannins and related polyphenols in forage legumes. *Journal of Animal Science* 73, 1516–1528.

Reyneri, A. (1989) Analisi della struttura epigea ed effetti della utilizzazione di graminacee foraggere sottoposte a pascolamento continuo e a rotazione. *Rivista di Agronomia* 23, 297–305.

Romney, D.L., Sendalo, D.S.C., Owen, E., Mtenga, L.A., Penning, P.D., Mayes, R.W. and Hendy, C.R.C. (1996). Effects of tethering management on feed intake and behaviour of Tanzanian goats. *Small Ruminant Research* 19, 113–120.

Rubino, R., Pizzillo, M. and Scardella, P. (1988) Confronto tra diversi sistemi di pascolamento in un modello intensivo con capre Maltesi. *Agricoltura Ricerca* 81, 79–86.

Rubino, R., Moioli, B., Fedele, V., Pizzillo, M. and Morand-Fehr, P. (1995) Milk production of goats grazing native pasture under different supplementation regimes in Southern Italy. *Small Ruminant Research* 17, 213–221.

Tisserand, J.L. (2003) Comparison of the blood urea recycling in sheep and goat. In: *Book of Abstracts of the 54th Annual Meeting of the European Association for Animal Production, No. 9*, Rome, p. 150. Wageningen Academic Publishers, Wageningen, The Netherlands.

Tsiouvaras, C.N., Nastis, A., Papachristou, T., Platis, P. and Yiakoulaki, M. (1999) Kermes oak shrubland availability and grazing pressure responses as influenced by stocking rate and grazing system. *Options Méditerranéennes* 27, 155–164.

Van Quackebeke, E., Lefrileux, Y. and Pommaret, A. (1997) Comparison de deux modes d'exploitation de prairies à base de graminées au printemps par les chévres laitières. In: *4èmes Rencontres Recherches Ruminants*. Institut National de la Recherche Agronomique, Paris, p. 154 (Abstract).

10 Feeding Management of Dairy Goats in Intensive Systems

L. Rapetti and L. Bava

Dipartimento di Scienze Animali, Università degli Studi di Milano, Via Celoria 2, 20133 Milan, Italy

Introduction

Goats change their feeding behaviour according to the availability of forages and concentrates and they are more versatile than other domestic ruminants. For these reasons, goats can adapt to different conditions, ranging from desert to grassland.

The mouth of goats is similar to that of wild intermediate feeders, so they are efficient at picking their food, since the upper lip is mobile and this helps the tongue activity. Daily ruminal activity of goats is intermediate between that of wild selective ruminants, which have a simpler rumen and eat and ruminate frequently, and grass eaters, which spend less time feeding (Van Soest, 1994).

Goats have a greater ability to reduce feed particle size during chewing in comparison with sheep. Indeed, a comparative study showed that the total number of ingestive jaw movements (eating) did not differ between species but goats had finer particle size in their ingestive boli than sheep (Hadjigeorgiou *et al.*, 2003).

The chewing activity involves a reduction of feed particle size and, consequently, an increase in the surface area for rumen microbial attack.

In the rumen of goats there is often high ammonia concentration even when diets poor in proteins are used; this may be due to: (i) low water intake, which limits ammonia dilution; and (ii) good efficiency in recycling urea from the blood to the rumen, due to the high permeability of rumen epithelium to urea. The amount of recycled urea varies from 0.4 to 9.7 g/day, depending on N intake and animal physiological status, and can provide 20% of the rumen-degradable protein requirements of goats (Brun-Bellut *et al.*, 1991).

Goats' feeding activity is highly selective. Goats when grazing grassland prefer plants with high dry matter (DM) and protein content and when fed in mangers are able to be very selective, tasting the feed and choosing first the most palatable parts. In particular, goats have a preference for concentrates, thus decreasing forage intake when concentrates are available and showing great ability in avoiding rumen acidosis (Abijaoudé *et al.*, 2000).

It is well known that goats have a great capacity to utilize the fibre components of diets. When fed low-quality diets (high in fibre and low in protein content), they have higher DM intake and fibre digestibility than sheep (Domingue *et al.*, 1991). The differences in digestive capacity between goats and sheep narrow when diets are characterized by high content of good-quality forage and good nutrient balance.

In conclusion, when considering the goat as an 'intermediate feeder', goats: (i) are good selective feeders; (ii) have an efficient chewing and rumen activity; (iii) are able to efficiently utilize diets rich in fibre but also diets rich in concentrates and starch, being quite resistant to rumen acidosis; and (iv) can tolerate low water intake. In addition, goats can adapt to either poor pastures or rich and balanced diets. For these reasons, goats can be reared not only in extensive systems but also in intensive farm systems, where usually specialized dairy goat breeds are utilized.

All of these characteristics of goats should be taken in consideration when formulating diets: goats should not be considered as 'small cattle'.

Feeding Strategies

The choice among the different feeding strategies for goats in intensive farms is based on flock size and level of intensification of the farm. It is necessary to reach a compromise between managerial choices and nutrient requirements of goats. For example, utilization of a total mixed ration (TMR) can be the best choice to satisfy the nutrient requirements of highly productive flocks. However, it can be too expensive for small and medium-sized farms, especially due to the high cost of the TMR mixer.

The aim of the different feeding systems for goats is to satisfy nutrient requirements and, consequently, to guarantee good health, reproduction and production of the animals. However, in order to achieve that, many problems have to be solved, especially regarding feed supply techniques, considering the economic and managerial aspects.

Two different situations are discussed: (i) small farms where goat diets are composed of hay and concentrates; and (ii) large farms where the feeding system is based on a TMR (Fig. 10.1).

Traditional feeding systems

In this case, the techniques and the timing used to supply hay, silages and concentrates and the quality of the feeds have a fundamental effect on the response of the animals.

In this system it is common that concentrate and raw materials come from the market while hay and silage are normally produced on the farm. In many cases it is possible that the nutrient composition of concentrate does not

Fig. 10.1. Example of a total mixed ration with maize silage.

complement that of hay properly. As a consequence, the final ration does not fully satisfy the nutrient requirements of goats. Since the animal feed industry offers different types of concentrate, especially with different crude protein (CP) contents, it is not difficult to choose proper feedstuffs for the farm but it is important to adapt the protein content of the ration to the protein requirements of goats. If the N content of the ration is too high, milk urea and urine N contents increase, without increasing milk protein, as demonstrated by Bava *et al.* (2001) and Arieli *et al.* (2005).

Hay should be fed before concentrates, in order to avoid an excessive decrease of rumen pH, due to fast rumen degradability of non-fibre carbohydrates (NFC) of the concentrates. Moreover, it is better to offer small daily meals of concentrate, to avoid a drop in pH, especially when low-quality forages are offered and when high doses of concentrate need to be given to achieve high milk production. It is advisable to use concentrates as pellets, in order to avoid selection and pulmonary disease caused by dust that can occur when the particle size of the feedstuffs is too small. It could be useful to include in the diet a pellet formulated for protein, mineral and vitamin supplementation, and maize as an energy source to reduce the cost of the ration. Maize, as reported below, can be given as whole grain without a detrimental effect on DM digestibility.

Among the forages, the utilization of lucerne as hay or pellets is very common in temperate areas. This forage is characterized by higher protein content (more than 16% CP on a DM basis) and lower neutral-detergent fibre (NDF) concentration than permanent pasture hay. Moreover, it has a high content of soluble fibre with high degradability, which, together with its low NDF content, has a positive effect on rumen fill and DM intake. For all these reasons, lucerne is considered a valuable forage for feeding goats.

Total mixed rations

The well-known advantages of TMRs are that: (i) all ingredients are weighed and mixed together; (ii) the ration can be easily balanced to cover the nutrient requirements of the flock and actual intake and refusals can be easily monitored; (iii) the composition of the ration tends to be constant in time; and (iv) selection by the animals and dustiness are reduced when silages or wet raw material are included. In the case of TMRs made only of dry ingredients, water could be added to avoid the dustiness; if the ration is not prepared every day, water can be replaced by molasses, useful for palatability and also to increase the sugar content of the diet without increasing excessively the humidity and guaranteeing an optimal conservation of the TMR.

Complete diets ensure that goats receive the proper proportion of carbohydrates and N at each meal (Giger-Reverdin and Sauvant, 1991). Moreover, the TMR masks unpalatable feeds, such as certain by-products, it is easily managed and its preparation can be completely mechanized.

Maltz *et al.* (1991) reported that a TMR feeding system was superior to feeding the different dietary ingredients separately in terms of daily milk yield, milk fat and milk protein content, and feed conversion ratio of dairy goats.

Finally, a TMR can improve rumen function, due to the advantages associated with simultaneous feeding of roughage and concentrate. In fact, when TMR and traditional diets were compared, Saanen lactating goats fed TMR had higher rumen acetate/propionate ratio than animals fed traditional forage plus concentrate diets (acetate/propionate ratio: 3.5 versus 2.7, respectively). Moreover, ruminal pH was significantly higher for the TMR diet than for the traditional one in the samplings performed before feeding and 2 and 5 h after feeding: 6.67, 6.5, 6.55 versus 6.32, 5.98, 5.95 for TMR and traditional diet, respectively (Rapetti, unpublished data).

The use of a TMR system normally requires high investments for mixer wagon and distribution equipment (e.g. tractor, silos and feed driveway). These investments are justified only on large farms where many goats are reared.

Usually the principal ingredient of TMR is silage, for both cattle and goat rations.

When the TMR is a mixture of dry ingredients without silages, it can be purchased directly from the market or prepared by the farmer. The TMRs available on the market might be convenient for farms where not enough forages can be cultivated. The use of commercial TMRs can encourage the implementation of this advantageous feeding system without requiring high additional investments in machinery, labour and so on. However, the disadvantage is the high and variable price of TMR diets purchased directly from the market.

With the use of dry TMRs, special attention must be paid to the particle size of the feeds used, especially for hay. Goats are able to select large particles (grain, pellets and forages) and less-small particles (dust, meals and broken pellets), in particular when the diet offered is abundant (Fernàndez and Sànchez-Seiquer, 2003).

Roughage for Dairy Goats in Intensive Systems

Like all ruminants, the goat is primarily a forage consumer. Forages are the principal source of fibre but they might not completely satisfy the nutrient requirements of high-producing lactating goats. Many authors have indicated that high-quality forages could satisfy the needs for production to 2.5–3 kg of daily milk yield. To support milk yield over 3 kg/day it is necessary to complement a forage-based diet with proper amounts of concentrates (Masson *et al.*, 1991), mainly to cover energy requirements but also protein, mineral and vitamin requirements.

The forage/concentrate ratio is an important factor when balancing diets. As an example, the effects of different forage/concentrate ratios on DM intake and milk yield and quality of Saanen goats in mid-lactation are shown in Table 10.1 (Rapetti *et al.*, 1997). Diets with the following forage/concentrate ratios were fed *ad libitum* in a Latin square design: 70:30 (F70), 50:50 (F50), 30:70 (F30). Permanent pasture hay was produced on the farm, while concentrate was purchased from the market. Dietary NDF concentration was 41.5, 47.9 and 54.4% on a DM basis for the diets F30, F50 and F70, respectively. The DM intake of the high-concentrate diet (F30) was about 11 and 21% higher than that of F50 and F70 diets, respectively. The higher intake of concentrate increased milk production but did not reduce milk fat content. These data confirm that goats are less sensitive than cows to diets with low fibre content in terms of milk fat depression (Calderon *et al.*, 1984). These results are also in agreement with those of Sauvant *et al.* (1987), who concluded that for diets with forage/concentrate ratio higher than 20:80, the goat's nutritional energy status is a more important factor than the relative proportion of the forage in the ration. It is important to highlight that the milk fat/protein ratio was lower than one for all tested diets (Table 10.1). This was probably due to the genetic characteristics of the goats used and the climatic conditions (beginning of the summer) during the trial, characterized by high temperatures and high humidity, which decrease milk fat production. The renneting properties of the milk and curd firmness were not significantly affected by the forage/concentrate ratio of the diets.

Milk efficiency of the diets, expressed as the ratio of milk yield to DM intake, was highest for the F50 diet (1.40) and similar between F70 and F30 diets (1.37 and 1.38, respectively). The milk efficiency of the F70 diet was mainly due to the low energy concentration of this diet, while that of the F30 diet was due to the conversion of part of the metabolizable energy into body fat instead of milk, as resulted from the energy balances determined by means of indirect calorimetry. The results reported in Table 10.1 are partially in agreement with those of the experiment reported by Goetsch *et al.* (2001), in which late-lactation Alpine goats, fed diets with increasing amounts of concentrates (from 20 to 65% of concentrates), had higher milk yields when fed a diet with 50% concentrates.

Table 10.1. Effects of different forage/concentrate ratios on dry matter (DM) intake, milk yield, milk quality and renneting properties of Saanen goats. (Adapted from Rapetti *et al.*, 1997.)

Item	Diet[a]			SE
	F30	F50	F70	
DM intake (g/day)	2535[A]	2318[B]	2091[C]	19.6
Milk yield (g/day)	3508[A]	3248[AB]	2868[B]	46.0
Milk fat content (%)	2.62	2.66	2.55	0.04
Milk fat yield (g/day)	85.0[A]	78.7[AB]	69.5[B]	1.1
Milk protein content (%)	2.90	2.81	2.80	0.07
Milk protein yield (g/day)	92.5[A]	85.6[AB]	75.6[B]	1.2
Coagulation time (min)	15.7	14.4	15.9	1.0
Renneting time (min)	7.7	7.1	8.2	0.5
Curd firmness (min)	21.8	22.7	21.9	0.3

SE, standard error.
[A,B,C]Mean values in a row with unlike superscript letters were significantly different ($P \leq 0.01$).
[a]Forage/concentrate ratio of the diets: 30:70 (F30); 50:50 (F50); 70:30 (F70).

Based on the results of both trials reported above (Rapetti *et al.*, 1997; Goetsch *et al.*, 2001), it is possible to conclude that in mid–late lactation diets too rich in concentrates may not influence milk production positively, and may also cause an excessive fattening of dairy goats, with the risk of metabolic disorders during the transition period. Therefore, since it is important to monitor body fattening during productive and reproductive stages of the animals, the evaluation of body condition score should be more often practised in goat farms.

The quality of forages is very important in intensive systems, and should be carefully evaluated especially when they are purchased from the market. If forage quality is poor, a large amount of concentrate needs to be supplied in the diet. This increases not only feeding costs, but also the risk of metabolic disorders.

The importance of forage quality in intensive feeding systems was studied in a trial in which maize silage and Italian ryegrass silage were compared in lactating Saanen goats (Crovetto *et al.*, 1994). Intake of ryegrass silage was lower than that of maize silage, probably due to its high humidity content. However, since fibre digestibility of ryegrass silage was higher than that of maize silage, its nutritive value was high. As a consequence, the two diets caused similar effects on milk production.

Also, autumn–winter cereals silages could be good products for dairy goats because of their high productivity in terms of quantity and quality. It is important to remember that the correct stage to harvest whole-crop silage is when the grain has reached its full size and weight, but before it becomes hard, and this coincides with dough and milk-dough stages (D'Urso *et al.*, 1989). At these stages the DM of plants is greater than 30%; by contrast, if the DM is lower than 30%, the forage needs to be pre-wilted before ensiling.

Importance of Feed Physical Form

The physical form of dietary particles is very important for the digestive function of ruminants, because it influences: (i) chewing activity and salivary production; (ii) rumen microbial activity; (iii) together with particle density, rumen feed passage rate; and (iv) milk production and milk fat percentage.

Fibre sources, like roughages, grains or vegetable by-products, influence the particle size of the diet. To underline the importance of fibre functions some authors have defined the effectiveness of a dietary fibre source as the ability to stimulate chewing activity and to maintain milk fat percentage and fat-corrected milk production (Grant, 1997).

Particle size is defined by the length and diameter of particles. Length plays a role in rumination and feed intake. Welch (1982) demonstrated that particles that were too long did not pass from the rumen and reduced hay intake permanently. By contrast, finely ground forage particles, from hay or silage, decreased rumination time and ruminal pH and increased ruminal propionate production (Grant *et al.*, 1990), due to a reduction of chewing activity (Kuehn *et al.*, 1997; Andrighetto *et al.*, 1998).

Particle size can be measured by either dry-sieving or wet-sieving (sieving associated with water sprayed on the sieves), through calibrated screens arranged by decreasing mesh size (screen size) from top to bottom. Wet-sieving tends to sort particles according to length, while the dry method tends to sort them according to cross-sectional diameter. Since the rumen is a wet system and due to the fact that soaking the feed in water dissolves most of the finely divided non-cell-wall matter, wet methods of sizing particles in feeds should be favoured (Van Soest, 1994).

Feed particle size can also be measured *in vivo* by quantifying the chewing activity of the animals with: (i) the visual observation method, with fixed observation intervals (e.g. at 5 min); and (ii) electronic measurements. The automatic electronic devices utilize a bit recorder that identifies different jaw movements: eating, ruminating, swallowing and accessory movements. By using specific software, it is then possible to calculate the number of chews, the total time for each activity (eating and rumination), the number of boli, the number of remastications per bolus, etc. A comparative study between goats and sheep, in which chewing activity was measured using electronic equipment, showed that goats were more able to reduce long forage particles into smaller particles than sheep. Moreover, while both species had similar faecal particle size, goats spent less time per day masticating and performed fewer chews, during both ingestion and rumination, than sheep, demonstrating goats to have a more efficient chewing activity (Hadjigeorgiou *et al.*, 2003).

Unprocessed cereal grains and by-products

Besides roughages, particle size is also influenced, to a lesser extent, by whole grains and vegetable by-products, defined as non-forage fibre sources by Grant (1997).

Cereal grains are the main source of starch in ruminant diets and can be supplied in different physical forms (i.e. meal, rolled, flaked, expanded) depending on the mechanical and thermal treatments they receive. The aim of these treatments is to increase rumen microbial degradation rate on the starch present inside the kernels, which is the main energy source for microbes in cereal grains. It is well known that if treated cereal grains are supplied in large amounts, there is an increased risk of rumen acidosis, with negative consequences such as: (i) growth of lactic bacteria; (ii) inhibition of cellulolytic bacteria; (iii) decrease of rumen fibre degradability; (iv) reduction of volatile fatty acids production; and (v) decrease of milk fat content. Therefore, it is important to regulate the inclusion of treated grains in the diets and moderate the physical treatments.

A good and cheaper alternative could be to include whole grains instead of processed grains in goat rations. Some studies (Nicholson *et al.*, 1971; Weston, 1974) indicate that cattle utilize whole grains only partially: their chewing activity is insufficient to break all kernels, so a great amount of them pass into the reticulo-rumen unfermented and are found in the faeces as whole grain. However, goats and sheep are more able to utilize whole grains compared with cattle (Ørskov, 1979).

To evaluate the importance of the physical form of grains, a digestibility trial was conducted in which lactating Saanen goats were fed either a diet with whole maize grain 800 g/day per head or a diet containing the same quantity of maize meal (Rapetti and Bava, 2004). Starch digestibility was lower for whole grain than for ground grain diets (94.2 versus 97.8%, $P < 0.05$), while DM intake and milk yield were higher for the whole grain diet (2295 versus 2162 g DM/day, $P < 0.10$, and 4570 versus 4075 g milk/day, $P < 0.05$, for whole grain and meal grain diets, respectively). Moreover, only a few whole kernels were found in the faeces of goats fed whole maize grain, probably due to the higher ruminating activity of these goats compared with the group fed the maize meal diet (328 versus 262 min/day, $P = 0.079$; Rapetti, unpublished data, 2003). In addition, the decrease of rumen pH observed 4 h after feeding, in relation to that measured at feeding time, was lower for the whole maize (–0.10) than for the maize meal (–0.52) based diets. The NDF digestibility was positively affected by the inclusion of the whole maize (44.8 versus 39.7%, for whole grain and meal diets, respectively, $P < 0.10$), confirming the results obtained by Ørskov and Fraser (1975) in sheep fed supplements made of either whole or processed barley.

Even if in many cases by-products are included in the ration to complement the supply of energy and protein given by roughages, they often have high fibre concentration. By-products, such as sugarbeet pulp or soybean hulls, have a slightly lower content of digestible energy than cereals, but due to their high NDF concentration, they positively influence rumen fermentations and pH, especially when diets rich in starch are used (Bava, 2000). By-products are convenient because they: (i) can be easily purchased from the market; (ii) are easy to store, move and supply; and (iii) generally do not require particular machinery or equipment.

Despite the similarity in NDF concentration, forages and by-products provide different amounts of effective fibre. In particular, by-products have lower particle size than forages, and this influences chewing activity. In addition, the fibre of certain by-products (beet pulps, soy hulls) has a very high degradation rate (Van Soest, 1994). Thus, diets with a high proportion of these by-products and low content of forages can be rich in fibre with high degradation rate and of small particle size. This would induce, at the same time, high passage rates, and consequently limited rumination and salivary production. In this regard, the literature on dairy cattle shows that it is possible to substitute up to 30–40% of total dietary NDF content with NDF from by-products (Beauchemin *et al.*, 1991; Grant, 1997; Zhu *et al.*, 1997).

The utilization of by-products or concentrates rich in fibre in substitution of roughages could be useful also for dairy goats, considering the high adaptability of this ruminant species to different diets. This hypothesis was tested by our research group on lactating Saanen goats in two experiments (Bava *et al.*, 2001; Rapetti *et al.*, 2005). In the first experiment, goats were fed two diets throughout the entire lactation: a traditional silage-based diet and a non-forage diet. In the second trial, goats were fed three diets during mid-lactation: non-forage diet, grass-based diet and hay-based diet. Feed particle size was determined with the wet-sieving method in both trials, and chewing activity was measured continuously with an electronic instrument in the second one. Tables 10.2 and 10.3 report the chemical analysis of diets used in the two experiments. The two non-forage diets contained by-products (cottonseed expeller, coconut meal, sugarcane molasses, sugarbeet pulp, grapecake skins and palmkernel meal) and whole seeds (cottonseed and maize grain). In experiment 1, the mean particle size of the non-forage diets was smaller than that of control diets (Fig. 10.2). In fact, only 22% of particles of this diet had diameters greater than 4 mm (Fig. 10.2). In the second trial the physically effective NDF was lower for the non-forage diet in comparison with the other diets: 31.1, 27.7 and 24.8% on DM basis for hay-based diet, fresh grass-based and non-forage diet respectively. Indeed, in this experiment the eating and ruminating activity were not significantly different between the non-forage diet and forage-based diets, probably due to the inclusion of whole seeds, sugarbeet pulp and cracked carob beans in the non-forage diets. In the first trial, the non-forage diet had a higher DM intake and lower digestibility than the control diet, probably due to its high content of fibre sources with small particle size (Table 10.2). By contrast, in the second study, the non-forage diet, characterized by a high NFC concentration, had a lower DM intake but higher digestibility than the hay-based diet (Table 10.4).

In the first trial, milk fat content was significantly higher for the non-forage diet (4.21%) than for the forage diet (3.63%) (average for the whole lactation), while milk yield was not significantly influenced by treatments (3548 versus 3357 g/day for forage and non-forage diet, respectively). In this case, the high milk fat concentration of the non-forage diet could partially be explained by its higher dietary fat concentration. By contrast, in the second experiment the milk fat content was lower for the non-forage than for the grass-based diet, with the hay-based diet having intermediate values (Table 10.4).

The results obtained in these experiments suggest that, in rations for high-producing lactating goats, part of forage fibre can be substituted with by-product fibre without negative effects on chewing and rumen activity and milk yield. These data are in agreement with those reported by other authors

Table 10.2. Experiment 1 on non-forage diets. Ingredients and chemical composition of non-forage and control diets used for an experiment carried out with Saanen goats during the whole lactation. (Adapted from Bava *et al.*, 2001.)

Item	Non-forage diet	Control diet
Ingredient (% of DM)		
Italian ryegrass silage		33.7
Maize silage		22.2
Soybean meal, solvent extract		14.6
Maize meal		29.3
Sunflower meal	33.0	
Cassava	20.0	
Coconut meal	16.9	
Whole cottonseed	8.5	
Cottonseed expeller	7.0	
Grapecake skins	6.0	
Palmkernel meal	3.0	
Sugarcane molasses	1.6	
Vitamin–mineral supplement	4.0	1.2
Chemical composition		
DM (% of as-fed)	91.5	46.1
CP (% of DM)	21.4	14.8
Ether extract (% of DM)	5.6	3.2
NDF (% of DM)	36.8	33.0
ADF (% of DM)	26.1	17.8
ADL (% of DM)	9.6	2.1
NFC (% of DM)	28.8	40.6

DM, dry matter; CP, crude protein; NDF, neutral-detergent fibre; ADF, acid-detergent fibre; ADL, acid-detergent lignin; NFC, non-fibre carbohydrates.

(Sanz Sampelayo *et al.*, 1998). In conclusion, when choosing forage substitutes for goats' diets, their particle size should be large enough to stimulate chewing activity and their acid-detergent fibre and acid-detergent lignin dietary contents should not be excessively high, in order to avoid negative effects on diet digestibility.

Nitrogen Feeding of Dairy Goats

Nitrogen metabolism in ruminants is different from that of other mammals, because of the essential contribution of ruminal microbes. In fact, microbial synthesis in the rumen provides most proteins supplied to the small intestine of ruminants, accounting for more than 50% of the total metabolizable protein supply. Moreover, microbial proteins are the main source of limiting essential amino acids, such as lysine and methionine, for the animal. The total amount of microbial protein flowing to the small intestine depends on the availability and efficiency of use of nutrients by ruminal bacteria (Bach *et al.*, 2005). Rumen N

Table 10.3. Experiment 2 on non-forage diets. Ingredients and chemical composition of grass-based, hay-based and non-forage diets of an experiment carried out with Saanen goats during mid-lactation. (Adapted from Rapetti *et al.*, 2005.)

Item	Grass-based diet	Hay-based diet	Non-forage diet
Ingredient (% of DM)			
Grass fresh	55.3		
Hay		55.2	
Soybean meal, solvent extract	10.3	10.3	12.0
Whole maize grain			9.6
Maize meal	29.0	29.1	9.8
Maize gluten meal	2.1	2.1	2.0
Whole cottonseed			11.3
Sugarbeet pulp			26.5
Cracked carob beans			25.8
Vitamin–mineral supplement	3.2	3.2	3.0
Chemical composition			
DM (% of as-fed)	54.1	89.2	90.0
CP (% of DM)	17.5	18.7	16.6
Ether extract (% of DM)	1.8	2.3	2.9
NDF (% of DM)	34.2	31.5	30.2
ADF (% of DM)	22.5	19.4	21.6
ADL (% of DM)	2.8	2.1	5.4
NFC (% of DM)	38.8	38.7	44.7

DM, dry matter; CP, crude protein; NDF, neutral-detergent fibre; ADF, acid-detergent fibre; ADL, acid-detergent lignin; NFC, non-fibre carbohydrates.

availability depends on the degradation rate of proteins, while the efficiency of use of N by bacteria depends on energy availability. If sufficient energy is available in the rumen, amino acids (derived from protein degradation) can be transaminated or used directly for microbial synthesis; if energy is limiting, amino acids are deaminated, and their carbon skeleton is then fermented (Bach *et al.*, 2005). Thus, ruminal microbial protein synthesis is largely affected by the supply of adequate amounts of rumen-fermentable carbohydrates to be used as an energy source. However, an excessive amount of carbohydrates in the diet can decrease rumen pH and negatively affect the growth of cellulolytic bacteria, dietary DM intake and milk production.

In highly producing ruminants, microbial protein can cover only part of the amino acid requirements. For this reason, as milk yield increases, it is necessary to increase the intestinal supply of amino acids from rumen-undegradable protein (RUP).

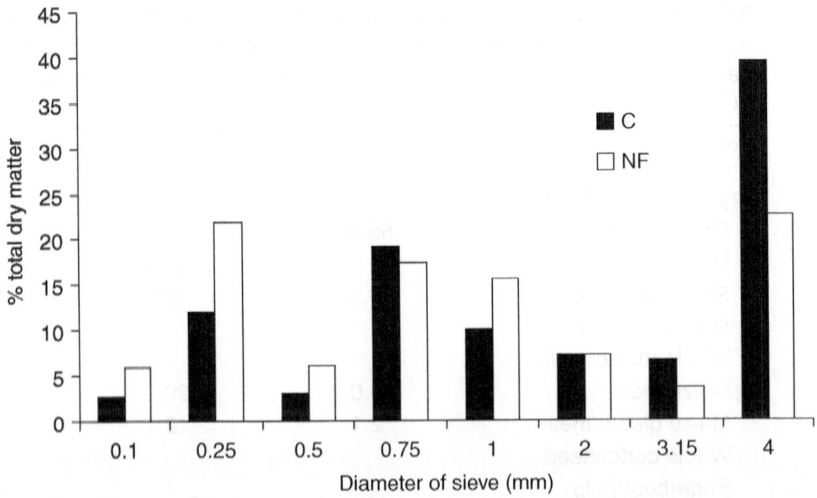

Fig. 10.2. Particle size of the diets (reported in Table 10.2) used in experiment 1 on non-forage diets in lactating Saanen goats (C, forage-based diet; NF, non-forage diet). (Adapted from Bava *et al.*, 2001.)

Table 10.4. Main results of the experiment 2 on non-forage diets (reported in Table 10.3). (Adapted from Rapetti *et al.*, 2005.)

Item	Grass-based diet	Hay-based diet	Non-forage diet	SE
DM intake (g/day)	2054	2354	2101	159.7
DM digestibility (%)	69.7	70.5	74.1	0.91
Milk yield (g/day)	3011[b]	3688[a]	3212[ab]	182.1
Milk fat (%)	3.37[a]	3.24[ab]	2.96[b]	0.09
Milk protein (%)	3.11	3.32	3.29	0.09
Casein (% of total N)	68.4[b]	70.7[ab]	73.7[a]	1.88
Milk urea N (mg/100 ml)	18.8[a]	18.6[a]	12.7[b]	0.82

SE, standard error; DM, dry matter.
[a,b]Least squares means within a row with different superscript letters were significantly different ($P < 0.05$).

However, increasing the supply of RUP does not assure an increase in animal performance, because it will be highly affected by the concentration, type and digestibility of the essential amino acids supplied by the RUP source.

Many studies have described the effects of the substitution of soybean meal with high-RUP sources, such as heat and chemically treated soybean meal, maize gluten meal, distillers' grains and brewers' grains, on the production of milk by dairy cows, as summarized by Santos *et al.* (1998). Their review showed that milk yield of RUP-rich diets was significantly higher than that of the soybean meal-based diet in only 17% of the comparisons. Positive effects of RUP on milk yield were obtained mostly using fishmeal and treated soybean meal.

Thus, the increase of RUP content of the diet might not have positive effects on milk yield in goats either, as demonstrated by some researchers (Brun-Bellut *et al.*, 1990; Lu *et al.*, 1990a,b) who used meat or bone meal as RUP sources. Similar results were registered in another experiment, carried out on Saanen goats in mid-lactation (97 days in milk, on average), with two diets with different RUP content (35 and 40% of CP), obtained with a partial substitution of soybean meal with treated canola meal. Milk yield (average 3454 g/day), milk CP (average 2.97%) and casein N (average 72.8% of total N) content did not differ between the two diets (Rapetti, unpublished data).

Information regarding the amino acid requirements of dairy goats is scarce. Semptey (1996), cited by Rousselot (1997), suggested including 2.2% of digestible methionine and 6.8% of digestible lysine, on the basis of total digestible protein (PDI), in the diets of goats. These values are lower than those reported for cows (2.5 and 7.3% for digestible methionine and lysine, respectively) (Rulquin *et al.*, 1993).

To reach high levels of digestible methionine and lysine in the diets for goats, in most cases it would be necessary to include rumen-protected amino acids. However, the addition of ruminal-protected amino acids does not guarantee positive effects, as observed in a specific experiment in which the addition of rumen-protected lysine and methionine did not influence milk yield and quality of Saanen goats (Tables 10.5 and 10.6) (Rapetti *et al.*, 1999). A possible

Table 10.5. Ingredients and chemical composition of a control diet (C) and a diet with lysine and methionine (C+AA) fed to Saanen goats. (Adapted from Rapetti *et al.*, 1999.)

	Diet	
Item	C	C+AA
Ingredients (% of DM)		
Italian ryegrass silage	48.4	48.1
Maize meal	26.7	26.5
Barley meal	11.1	11.0
Soybean meal	12.0	11.9
Rumen-protected methionine[a]	–	0.26
Rumen-protected lysine[a]	–	0.38
Mineral–vitamin supplement	1.8	1.8
Chemical composition		
CP (% of DM)	15.4	15.4
Ether extract (% of DM)	3.2	3.7
NDF (% of DM)	35.5	35.2
PDI (% of DM)	10.5	10.5
Intestinal digestible methionine (% of PDI)	1.82	2.35
Intestinal digestible lysine (% of PDI)	6.55	6.78

DM, dry matter; CP, crude protein; NDF, neutral-detergent fibre; PDI, intestinal digestible protein.
[a]The rumen-protected methionine and lysine supplements were constituted by 65% of a lipid matrix and 35% of amino acid. Effective rumen degradability (with a passage rate of 6%/h) was 22.2 and 65.5% for methionine and lysine, respectively.

Table 10.6. Dry matter (DM) intake and milk production of Saanen goats fed a control diet (C) and a diet with lysine and methionine (C+AA). (Adapted from Rapetti *et al.*, 1999.)

	Diet		
Item	C	C+AA	SE
DM intake (g/day)	2258	2231	27.7
Milk yield (g/day)	3754	3697	43.0
Fat (%)	2.58	2.65	0.04
Protein (%)	2.80	2.85	0.03
Casein N (% of total N)	74.6	75.0	0.29
NPN (% of total N)	10.7	10.8	0.32

SE, standard error; NPN, non-protein N.

explanation of these results could be related to the fact that mammary uptake of lysine and methionine seem to be affected not only by supply but also by the mammary synthetic capacity, since mammary tissue is capable of extracting these amino acids according to its needs independently of arterial supply (Madsen *et al.*, 2005).

Conclusions

Goats can easily adapt to intensive feeding systems. They can tolerate high amounts of concentrate rich in starch but also diets with high amount of forages, due to their efficiency in chewing and selecting the feeds of the diet. In intensive feeding systems, TMRs are advantageous to balance nutrient supply and to reduce feed selection. Moreover, goats are able to eat and efficiently utilize diets without forages, as long as the particle size of the ration and its fibre level are carefully balanced.

References

Abijaoudé, J.A., Morand-Fehr, P., Tessier, J., Schmidely, P. and Sauvant, D. (2000) Influence of forage:concentrate ratio and type of starch in the diet on feeding behaviour, dietary preferences, digestion, metabolism and performance of dairy goats in mid lactation. *Animal Science* 71, 359–368.

Andrighetto, I., Berzaghi, P. and Cozzi, G. (1998) Valorizzare le caratteristiche foraggiere del silomais mediante la scelta della lunghezza di trinciatura. *Farmer News* (special number), 44–46.

Arieli, A., Sasson-Rath, R., Zamwel, S. and Mabjeesh S.J. (2005) Effect of dietary protein and rumen degradable organic matter on milk production and efficiency in heat-stressed goats. *Livestock Production Science* 9, 215–223.

Bach, A., Calsamiglia, S. and Stern, M.D. (2005) Nitrogen metabolism in the rumen. *Journal of Dairy Science* 88(E Suppl.), E9–E21.

Bava, L. (2000) Studio di alcuni aspetti del metabolismo di capre in lattazione alimentate con una dieta priva di foraggi. PhD thesis, University of Pisa, Pisa, Italy.

Bava, L., Rapetti, L., Crovetto, G.M., Tamburini, A., Sandrucci, A., Galassi, G. and Succi, G. (2001) Effects of a nonforage diet on milk production, energy, and nitrogen metabolism in dairy goats throughout lactation. *Journal of Dairy Science* 84, 2450–2459.

Beauchemin, K.A., Farr, B.I. and Rode, L.M. (1991) Enhancement of the effective fiber content of barley-based concentrates fed to dairy cows. *Journal of Dairy Science* 74, 3128–3139.

Brun-Bellut, J., Blanchart, G. and Vignon, B. (1990) Effects of rumen-degradable protein concentration in diets on digestion, nitrogen utilization and milk yield by dairy goats. *Small Ruminant Research* 3, 575–581.

Brun-Bellut, J., Lindberg, J.E. and Hadjipanayiotou, M. (1991) Protein nutrition and requirements of adult dairy goats. In: Mohrand-Fehr, P. (ed.) *Goat Nutrition*. Pudoc, Wageningen, The Netherlands, pp. 82–93.

Calderon, I., De Peeters, E.J., Smith, N.E. and Franke, A.A. (1984) Composition of goat milk: changes within milking and effects of high concentrate diet. *Journal of Dairy Science* 67, 1905–1911.

Crovetto, G.M., Rapetti, L., Tamburini, A., Succi, G. and Galassi, G. (1994) Energy utilization of diets with hay, maize silage or Italian ryegrass silage as main forage in lactating goats. In: Aguilera, J.F. (ed.) *Energy Metabolism of Farm Animals*. Consejo Superior de Investigaciones Cientificas, Servicio de Publicaciones, Mojacar, Spain, pp. 175–178.

Domingue, B.M.F., Dellow, D.W. and Barry, T.N. (1991) Voluntary intake and rumen digestion of a low-quality roughage by goats and sheep. *Journal of Agricultural Science* 117, 111–120.

D'Urso, G., Avondo, M., Licita, G. and Sinatra, M.C. (1989) Composizione chimica, digeribilità e caratteristiche fermentative del triticale insilato a differenti stadi di vegetazione. *Agricoltura Mediterranea* 119, 470–477.

Fernàndez, C. and Sànchez-Seiquer, P. (2003) Feed intake and digestibility of total mixed ration fed Murciano-Granadina dairy goats. *Pakistan Journal of Nutrition* 2, 25–32.

Giger-Reverdin, S. and Sauvant, D. (1991) Evaluation and utilization of concentrates in goats. In: Mohrand-Fehr, P. (ed.) *Goat Nutrition*. Pudoc, Wageningen, The Netherlands, pp. 172–183.

Goetsch, A.L., Detweiler, G., Sahlu, T., Puchala, R. and Dawson, L.J. (2001) Dairy goat performance with different dietary concentrate levels in late lactation. *Small Ruminant Research* 41, 117–125.

Grant, R.J. (1997) Interactions among forages and nonforage fiber sources. *Journal of Dairy Science* 80, 1438–1446.

Grant, R.J., Colenbrander, V.F. and Mertens, D.R. (1990) Milk fat depression in dairy cows: role of particle size of alfalfa hay. *Journal of Dairy Science* 73, 1823–1833.

Hadjigeorgiou, I.E., Gordon, I.J. and Milne, J.A. (2003) Intake, digestion and selection of roughage with different staple lengths by sheep and goats. *Small Ruminant Research* 47, 117–132.

Kuehn, C.S., Linn, J.G. and Jung, H.G. (1997) Effect of corn silage chop length on intake, milk production, and rumen function in lactating dairy cows. In: *1997 Research Summaries*, pp. 83–84. US Dairy Forage Research Center, Madison, Wisconsin; available at http://www.dfrc.wisc.edu:16080/RS97_pdfs/FU6.PDF (accessed July 2007).

Lu, C.D., Potchoiba, M.J., Sahlu, T. and Fernandez, J.M. (1990a) Performance of dairy goats fed isonitrogenous diets containing soybean meal or hydrolyzed feather meal during early lactation. *Small Ruminant Research* 3, 425–434.

Lu, C.D., Potchoiba, M.J., Sahlu, T. and Kawas, J.R. (1990b) Performance of dairy goats fed soybean meal or meat and bone meal with or without urea during early lactation. *Journal of Dairy Science* 73, 726–734.

Madsen, T.G., Nielsen, L. and Nielsen, L.O. (2005) Mammary nutrient uptake in response to dietary supplementation of rumen protected lysine and methionine in late and early lactating dairy goats. *Small Ruminant Research* 56, 151–164.

Maltz, E., Silanikove, N., Karaso, Y., Shefet, G., Meltzer, A. and Barak, M. (1991) A note on the effects of feeding total mixed ration on performance of dairy goats in late lactation. *Animal Feed Science and Technology* 35, 15–20.

Masson, C., Rubino, R. and Fedele, V. (1991) Forage utilization in goats. In: Mohrand-Fehr, P. (ed.) *Goat Nutrition*. Pudoc, Wageningen, The Netherlands, pp. 145–159.

Nicholson, J.W.G., Garril, A.D. and Burgess, P.L. (1971) Loss digestible nutrients when ensiled barley is fed whole. *Canadian Journal of Animal Science* 51, 697–700.

Ørskov, E.R. (1979) Recent information on processing of grain for ruminants. *Livestock Production Science* 6, 335–347.

Ørskov, E.R. and Fraser, C. (1975) The effects of processing and barley-based supplements on rumen pH, rate of digestion and voluntary intake of dried grass in sheep. *British Journal of Nutrition* 34, 493–500.

Rapetti, L. and Bava, L. (2004) Effect of grinding of maize and level of starch on digestibility and lactation performance of Saanen goats. *South African Journal of Animal Science* 34(Suppl. 1), 85–88.

Rapetti, L., Tamburini, A., Crovetto, G.M., Galassi, G. and Succi, G. (1997) Energy utilization of diets with different hay proportions in lactating goats. *Zootecnica e Nutrizione Animale* 23, 317–328.

Rapetti, L., Bava, L., Sandrucci, A., Tamburini, A., Galassi, G. and Crovetto, G.M. (1999) Effect of rumen-protected methionine and lysine in diets with different protein content on N balance of lactating goats. In: *Proceedings of the 50th Annual Meeting of the European Association for Animal Production*. Wageningen Pers, Zurich, Switzerland, p. 257 (Abstract).

Rapetti, L., Bava, L., Tamburini, A. and Crovetto, G.M. (2005) Feeding behaviour, digestibility, energy balance and productive performance of lactating goats fed forage-based and forage-free diets. *Italian Journal of Animal Science* 4, 71–83.

Rousselot, M.C. (1997) Maîtrise et amélioration de la qualité nutritionnelle du lait de chèvre par l'alimentation. In: *Intérêts nutritionnel et diététique du lait de chèvre*, 7 novembre 1996. Les Colloques No. 81. Institut National de la Recherche Agronomique, Paris, pp. 151–162.

Rulquin, H., Pisulewski, P.M., Veritè, R. and Guinard, J. (1993) Milk production and composition as a function of postruminal lysine and methionine supply: a nutrient-response approach. *Livestock Production Science* 37, 69–90.

Santos, F.A.P., Santos, J.E.P., Theurer, C.B. and Huber, J.T. (1998) Effects of rumen-undegradable protein on dairy cow performance: a 12-year literature review. *Journal of Dairy Science* 81, 3182–3213.

Sanz Sampelayo, M.R., Perez, L., Boza, J. and Amigo, J. (1998) Forage of different physical forms in the diets of lactating Granadina goats: nutrient digestibility and milk production and composition. *Journal of Dairy Science* 81, 492–498.

Sauvant, D., Hervieu, J., Giger, S., Ternois, F., Mandran, N. and Morand-Fehr, P. (1987) Influence of dietary organic matter digestibility on goat nutrition and production at the onset of lactation. *Annales de Zootechnie* 36, 335–336.

Van Soest, P.J. (1994) *Nutritional Ecology of the Ruminant*, 2nd edn. Cornell University, Ithaca, New York, pp. 140–150.

Welch, J.G. (1982) Rumination, particle size and passage from the rumen. *Journal of Animal Science* 54, 885–894.

Weston, R.H. (1974) Factors limiting the intake of feed sheep. *Agricultural Research* 25, 349–362.

Zhu, J.S., Stokes, S.R. and Murphy, M.R. (1997) Substitution of neutral detergent fiber from forage with neutral detergent fiber from by-products in the diets of lactating cows. *Journal of Dairy Science* 80, 2901–2906.

11 Goat Nutrition for Fibre Production

A. Di Trana[1] and L. Sepe[2]

[1]*Università degli Studi della Basilicata, Dipartimento di Scienze delle Produzioni Animali, Viale dell'Ateneo Lucano 10, Campus di Macchia Romana, 85100 Potenza, Italy;* [2]*CRA – Unità di Ricerca per la Zootecnia Estensiva, Via Appia-Bella Scalo, Muro Lucano (PZ), 85054 Potenza, Italy*

Introduction

The primary products of most goat breeds are milk and meat. Nevertheless, goat breeds capable of producing a highly insulating and valuable fleece have evolved in low-temperature environments. In such environments, goats are an important resource for local animal production, since they produce milk, excellent fibre and also meat from kids.

Even though goats for fibre represent about 20% of the world's goat production, consumers do not associate cashmere or mohair with goats as immediately as they relate wool to sheep.

Goats for fibre are subdivided into two large groups: (i) Cashmere goats, to which different breeds and populations belong; and (ii) Angora goats, from which mohair is obtained. Cashmere goats can have a double or triple aptitude, without showing the strong competition between fibre and milk for nutrients found in Angora goats and sheep.

After providing a description of goats for fibre, focusing on the specific nature of their fibres, information on their nutrition is given in this chapter.

Cashmere Goats

Almost all goat breeds can produce two types of fibre. The so-called secondary fibre or down is extremely fine, generally shorter than the main fleece (called primary fibre or guard hair), and is visible especially in young goats. Figure 11.1 shows the presence of smaller fibres produced by secondary follicles (SFs) in dairy goats, in comparison to fibre goats.

In Cashmere goats these fibres are far more developed, and therefore provide efficient thermal insulation. Goats are classified as Cashmere when their

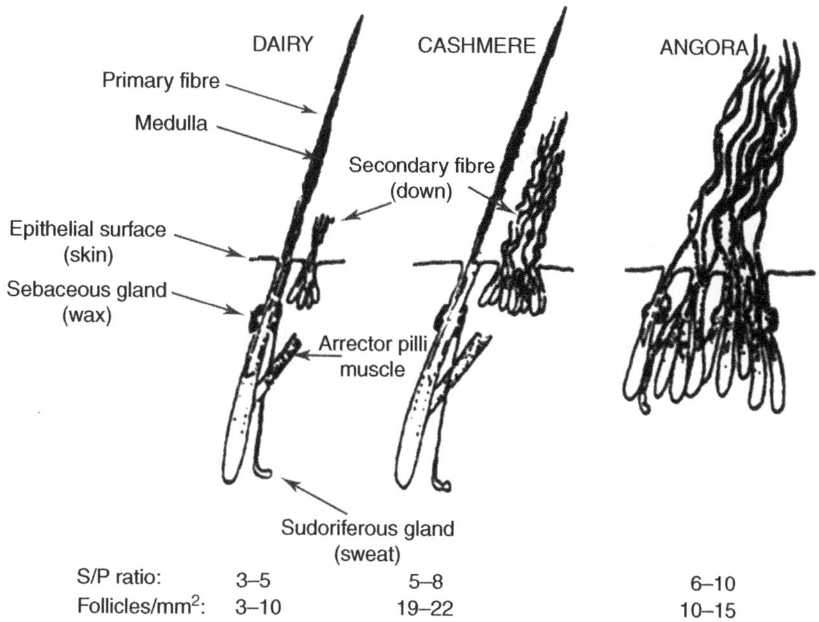

Fig. 11.1. Comparison of the organization of the secondary follicles (S) and primary follicles (P) in dairy and fibre goats. (Adapted from Millar, 1986.)

down fibre, or cashmere, is composed of fibres which have a precise diameter and length, with features of elasticity and lightness.

This fibre owes its name to Kashmir, an area situated to the west of the Himalayas. This fibre is also known as *cachemire* in French or as pashmina, the Indian–Pakistan equivalent (which, however, has not been recognized by the official nomenclature).

Several indigenous fibre goats are reared in Spain. Recently, a fibre aptitude has been observed in certain Italian goat populations (Rubino *et al.*, 2000), such as the Nicastrese goat (Di Trana *et al.*, 2004). Moreover, the presence of a modest quantity of fibre has been found in the Bionda dell'Adamello, an autochthonous goat breed from the Lombardy Alps.

Cashmere Fibre

Cashmere is a fibre without medulla, produced by secondary piliferous follicles, characterized by the features described in Table 11.1.

Since diameter is the main characteristic for distinguishing cashmere from other animal fibres, different industrial textile institutions have established accepted upper limits for cashmere diameter (Table 11.2) (Phan and Wortmann, 2000).

The growth of cashmere is seasonal, as opposed to wool, which has a continuous growth. The development cycle of cashmere fibre is greatly influenced

Table 11.1. Cashmere fibre characteristics.

Parameter	Range	Merchandise considerations
Average diameter (μm)	13–19	The smaller the diameter, the smoother is the product to the touch
Length (mm)	5–90	When shorter than 20 mm, product quality decreases
Colour	white, cream, grey, brown	White is more appreciated because of minor coloration costs
Yield (%)	30–85	Combed cashmere has a higher price than sheared cashmere, due to its higher yield and length
Crimps	from tight to loose	They give elasticity and softness. When closing an amount of cashmere in the hand, the mass returns to its original volume when the hand is opened

Table 11.2. Upper limits for the diameter accepted by the main international textile institutions. (Adapted from Phan and Wortmann, 2000.)

	International institutions			
	Chinese International Standard (China)	CCMI (Europe, Japan, USA)	AATCC (USA)	ASTM (USA)
Upper limit of cashmere diameter (μm)	16.0 ± 0.5	18.5 ± 0.5	18.5	19.0

CCMI, Cashmere and Camel Hair Institute; AATCC, American Association of Textile Chemists and Colourists; ASTM, American Society for Testing and Materials.

by the photoperiod, and, as a consequence, by the concentration of the melatonin and prolactin hormones in plasma. The growth phase (anagen) starts between the spring equinox and the summer solstice and finishes between the autumn equinox and the winter solstice. This is followed by a regressive phase of the piliferous follicle (catagen) and a final phase of rest (telogen), which is associated with a natural loss of hair or moulting (Lynch, 1990). The latter phenomenon occurs between the winter solstice and the spring equinox. Since photoperiod is related to latitude, at lower latitudes the telogen comes about earlier (in the middle of January) compared with higher latitudes (in February), and mostly coincides with the kidding period (McGregor, 1996a). According to the Chinese point of view, the highest-quality cashmere is that produced on the 40th parallel in Inner Mongolia (Petrie, 1995).

Fibre harvesting is carried out by two techniques: (i) combing; and (ii) total shearing. Combing is done using combs which have long, closely set teeth. In this way, a mass with a high yield (78–83%) is obtained, and is therefore much appreciated by textile industries because of the reduced guard hair percentage. Combing is carried out when the animal is 'ready', i.e. when a tuft of down fibre comes away very easily.

In areas where drastic changes in temperature are unlikely (e.g. Australia and Texas), total shearing is done. Sheared fleece has a lower price on the market, because of the lower yield of dehairing (the elimination of the guard hair from the mass), and the shorter cashmere produced.

Due to its high thermal insulation properties (three times higher than wool of the same weight) and its extreme lightness, as well as its limited production and high processing costs (due to the laborious dehairing process), cashmere is considered a luxury textile fibre and is utilized mostly in the clothing sector. Based on diameter, cashmere is classified into two typologies: (i) hosiery (< 16 μm), utilized for knitwear and luxury items of clothing such as stockings and ties; and (ii) weaving (16–18.5 μm), for clothing and furnishings.

Since the demand from industry is fairly constant, prices vary depending on the availability of the Chinese fibre on the market, which has been run by private companies since 1991. Naturally, this has influenced the supply and price of raw cashmere for the European industries (i.e. Italy and Scotland).

The diameter and length of cashmere are features which have very high hereditability ($h^2 = 0.47$ and 0.70 for down diameter and down length, respectively), yet at the same time they are influenced by many different factors, such as sex, age, nutritional status, health and environment. For example, female goats produce better-quality cashmere, with a mean fibre diameter 1–2 μm lower than that of male goats (Ryder, 1989; Li *et al.*, 1996). Peak production is reached at 2 years of age and drops, progressively, up to the sixth to seventh year (Zhang and Shi, 1996). The nutritional, environmental and physiological factors which influence fibre production are analysed as follows.

Nutritional Requirements of Cashmere Goats

The protein and energy requirements for the production of cashmere fibre are estimated according to the chemical composition and energy content of the fleece (Table 11.3) (Russel, 1990). Protein and energy are also required for maintenance, milk production, pregnancy, grazing and thermoregulation (see Cannas *et al.*, Chapter 6, this volume).

Like other animal fibres, cashmere fibre and guard hair are composed of proteins belonging to the α-keratin family. These proteins are rich in sulphurated amino acids such as cystine and cysteine, as a result of the trans-sulphuration of methionine. Apart from the two already-mentioned types of hair, the fleece contains small quantities of sebum and sweat, produced by the sebaceous glands

Table 11.3. Chemical composition and net energy content of cashmere fibre, guard hair and their associated secretions, and nutrient and energy retention in cashmere and guard hair production, considering a growth period of 6 months. (Adapted from Russel, 1990.)

	N	S	Ca	P	Mg	K	Na	Energy
	Chemical composition (g/kg)							Content (MJ/kg)
Fibre	165	33	1	0.1	0.1	0.05	0.3	23.5
Sebum	1.5	–	0.3	0.8	0.1	6.8	0.3	40.8
Sweat	27	4	7	1.2	2	200	10	–
	Nutrient retention (mg/day)							Retention (kJ/day)
Fibre growth of 3.4 g/day	573	114	5	1	1	56	4	84

and sweat glands, respectively. These glands are associated with the primary follicles (PFs).

The retention of nutrients in the fleece, such as cashmere, guard hair and secretions, is evaluated on the basis of an annual production of 250 g of cashmere per head, which represents 30–40% of the fleece, and a growth period of 6 months (Table 11.3) (Russel, 1990).

The energy and protein requirements for the production of fleece in Cashmere goats are determined by the protein and energy retained in the fleece during the growth period. Over the period during which the fibre is growing, the proportion of the energy consumed that is stored in the fibre, or used to produce sebum and sweat, is about 0.006 (0.6% of total energy consumed), or, on an annual basis, about 0.003 (0.3% of total energy consumed) (Russel, 1990). The proportion of protein ingested retained is about 0.02 (2% of protein ingested) over 6 months or 0.01 (1% of protein ingested) over the whole year (Russel, 1990).

Fleece weight of Scottish Cashmere goats varies from 300 to 700 g per head, of which about 30–40% is cashmere (Table 11.4). The efficiency of utilization of metabolizable energy (ME) to fibre energy is not known precisely, but it is estimated to be 0.18 (McDonald *et al.*, 1992). Therefore, a goat which produces a fleece of 700 g with 280 g of cashmere and which retains 96.06 kJ of net energy per day (Table 11.4) would require 534 kJ (i.e. 96.06/0.18) of ME per day (for 6 months) for fleece growth only. This energy requirement for fibre production could be met by supplying daily 63 g of medium-quality hay or 42 g of barley grain.

Protein requirements are related to the quantity of energy supplied. On average, 9 g of rumen-degradable protein per megajoule of ME should be supplied (Russel, 1990). For most foodstuffs fed at normal levels, the degradability of the protein fraction could be assumed to be approximately 0.75 and thus the requirement for crude protein (CP) would be about 12 g CP/MJ ME (Russel, 1990). Therefore, for the production of 700 g of fleece with 280 g of cashmere, 6.4 g CP/day (for 6 months) are required, which could be supplied with 56 g of barley grains or 24 g of broad beans.

Table 11.4. Energy and protein retained in the fleece of Cashmere goats at different productive levels, considering an average period of fleece growth of 6 months.

Fleece weight (g/year)	Cashmere weight (g/year) 30% yield	40% yield	Energy retained in the fleece (kJ/day)	Protein retained in the fleece (g/day)
300	90	120	41.17	1.76
350	105	140	48.03	2.05
400	120	160	54.89	2.34
450	135	180	61.75	2.63
500	150	200	68.61	2.93
550	165	220	75.47	3.22
600	180	240	82.33	3.51
650	195	260	89.19	3.80
700	210	280	96.06	4.10

Nutrition of Cashmere Goats

Proper nutrition of Cashmere goats is needed to produce good-quality fibre, one or two kids and sufficient milk, as well as to ensure high reproductive efficiency and a marketable carcass.

Feed intake

The intake level of a Cashmere dry goat with a body weight (BW) of 40 kg, fed natural pasture hay, was 40.9 g dry matter (DM)/kg $BW^{0.75}$ per day (Table 11.5) (Sepe *et al.*, 1996). This value is within the wide range of values (40–140 g DM/kg $BW^{0.75}$) indicated for dairy goats and for French meat goats (Sauvant *et al.*, 1991). Moreover, an intake of 40.9 g DM/kg $BW^{0.75}$ per day is lower than the intake level of dry dairy goats reared in southern Italy (Fedele *et al.*, 2002).

Feed quality has a major impact on DM intake. For example, as organic matter digestibility of hay increased from 0.47 to 0.63, an increase in intake from 42.8 to 77.7 g DM/kg $BW^{0.75}$ occurred in Scottish Cashmere castrated male goats (Hadjigeorgiou *et al.*, 2001).

The herbage intake of Cashmere dry goats grazing natural Mediterranean pastures, characterized by high botanical variability and seasonal changes in quality, averaged 47 g DM/kg $BW^{0.75}$ on a season basis. A supplement of commercial concentrate with 16% CP, at a level of 12 g/kg $BW^{0.75}$, reduced herbage intake by 9 g DM/kg $BW^{0.75}$, for a total DM intake only slightly higher than that of the unsupplemented goats (Table 11.5) (Sepe *et al.*, 1996). Intake levels varied between 57 and 140 g DM/kg $BW^{0.75}$ in Cashmere goats reared in northern European pastures (Merchant and Riach, 1994).

Table 11.5. Level of dry matter intake of Cashmere goats fed indoors (natural pasture hay plus commercial concentrate 200 g/day) or grazing on hill pasture (360 m above sea level) in late spring. (Adapted from Sepe *et al.*, 1996.)

	DM intake (g/kg $BW^{0.75}$)	
	Mean	SD
Housed goats		
Dry	40.88^{b}	16.93
Lactation	68.89^{a}	21.22
Grazing goats		
With supplement	49.92^{a}	19.45
Without supplement	46.87^{a}	18.95

DM, dry matter; BW, body weight; SD, standard deviation.
[a,b]Mean values in a column with different superscript letters were significantly different ($P < 0.05$).

Proteins

In contrast to sheep and Angora goats, an increase in the amount of dietary protein did not have any positive effect on down length and production of Cashmere goats (Russel, 1995). It should be noted, however, that such an increase led to a greater BW, guard hair growth and thickening of cashmere.

Once production (maintenance and fleece) requirements are satisfied, no additional improvements in quality and quantity of the produced cashmere are determined by a further increase in dietary protein, by supplementation with feeds rich in sulphurated amino acids (casein, cotton, sunflower, lupin and lucerne), or with rumen-protected protein. This was observed in goats fed either in housed conditions or on pasture, as well as for different genetic types, with various production levels of cashmere (McGregor, 1998).

However, when animals fed on a diet lacking in protein were given a protein supplement, both cashmere growth and diameter thickening of the fibre increased (McGregor, 1998).

In contrast to cashmere, the guard hair, originating from the PF adjacent to the SF, is sensitive to increasing N content in the diet, even if these two types of fibre have similar chemical composition. This suggests that SFs producing cashmere have a reduced activity of the enzymes involved in the process of trans-sulphuration of the methionine in cysteine (Russel, 1990). Follicles which produce cashmere have fewer cells able to synthesize fibre proteins, and less intrinsic capacity for protein synthesis, as far as single cells are concerned, than mohair follicles (Lee *et al.*, 1994). Such a hypothesis might be extended to guard hair, based on the similar responses of both guard hair and mohair to protein supplements.

Table 11.6 reports the productive responses of Scottish Cashmere goats reared in mountain environments (1180 m above sea level (asl)) in southern Italy with a feeding regimen consisting of natural pasture (40% grasses, 30% legumes and 30% forbs), or of natural pasture (CP content of 15 and 12% of DM in spring and summer, respectively) plus supplements of 13 or 18% CP (Di Trana and Sepe, 2007). The use of a supplement increased the length of guard hair and the diameter of the down. The latter, however, was not advantageous for production, since it reduced cashmere quality. The highest dietary protein concentration increased guard hair yield and reduced cashmere yield (percentage of down fibre on the total fleece). This may also be related to the shorter growth period of cashmere fleece, caused by an earlier start of moult of cashmere in comparison to the other two diets (Di Trana and Sepe, 2007).

On the basis of these considerations, the most advantageous diet for a better cashmere output seems to be the one with a CP level of 13%, even if a thicker down was obtained compared with goats which have utilized pasture only. Feeding regimen based only on pasture reduced fibre length, without worsening significantly the cashmere yield with respect to the groups fed a supplement. Actually, the use of pasture alone improved the quality of cashmere, which became 2 and 4% thinner than that of goats fed 18% CP and 13% CP, respectively. Moreover, the availability of natural pasture caused a delay in moulting of about 1 month

Table 11.6. Effect of feeding regimen on fibre production traits of Scottish Cashmere goats grazing on natural pastures (CP content of 15 and 12% of DM in spring and summer, respectively) with or without supplements. (Adapted from Di Trana and Sepe, 2007.)

	Feeding regimen		
		Grazing plus supplement	
Parameter	Grazing natural pasture	CP 13%	CP 18%
Cashmere length (mm)	49	55	54
Maximum guard hair length (mm)	59	63	75
Cashmere growth rate (mm/day)	0.28	0.30	0.36
Growth period of cashmere (days)	182[a]	180[a]	148[b]
Cashmere diameter (μm)	16.8[b]	17.5[a]	17.2[ab]
Cashmere production (g/kg BW)	5.20	5.92	5.10
Cashmere yield (%)	33.5	34.1	32.0
Guard hair yield (%)	66.5	65.9	68.0
Start of moult (days)	393[a]	361[b]	350[b]

CP, crude protein; DM, dry matter; BW, body weight.
[a,b]Within rows, values with different superscript letters were significantly different ($P < 0.05$).

compared with the feeding regimen at 13% CP, and a 6-week delay compared with the feeding regimen of 18% CP (Table 11.6) (Di Trana and Sepe, 2007).

Ivey *et al.* (2000) studied the effects of diets differing in concentrations of CP (10 and 15%) and ME (2.00, 2.35 and 2.70 Mcal/kg; DM basis), fed *ad libitum*, on growth and cashmere fibre production of cashmere-producing Spanish wether goats, in an 84-day trial during autumn. Guard hair weight and cashmere length were not affected by dietary treatments. Cashmere fibre diameter was not altered by dietary ME level but was greater for 15% CP than for 10% CP (16.92 versus 16.06 μm; $P < 0.05$). Cashmere weight was influenced by an interaction between CP and ME levels; cashmere fibre weight with 10% CP was 92, 82 and 95 g, and with 15% CP was 63, 115 and 99 g for diets with 2.00, 2.35 and 2.70 Mcal ME/kg, respectively.

No cashmere growth responses have been observed by supplying protected protein to low-production goats or to high-production goats when the basal diet appeared to contain sufficient protein. When high-protein legume grains were fed to cashmere goats consuming a basal diet deficient in protein, large increases in both cashmere growth and mean fibre diameter were obtained (McGregor, 1998).

Amino acids

In Australia, cashmere growth of fully grown male Cashmere goats was not influenced by integration with methionine (Ash and Norton, 1987). However, other studies on Scottish Cashmere yearling goats showed that the inclusion of methionine (2.5 g/day), in addition to a basal diet containing 9.9 MJ ME and 107 g CP (per kg DM), increased cashmere yield and diameter, while the guard hair was not influenced (Souri *et al.*, 1998; Galbraith, 2000). These opposing results could be attributed to a different genetic potential for cashmere production of the animals utilized, and to an inadequate supply of the amino acids required in the non-supplemented group.

Considering that the annual production of cashmere is seldom higher than 300 g per head, the requirement for amino acids (< 1 g protein/day) for cashmere growth can be easily satisfied by the circulating amino acid pool, without being limited by the availability of protein in the diet (Galbraith, 2000).

Energy

Fibre production is less affected by energy supply than by protein supply (Reis, 1989). For the Cashmere goat, neither the timing nor increasing levels of feeding above maintenance increased the growth of down fibre, although these same feeding regimes promoted significant increases in both guard hair growth and BW gain in the same goats (Galbraith *et al.*, 2000).

In grazing goats, neither decreased stocking rate (Norton *et al.*, 1990) nor supplementary grain feeding (McGregor, 1996b) increased cashmere production. On the other hand, undernourishment, i.e. when maintenance requirements are not satisfied, reduced animal weight and cashmere growth in goats. This phenomenon was more evident when undernourishment occurred during the anagen phase of fibre growth (McGregor, 1998).

In castrated goats reared in Australia, undernourishment (at 0.73 of maintenance level), between the summer solstice and winter solstice, caused a 16% BW loss, a 24% reduction in cashmere production and a 2% reduction in fibre diameter, as compared with control goats fed at maintenance levels (McGregor, 1988). Similarly, when Australian cashmere does of two genotypes (a Random line, made of goats selected from a population mated in a random manner, under no selection pressure; and a Cashmere Plus line, selected for down production) were fed at 0.7 × maintenance, weight loss and reduction in cashmere production occurred (Restall *et al.*, 1994).

It should be pointed out that the response to undernourishment varies according to the origin of Cashmere goats. In fact, in selected breeds of Australian origin, probably having Angora blood, weight loss and reduction in the growth and diameter of cashmere were observed, while in non-selected types the responses to an inadequate dietary supply were not clear (Restall *et al.*, 1994).

Therefore, in goats of Australian origin, the most convenient feeding treatment, which increased cashmere production without decreasing its quality, led to an increase of 1–2 kg in BW during the period of maximum growth of the down fibre (i.e. between summer and autumn) (McGregor, 1998). Feeding goats to gain 4 kg in BW (Fig. 11.2a) (McGregor, 1998) over summer and autumn gave maximum cashmere growth. Maximum production was achieved at energy intakes of about 1.4 × maintenance. On average, cashmere production is greatest in heavier goats, which increase BW, and lowest in goats of inferior size, which lose weight (Fig. 11.2b) (McGregor, 1998).

In conclusion, the convenience of feed integration depends on several factors: (i) the cost of integration; (ii) seasonal conditions, such as dry or wet years, and duration of feed supplementation; (iii) goat productivity; (iv) the level of response to better nutrition; and (v) the quality of the cashmere produced.

Effects of Energy Supply on Cashmere Moulting

Nutrient supply has indirect effects on fibre moulting because of its relationship with the concentration of hormones which regulate nutrient distribution and competition between tissues. The mechanisms that regulate fibre shedding are still not well understood, but a large body of evidence indicates that this event is related to seasonal changes in hormonal concentration of prolactin (Dicks, 1994). Growth hormone indirectly stimulates wool growth in sheep (Wynn *et al.*, 1988), while no evident influence of this hormone on the cashmere growth cycle was found (Klören *et al.*, 1993). In a more recent study, insulin-like growth factor-I (IGF-I) receptors have been identified in hair follicles (Dicks *et al.*, 1996), suggesting a direct role of IGF-I in the control of hair follicle activity. Thyroid hormones also seemed to be involved in the regulation of seasonal pelage exchange in goats (Rhind and McMillen, 1995).

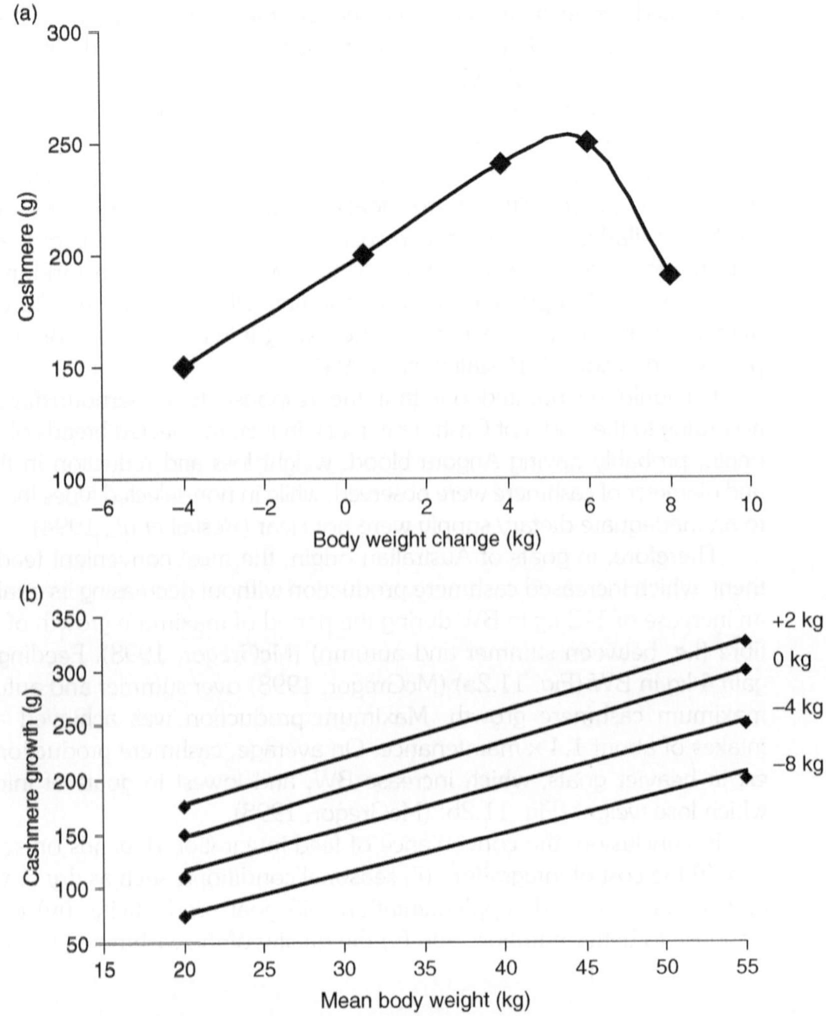

Fig. 11.2. (a) Relationship between the annual production of cashmere and the body weight change in Australian Cashmere goats; (b) relationship between cashmere growth and mean body weight for goats with different body weight changes during the cashmere growing season. (Adapted from McGregor, 1998.)

Energy restrictions at levels lower than maintenance (0.8 × maintenance), during the 6 months following the winter solstice, caused weight loss and delay in moulting of about 1 month. Energy levels higher than maintenance (1.2 and 2 × maintenance) anticipated moulting (Merchant and Riach, 2002). These authors have proposed that undernutrition also delayed the onset of regrowth of the new coat. At latitudes of 40–41°N, where Cashmere goats were reared either in mountain environments (1200 m asl) or on hills (360 m asl), the moulting peak was reached earlier in the mountains. In fact, during February, 79% of the cashmere is harvested in the mountain environment and 67% in the hilly areas. These results are also related to the different dietary regimens of the two environments. From autumn until the beginning of winter, goats reared on the hills grazed pastures for a longer period of time, during which the amount and quality of the available biomass may not have satisfied nutrient requirements. By contrast, in mountain environments, due to the bad weather conditions of that period, goats were fed indoors, thus assuring a more constant supply of nutrients (Di Trana *et al.*, 1998).

At latitudes of 40–41°N, for Cashmere kids at first moult, maximal fibre shedding occurred a month earlier in males than in females. This was due to the higher prolactin and growth hormone concentration in males (Celi *et al.*, 2003).

Environmental and Physiological Factors

Cashmere goats reared in different altitudes and weather conditions show different productive capacities. At higher altitudes (1180 m asl) and low temperature humidity index (THI), the percentage of active SFs, fibre length, cashmere yield and growth period increased in comparison to animals at lower altitudes (50 m asl) and a higher THI, while diameter remained the same (Celi *et al.*, 2000, 2001; Di Trana *et al.*, 2001, 2003).

The effects of latitude on down production are similar for high- and low-producing Cashmere goats. The down produced by Cashmere and Spanish goats, reared at a latitude above the 40th parallel, was higher and finer by 59 and 83%, respectively, compared with that of the same breeds reared at a latitude near the 30th parallel (Lupton *et al.*, 2000).

In the Cashmere goat where the anagen phase is shorter and more concentrated in certain months of the year, the overlap between lactation and the production cycle of cashmere fibre reduces the activity of the SF. This was limited to the first 4 months of lactation (between June and September), with negligible effects on yield, the total annual production and diameter of cashmere (Celi *et al.*, 2002).

At Italian latitudes (40–41°N), the overlap between the period of pregnancy and the beginning of lactation with that of the activity of SFs (May–December) decreased production, length and yield of cashmere, without influencing its diameter (Celi *et al.*, 2005). Similar behaviour was observed in Cashmere goats of Australian origin (Klören and Norton, 1993).

Milk Production and Quality in Cashmere Goats

Information regarding milk quantity and quality produced by Cashmere goats is scarce. Table 11.7 reports the milk production and fat and protein content, at the 4th week after the colostrum phase, of Scottish feral Cashmere goats that had previously been reared in extensive systems until 14 or 9 weeks before delivery. Following housing the does were fed according to BW and were turned out to graze following kidding (Galbraith *et al.*, 1992).

In extensive systems, individual milk production increases according to the availability of the fodder biomass.

Table 11.7. The production and quality of milk produced by Scottish feral Cashmere goats reared in extensive systems, housed before delivery and turned out to graze following kidding. (Adapted from Galbraith *et al.*, 1992.)

	Housed before delivery for		
Values at 4 weeks postpartum	14 weeks	9 weeks	SED
Milk production (g/day)	200–230	225–379	66.9
Fat (%)	2.5–3.0	4.1–6.4	1.0
Protein (%)	4.0–4.1	3.8–3.9	0.3

SED, standard error of the difference.

Table 11.8. Milk production and chemical composition of Cashmere goat's milk. (Adapted from Di Trana and Sepe, 2000.)

	Days after delivery						
Item	45	60	75	90	105	120	SEM
Milk (g/day)	1079^{a}	1028^{a}	943^{a}	725^{b}	558^{c}	534^{c}	59.2
Fat (%)	3.5^{ab}	3.6^{ab}	3.2^{a}	3.5^{ab}	3.7^{b}	4.3^{c}	0.2
Protein (%)	3.9	3.9	3.7	3.8	3.8	3.8	0.1
Lactose (%)	4.8	4.8	4.9	4.6	4.5	4.3	0.1
Casein N (mg/dl)	497	457	445	462	464	465	22
Whey N (mg/dl)	100	97	105	95	94	94	8
Non-protein N (mg/dl)	44	45	42	46	47	44	2

SEM, standard error of the mean.
[a,b,c]Mean values in rows with different superscript letters were significantly different ($P < 0.05$).

From Cashmere goats of Scottish origin, introduced into a Mediterranean environment and reared in an extensive system, a fair amount of milk is obtained together with the production of fibre of excellent quality. During the 75 days of lactation after the sale of kids, pluriparous goats kidding at the beginning of March produced 60.9 kg of milk per head. Contents of milk fat, protein and its nitrogenous fraction (Table 11.8) (Di Trana and Sepe, 2000) were similar to those of autochthonous goat breeds in southern Italy (De Maria Ghionna *et al.*, 1984). This suggests that milk and its products could be a further source of profit for the fibre goat sector.

The Dahlem Cashmere breed, created in the 1980s by cross-breeding the Angora breed with milk breeds, has a high level of milk production with a total protein content of 3.6%, which is similar to that observed in the Scottish Cashmere goats. Since Scottish Cashmere goats showed lower milk production, their milk casein content was higher (2.9 g/100 g) (Di Trana and Sepe, 2000) than that of the Dahlem Cashmere breed (2.7 g/100 g) (Dimassi *et al.*, 2004).

The Angora Goat

Unlike Cashmere goats, the Angora goat experiences strong competition for nutrients between milk and fibre, similar to that found in sheep. The main difference between the Angora and the Cashmere goat is in the physiology of their fibre growth as well as the structure of the epidermic scales of the fibrous filament and the ratio between SFs and PFs (Fig. 11.1).

The Angora goat has been reared as a pure breed in Turkey and was first exported to South Africa in 1838; subsequently it reached Russia and Australia, where cross-breeding was carried out with local Cashmere breeds (Millar, 1986). This created a breed named Cashgora, with fibre which had intermediary features between cashmere and mohair.

Mohair Fibre

Mohair fibre is produced exclusively by Angora goats from the SF, and is characterized by the features described in Table 11.9 (Shelton, 1981; Epplestone and Moore, 1990; Antonini *et al.*, 1994b).

The fineness of the fibre is the main element which distinguishes mohair from other fibres. In Italy, this fibre is classified by diameter and length according to the South African method (Table 11.10) (Gallico, 1992).

In Angora goats a marked aptitude for fibre production is demonstrated by a predominant number of SFs compared with PFs (Fig. 11.1), attaining the SF/PF ratio of 8–10 (Shelton, 1981).

In contrast to cashmere, mohair has a continuous anagen all year long, even if it is faster in the summer. This kind of growth has a stronger influence on the

Table 11.9. Characteristics of mohair fibre. (Adapted from Shelton, 1981; Epplestone and Moore, 1990; Antonini *et al.*, 1994b.)

Item	Range	Merchandise considerations
Average diameter (μm)	25–39	The thinner the softer
Length (mm)	75–150	Turkish mohair is longer and, owing to this, more requested
Yield (%)	70–90	
Colour	white, grey, brown, black	Normally white; occasional black hairs depreciate the mass
Sheen	–	Main characteristic which makes mohair especially valuable
Kemp (fibres with medulla, short and dead) (%)	1–2	As kemp increases, flock quality decreases

Table 11.10. Classification of mohair used in Italy according to diameter and length; quality increases from adult to super kid and from E to A. (Adapted from Gallico, 1992.)

Type	Diameter (μm)	Class	Length (mm)
Super kid	24.0–26.5	A	150 and above
Kid	26.5–29.5	B	125–150
Young goat	29.5–34	C	100–125
Fine adult	34–36	D	75–100
Adult	36–39	E	< 75

nutritional requirements during the various physiological stages of Angora goats compared with that of fibre growth on Cashmere goats.

A well-reared and well-fed adult Angora goat has an average monthly growth of mohair of 2–5 cm (Shelton, 1981). Thanks to its particular structure, mohair has a characteristic silken sheen and a mean tenacity about 30% higher than that of wool.

The harvesting of the fleece takes place by shearing. The first shearing, carried out on 6-month-old kids, supplies the very best quality fibre. In general, two shearings are carried out annually, the first in April, when the fibre is 120 mm long, and the second in September. Turkish mohair is obtained from one single shearing per year, and is therefore longer. In this case, the annual production per head is reduced to 2.5 kg compared with the 3 kg of double shearing in Australia and the 4–5 kg from the South African goat population.

Mohair is considered a luxury fibre for both its quality and its limited availability, even if it is evaluated at a lower price than cashmere. This depends on the lower costs for transformation into yarn. There is a traditional use of mohair for rugs and curtains by Arab nomads. The use of this fibre for men's and women's clothing is widespread. Other than for luxury upholstery, the demand for mohair varies according to the requirements of the textile industry and the designers. Age is the determining factor for both the quality and quantity of mohair in a rearing system which satisfies the nutritional requirements of the Angora goat. The number of SFs increases from the moment of birth up to 2 years of age, bringing the SF/PF ratio from the initial 2–3 to 8–10. As kids grew, mohair diameter increased from 24 μm in a 6-month-old kid to 49 μm in an adult goat, so the mohair production and yield also increased (Shelton, 1981).

Female goats had a higher number of follicles per square millimetre, and lower diameter and percentage of medullated fibres than the males (Antonini *et al.*, 1994b). As confirmed by this information, the best-quality mohair comes from females, and this has led to the use of castration in male kids (Willingham *et al.*, 1988).

The production of kemp increases with age and has a high degree of hereditability ($h^2 = 0.43$); therefore, there is potential for genetic improvement of this trait.

Nutritional Requirements of Angora Goats

The amount of fibre produced by the Angora goat on a daily basis varies between 5 g/day (1.8 kg/year) and 16.7 g/day (6.0 kg/year), being higher than the amount produced by sheep.

Recently, a system for goats, with specific predictions for Angora goats, has been developed by the E(Kika) de la Garza Institute for Goat Research at Langston University (USA) to estimate the nutritional requirements for maintenance, fibre and milk production and also weight gain under different grazing and climatic conditions (Luo *et al.*, 2004a,b). A summary of these requirements is reported in Table 11.11.

Mineral and vitamin requirements

Fibre-producing goats do not have any particular requirements for minerals and vitamins. A mineral deficiency, as with other goat breeds, causes a decrease in production and threatens animal health. In areas where the soil–plant system does not guarantee the right amount of microelements (e.g. Se, Cu, Co, I and Zn), diet integration with these is necessary. Anyway, mineral and vitamin integration for Cashmere and Angora goats must be taken into consideration whenever the feeds utilized in their diet are deficient.

Specific nutritional requirements for Angora goats have been established for S, an essential microelement for the synthesis of sulphurated amino acids, microbial protein in the rumen and fibre proteins. An intake of S equal to 0.267% of DM and a ratio of 7.2:1 between N and S are considered optimal (Qi *et al.*, 1992).

Table 11.11. Daily total nutritional requirements and intake of the Angora goat at different fibre production levels grazing on hill pastures (body condition score was equal to 2, and average air temperature was 12°C). (Adapted from Luo *et al.*, 2004a,b.)

BW (kg)	Clean mohair production (g/day)	ME (MJ/day)	MP (g/day)	CP (g/day)	DM intake (% of BW)
30	5	6.85	51	76	2.70
	10	7.64	59	88	2.77
	15	8.42	68	101	2.84
40	5	8.22	61	92	2.45
	10	9.00	70	104	2.51
	15	9.79	78	116	2.56
50	5	9.50	71	106	2.28
	10	10.29	79	118	2.33
	15	11.07	88	131	2.37

BW, body weight; ME, metabolizable energy; MP, metabolizable protein; CP, crude protein; DM, dry matter.

By increasing the S in the diet, the production of mohair and the length and resistance of the flock increase, without any reduction in fibre quality.

A deficit in P and vitamin A reduced the keratinization of the skin and the activity of piliferous follicles (Morand-Fehr and Galbraith, 1992). Therefore, it is better to integrate these nutrients in the diet, especially when fresh forage availability is limited.

In Inner Mongolian White Cashmere goats, the optimum Cu level is 27.46 mg/kg of diet during the cashmere growing period (Zhang *et al.*, 2004).

Among the vitamins, biotin has an important influence on the growth of tegumental tissues and piliferous follicles. In fact, deficiency of this vitamin in kids led to a reduction of weight gain, an *in vivo* mohair loss and a reduction of *in vitro* vitality of the mohair follicles (Galbraith *et al.*, 2000).

Nutrition of Angora Goats

Feed intake

As for all ruminants, the intake of Angora goats differs according to the type of hay utilized. In fact, the daily intake was 716, 546 and 724 g of DM per head for hays of lucerne, poliphita lowland pasture and meadow (73.7% *Trifolium repens* and 26.3% *Lolium perenne*), respectively. Daily intake was 576 and 864 g of DM per head for sorghum and oat hays, respectively (Antonini *et al.*, 1994a).

As NDF concentration in the hay increased from 38.2 to 73.6% and up to 75.8%, organic matter intake was, respectively, 67.6, 45.5 and 34.6 g/kg $BW^{0.75}$ (Morand-Fehr, 1993). During lactation, increases in the CP level of the diet from 9 to 18% of DM caused a non-significant increase of DM intake from 1.54 to 1.56 kg/day ($P > 0.05$) (Sahlu *et al.*, 1992a).

Pasture intake is greater in periods of time when forage availability is high. As for other breeds (see Avondo *et al.*, Chapter 7, this volume), intake of Angora goats is influenced by animal factors, especially selective behaviour, as well as by the type of herbage utilized. Pasture intake is also related to stocking rate and individual production. Both the productivity and the economic efficiency of fibre-producing farms are influenced by stocking rates. The effects of high stocking rates (higher than 7.5 head/ha) on mohair features of grazing Angora goats are often negative and commercially relevant (Table 11.12) (McGregor, 1998).

On pastures in temperate areas, it is advisable to use a stocking rate of 7.5 head/ha (season average) for Angora goats. Such a stocking rate minimizes both gastrointestinal parasite infections and post-shearing weather stress. At high rates of stocking the mortality rate (primarily gastrointestinal parasitism and post-shearing weather stress) was twice that at lower rates of stocking (McGregor, 1998). Under such conditions, the goats increase in weight fairly quickly and produce more and longer mohair, which, however, is not as fine as that of goats reared at high stocking rates.

Table 11.12. The main effects of high stocking rates (> 7.5 head/ha) of Angora goats grazing annual pastures in temperate areas on mohair production and quality parameters. (Adapted from McGregor, 1998.)

Item	Effect of high stocking rate
Weight of mohair	20% reduction in fibre production per head
Fibre diameter	Thinner coat – diameter decrease by ≥ 5 μm
Kemp	Increase in the incidence of kemp
Clean mohair yield	Decrease by ≥ 5%
Mohair length	Reduction of fibre length

Energy

In Mediterranean environments, energy deficit usually occurs in summer. This is determined by scarce pasture resources and leads to a decrease in body condition score and BW of the animals. During this time, weight loss can be minimized by utilizing cereal grains (e.g. barley, oats and wheat), which are relatively low-cost energy sources that do not dirty the fleece remaining on the flocks. The use of energy supplements influences BW and mohair diameter. Goats losing weight produced less mohair (from 15 to 40%) and their fibre was thinner by 2–3 μm than goats at constant weight. On the other hand, goats fed for BW gain produced much more mohair (up to 100% increase) and fibre which was 2–3 μm thicker than animals with a constant weight (McGregor, 1998). When integration was supplied, an increase of 0.26 and 0.40 μm in diameter for each kilogram of weight gain was recorded, respectively, for animals on pasture and those housed (McGregor, 1998).

Determination of the economic convenience of using energy integration should take diameter increase into consideration, since a 5% depreciation of the price occurs for each micrometre increase (up to diameters of 32 μm) (van der Westhuysen, 1982). In young animals, a dietary supplement which increased weight reduced the incidence of medullated fibres when compared with goats which had an unvaried BW during the growth cycle of the fibre. In 6-month-old Angora kids, during a period of scarce herbage availability, supplementation actually doubled fibre production. The percentage of medullated fibre (7.6 and 0.9% in New Zealand and South African genotype, respectively) and kemp (4 and 0.5% in New Zealand and South African genotype, respectively) (Newman and Paterson, 1999) was not strongly influenced by the utilization of dietary supplement, but rather by genetic factors and age (Galbraith *et al.*, 2000).

Energy supplementation in pregnant or lactating goats is advisable when goat welfare is at risk (e.g. scarce pasture availability or pregnancy) or when animal reproductive and productive performance is to be improved. An increase in fleece value at the kid's first shearing is added to these benefits.

In extreme situations, such as bad weather conditions after shearing, or a lack of hay or pasture availability, a quick and high intake of cereal grains could induce rumen acidosis. In this case, it is advisable to supply rumen buffers together with cereal grains to limit the increase in rumen acidity or to supply feed as dried beet pulps, citrus pulp and soybean hulls.

Proteins

Protein supplements with high rumen escape and rich in sulphurated amino acids increase weight and diameter of mohair. The beneficial effects of this kind of integration depend on the quantity and the balance of nutrients in the basal diet (Galbraith *et al.*, 2000).

During the period of the year in which the lengthening of the mohair is more rapid, a substantial increase in fibre growth can be obtained with diets containing large amounts of sulphurated amino acids reaching the intestine.

The effects of two protein (108 and 180 g CP/kg DM) and energy (10.0 and 11.9 MJ ME/kg DM) levels on the production and the quality of the mohair were evaluated in castrated Angora goats. The results confirmed the fact that mohair growth can be significantly limited by an inadequate amount of dietary protein, and also that protein supplement stimulates yield (Shahjalal *et al.*, 1992; Galbraith, 2000). However, such supplement also increases fibre diameter, which is a commercially undesirable consequence. As an example, an increase in dietary protein from 12 to 18% increased the yield of the clean fleece by about 23% and the average fibre diameter by 5.2% (Galbraith *et al.*, 2000).

It is difficult to optimize protein supplements in the diet of goats to improve at the same time the quality and the quantity of the fibre. The highest production of mohair was reached by supplying 165 g CP/kg DM per day, while the lowest fibre diameter was achieved by a lower CP supply (102 g/kg DM) (Shahjalal *et al.*, 1991).

In 18-month-old Angora female goats, high mohair yield was reached with CP levels up to 190 g/kg DM with an isoenergetic diet (Sahlu *et al.*, 1992a,b). A dietary supplementation with horsebeans (*Vicia faba* var. *minor*) in 90- and 150-day-old kids increased fibre growth rate, follicle activity and fibre length, and decreased mohair diameter, as opposed to what happens with adult animals (Trabalza Marinucci *et al.*, 2003).

Protein sources with a high biological value or a high content of rumen-protected protein, such as cysteine treated with formaldehyde and heat-treated soybean, increased the growth of mohair fibre (Morand-Fehr and Galbraith, 1992; Galbraith *et al.*, 2000).

Methionine, cystine and cysteine are amino acids involved in the synthesis of the proteins deposited in mohair. A great amount of cystine and cysteine can be recuperated through the trans-sulphuration of methionine. This amino acid has a specific role in the stimulation of mohair growth. In fact, with 1-year-old Angora goats the inclusion of protected methionine (2.5 g/head per day), in a diet of 9.9 MJ ME/kg DM and 107 g CP/kg DM, increased both the yield and the diameter of the mohair produced in 2 months (Souri *et al.*, 1998; Galbraith, 2000). Protected methionine improved mohair production by 0.8 g/head per day for each gram of methionine supplied (Morand-Fehr and Galbraith, 1992).

In conclusion, Angora goats respond to a dietary supplementation of good-quality protein, or protected methionine, with an increase in mohair yield and a decrease in its quality.

Photoperiod

The seasonal photoperiod has an influence on mohair growth. In the winter, the lengthening of mohair is at a minimum whereas it reaches its maximum value in the summer. Castrated Angora goats kept at a constant BW had 55% higher fibre growth in the summer than in the winter. Medullated fibre and kemp showed a similar trend. In the winter, the percentage of inactive piliferous follicles was greater and increased as the animals aged (McGregor, 1998). The practical implications of these observations are: (i) an increase in the presence of kemp in the fleece collected in the autumn; and (ii) an indication for the most appropriate slaughtering age based on the reduction in the percentage of active follicles as the animal ages.

Conclusions

The Cashmere goat is recognized to have a large productive potential due to: (i) its limited nutritional requirements for fibre production; (ii) the reduced influence of protein or energy supplementation on the yield; (iii) its easy adaptation to various rearing environments; and (iv) its production, in addition to fibre, of kids and milk.

For these particular reasons, the rearing of Cashmere goats could give a real boost to farm activities or even to farm holidays utilizing market niches, so as to guarantee an added income for that type of enterprise. Another advantage would be that this goat could utilize and protect native pasture systems.

On the other hand, the Angora goat, having undergone a selective process, is characterized by nutritional requirements closely associated to its fleece production.

The productive response of the Angora goat is sensitive to both protein and energy supplementation, with an increase in mohair yield and a decrease in quality. Therefore, since the value of the product decreases as fibre diameter increases, an economic analysis of feed supplementation is necessary.

Because of its functional features and the strong competition for nutrients between milk and fibre, the Angora goat expresses a productive potential which is limited to luxury fibres.

References

Antonini, M., Abbadessa, V., Trabalza Marinucci, M. and Misiti, S. (1994a) Prime osservazioni sul comportamento alimentare di capre Angora importate in Italia: ingestione, selezione e digeribilità apparente di alcuni foraggi tipici. In: *Proceedings of the XI National Congress of Società Italiana di Patologia e di Allevamento degli Ovini e dei Caprini*. Università degli Studi di Perugia, Perugia, Italy, pp. 389–392.

Antonini, M., Carnevali, F., Castrignano, F., Misiti, S. and Pieramati, C. (1994b) Angora goat production in Italy. In: Laker, J.P. and Bishop, S.C. (eds) *Proceedings of the European Fine Fibre Network, Occasional Publication 1*. Macaulay Land Use Research Institute, Peebles, UK, pp. 65–75.

Ash, A.J. and Norton, B.W. (1987) Effect of D,L-methionine supplementation on fleece growth by Australian cashmere goats. *Journal Agricultural Science (Cambridge)* 109, 197–199.

Celi, R., Di Trana, A., Colonna, M.A., Toteda, F., Facciolongo, A.M. and Rubino, R. (2000) Influence of altitude on Cashmere production in goats reared in Southern Italy. In: Gruner, L. and Chabert, Y. (eds) *Proceedings of the 7th International Conference on Goats*. Institut de l'Elevage and Institut National de la Recherche Agronomique, Paris, pp. 652–654.

Celi, R., Labate, M., Di Trana, A., Desantis, S., Rubino, R. and Cirillo, F. (2001) Percentage of active primary and secondary follicles and fibre length in cashmere goats reared at different altitudes. In: Antongiovanni, M., Bozzi, R., Franci, O., Giorgetti, A., Gualtieri, M., Acciaioli, A., Parisi, G., Martini, A. and Biagioli, O. (eds) *Proceedings of the ASPA XIV Congress, Recent Progress in Animal Production Science*, Vol. 2. Dipartimento di Scienze Zootecniche, Università di Firenze, Firenze, Italy, pp. 523–525.

Celi, R., Di Trana, A. and Celi, P. (2002) The influence of lactation on the quantity and quality of cashmere production. *Italian Journal of Animal Science* 1, 79–86.

Celi, P., Seren, E., Celi, R., Parmeggiani, A. and Di Trana, A. (2003) Relationships between blood hormonal concentrations and secondary fibre shedding in young cashmere-bearing goats at their first moult. *Animal Science* 77, 371–381.

Celi, R., Di Trana, A., Celi, P., Marsico, G. and Forcelli, M. (2005) The influence of pregnancy and the beginning of lactation on pelage traits in cashmere goats. *Italian Journal of Animal Science* 4, 85–96.

De Maria Ghionna, C., Catillo, G., Angelucci, M., Zarriello, G. and Rubino, R. (1984) Indagine conoscitiva sulla produzione e qualità del latte di un gruppo di capre locali, allevate nelle aree interne della Lucania. *Annali Istituto Sperimentale per la Zootecnia* 17, 155–166.

Dicks, P. (1994) The role of prolactin and melatonin in regulating the timing of the spring moult in the cashmere goat. In: Laker, J.P. and Allain, D. (eds) *Proceedings of the European Fine Fibre Network, Occasional Publication 2*. Macaulay Land Use Research Institute, Aberdeen, UK, pp. 109–128.

Dicks, P., Morgan, C.J., Morgan, P.J., Kelly, D. and Williams, L.M. (1996) The localisation and characterisation of insulin-like growth factor-I receptors and the investigation of melatonin receptors on the hair follicles of seasonal and non-seasonal fibre-producing goats. *Journal of Endocrinology* 151, 55–63.

Dimassi, O., Herold, P. and Zárate Valle, A. (2004) Yield, composition and cheese making potential of Dahlem Cashmere goat milk. *South African Journal of Animal Science* 34(Suppl. 1), 159–161.

Di Trana, A. and Sepe, L. (2000) Milk production traits of Cashmere goats in Southern Italy. In: Gruner, L. and Chabert, Y. (eds) *Proceedings of the 7th International Conference on Goats*. Institut de l'Elevage and Institut National de la Recherche Agronomique, Paris, p. 612.

Di Trana, A. and Sepe, L. (2007) The effect of protein supplement on the quantity, quality and moult of cashmere fibre in Cashmere goats kept at pasture in Southern Italy. In: Rubino, R. and Sepe, L. (eds) *Proceedings of the International Symposium*

'The quality of goat products: models and tools for evaluation and promotion', Bella (Potenza), Italy. Consiglio per la Ricerca e sperimentazione in Agricoltura, Bella (Potenza) Italy, pp. 121–124.

Di Trana, A., Sepe, L. and Rubino, R. (1998) Influenza dell'ambiente di allevamento e della frequenza di raccolta sulla muta della fibra cashmere. In: *Proceedings of the XIII National Congress of Società Italiana di Patologia e di Allevamento degli Ovini e dei Caprini*. Università degli Studi di Palermo, Palermo, Italy, pp. 454–458.

Di Trana, A., Celi, R., Desantis, S., Celi, P. and Labate, G.M. (2001) Influence of environmental temperature and humidity on cashmere traits in goats reared in southern Italy. In: Antongiovanni, M., Bozzi, R., Franci, O., Giorgetti, A., Gualtieri, M., Acciaioli, A., Parisi, G., Martini, A. and Biagioli, O. (eds) *Proceedings of the ASPA XIV Congress, Recent Progress in Animal Production Science*, Vol. 2. Dipartimento di Scienze Zootecniche, Università di Firenze, Firenze, Italy, pp. 526–528.

Di Trana, A., Celi, P. and Celi, R. (2003) The effect of environmental temperature and humidity on cashmere yield, secondary active follicles and thyroid hormones in cashmere goats. In: Lacetera, N., Bernabucci, U., Khalifa, H.H., Ronchi, B. and Nardone, A. (eds) *Proceedings of 'Interaction between climate and animal production'*. European Association for Animal Production, Technical Series No. 7. Wageningen Academic Publishers, Wageningen, The Netherlands, p. 96.

Di Trana, A., Celi, P., Sepe, L., Desantis, S. and Rubino, R. (2004) Fleece and skin traits of goats of different genetic types reared in Southern Italy. *South African Journal of Animal Science* 34(Suppl. 1), 156–158.

Epplestone, J. and Moore, N.W. (1990) Fleece and skin characteristics of selected Australian Angora goats. *Small Ruminant Research* 3, 397–402.

Fedele, V., Claps, S., Rubino, R., Calandrelli, M. and Pilla, A.M. (2002) Effect of free-choice and traditional feeding systems on goat feeding behaviour and intake. *Livestock Production Science* 74, 19–31.

Galbraith, H. (2000) Protein and sulphur amino acid nutrition of hair fibre-producing Angora and Cashmere goats. *Livestock Production Science* 64, 81–93.

Galbraith, H., Bell, J., Scaife, J.R., Acamovic, T., MacDonald, D.C. and Roger, L.C. (1992) Milk production, vitamin and immunological status of cashmere does and kids differing in pre-partum housing date and buck genotype. In: *CIV Proceedings of the British Society of Animal Production*. BSAS, Edinburgh, pp. 472–473.

Galbraith, H., Norton, B.W. and Sahlu, T. (2000) Recent advances in the nutritional biology of Angora and Cashmere goats. In: Gruner, L. and Chabert, Y. (eds) *Proceedings of the 7th International Conference on Goats*. Institut de l'Elevage and Institut National de la Recherche Agronomique, Paris, pp. 59–65.

Gallico, L. (1992) La fibra tessile mohair. *Notiziario dell'ENEA e di RENAGRI* 22, 44–49.

Hadjigeorgiou, I.E., Gordon, I.J. and Milne, J.A. (2001) The intake and digestion of a range of temperate forages by sheep and fibre-producing goats. *Small Ruminant Research* 39, 167–179.

Ivey, D.S., Owens, F.N., Sahlu, T., Teh, T.H., Claypool, P.L. and Goetsch, A.L. (2000) Growth and cashmere production by Spanish goats consuming *ad libitum* diets differing in protein and energy levels. *Small Ruminant Research* 35, 133–139.

Klören, W.R.L. and Norton, B.W. (1993) Fleece growth in Australia cashmere goats. II. The effect of pregnancy and lactation. *Australian Journal of Agricultural Research* 44, 1023–1034.

Klören, W.R.L., Norton, B.W. and Waters, M. J. (1993) Fleece growth in Australian cashmere goats. III. The seasonal patterns of cashmere and hair growth, and the underlying seasonal patterns of growth hormone, prolactin and thyroxine concentration. *Australian Journal of Agricultural Research* 44, 1035–1050.

Lee, D.R., Galbraith, H. and Scaife, J.R. (1994) *In vitro* fibre production and protein synthesis in secondary hair follicles of the Cashmere and Angora goat. *Animal Production* 58, 483.

Li, Y., Ma, N., Song, Y., Luan, W. and Lou, Y. (1996) Effects of non-genetic factors on main economic traits in Liaoning cashmere goat. In: Holst, P.J., Rubino, R., Galal, E., Soedjana, T., Haenlein, G.F., Gatenby, R., Morand-Fehr, P., Lebbie, S.H. and Restall, B. (eds) *Proceedings of the VI International Conference on Goats*. International Academic Publishers, Beijing, pp. 202–204.

Luo, J., Goestch, A.L., Nsahlai, I.V., Sahlu, T., Ferrel, C.L., Owens, F.N., Galyeans, M.L., Moore, J.E. and Johnson, Z.B. (2004a) Prediction of metabolizable energy and protein requirements for maintenance, gain and fibre growth of Angora goats. *Small Ruminant Research* 53, 339–356.

Luo, J., Goestch, A.L., Nsahlai, I.V., Moore, J.E., Galyeans, M.L., Johnson, Z.B., Sahlu, T., Ferrel, C.L. and Owens, F.N. (2004b) Voluntary feed intake by lactating, Angora, growing and mature goats. *Small Ruminant Research* 53, 357–377.

Lupton, C.J., Dooling, A.R., Lankford, K., Huston, J.E. and Pfeiffer, F.A. (2000) Effect of location on fiber production by Cashmere goats. In: Gruner, L. and Chabert, Y. (eds) *Proceedings of the 7th International Conference on Goats*. Institut de l'Elevage and Institut National de la Recherche Agronomique, Paris, p. 655.

Lynch, P. (1990) Cashmere growth: its control and possible manipulation. In: Russel, A.J.F. (ed.) *Scottish Cashmere – The Viable Alternative*. D. & J. Croal Ltd, Haddington, UK, pp. 5–12.

McDonald, P., Edwards, R.A. and Greenhalgh, J.F.D. (1992) *Animal Nutrition*, 4th edn. Longman Group UK Limited, Harlow, UK.

McGregor, B.A. (1988) Effect of different nutritional regimens on the productivity of Australian cashmere goats and portioning of nutrients between cashmere and hair growth. *Australian Journal of Experimental Agriculture* 28, 459–467.

McGregor, B.A. (1996a) Environmental, nutritional and management influences on quality and production of mohair and cashmere. In: Holst, P.J., Rubino, R., Galal, E., Soedjana, T., Haenlein, G.F., Gatenby, R., Morand-Fehr, P., Lebbie, S.H. and Restall, B. (eds) *Proceedings of the VI International Conference on Goats*. International Academic Publishers, Beijing, pp. 285–296.

McGregor, B.A. (1996b) Lupin grain but not barley straw supplements allow cashmere buck kids to grow rapidly during winter. *Proceedings of the Australian Society of Animal Production* 21, 294–297.

McGregor, B.A. (1998) Nutrition, management and other environmental influences on the quality and production of mohair and cashmere with particular reference to Mediterranean and annual temperate climatic zones: a review. *Small Ruminant Research* 28, 199–215.

Merchant, M. and Riach, D.J. (1994) The intake and performance of cashmere goats grazing sown swards. *Grass and Forage Science* 49, 429–437.

Merchant, M. and Riach, D.J. (2002) The effect of plane of nutrition and shearing on the pattern of the moult in Scottish cashmere goats. *Animal Science* 74, 177–188.

Millar, P. (1986) The performance of cashmere goats. *Animal Breeding Abstracts* 54, 181–199.

Morand-Fehr, P. (1993) Nutritional characteristics and feeding strategies for Angora goats. In: Güney, O., Biçer, O. and Ranieri, M.S. (eds) *Production of Hides, Skins, Wool and Hair*. Pudoc Scientific Publishers, Wageningen, The Netherlands, pp. 97–107.

Morand-Fehr, P. and Galbraith, H. (1992) Nutritional characteristics and feeding strategies for fibre producing goats. In: Galbraith, H. (ed.) *New Developments in Goat*

Husbandry for Quality Fibre Production. Aberdeen University Studies Committee, Aberdeen, UK, pp. 40–66.

Newman, S.A.N. and Paterson, D.J. (1999) Variation in fibre and fleece characteristics between and within South African, New Zealand, and South African × New Zealand Angora goat genotypes. *New Zealand Journal of Agricultural Research* 42, 77–82.

Norton, B.J., Wile, C.A. and Hales, J.W. (1990) Grazing management studies with Australian cashmere goats: 1. Effect of stocking rate on the growth and fleece production of weaner goats grazing tropical pastures. *Australian Journal of Experimental Agriculture* 30, 769–775.

Petrie, O.J. (1995) *Harvesting of Textile Animal Fibres*. FAO Agricultural Services Bulletin No. 122. Food and Agriculture Organization of the United Nations, Rome.

Phan, K.H. and Wortmann, F.J. (2000) Quality assessment of goat hair for textile use. In: Gruner, L. and Chabert, Y. (eds) *Proceedings of the 7th International Conference on Goats*. Institut de l'Elevage and Institut National de la Recherche Agronomique, Paris, pp. 638–640.

Qi, K., Lu, C.D., Owens, F.S. and Lupton, C.J. (1992) Sulphate supplementation of Angora goats: metabolic and mohair responses. *Journal of Animal Science* 70, 2828–2837.

Reis, P.J. (1989) The influence of absorbed nutrients on wool growth. In: Rogers, G.E., Reis, P.J., Ward, K.A. and Marshall, R.C. (eds) *The Biology of Wool and Hair*. Chapman and Hall, London, pp. 185–204.

Restall, A.J.F., Restall, H., Restall, M. and Parry, A. (1994) Seasonal production of cashmere and environmental modification in Australian cashmere goats. In: Laker, J.P. and Allain, D. (eds) *Proceedings of the European Fine Fibre Network, Occasional Publication 2*. Macaulay Land Use Research Institute, Aberdeen, UK, pp. 63–74.

Rhind, S.M. and McMillen, S.R. (1995) Seasonal changes in systemic hormone profiles and their relationship to patterns of fibre growth and moulting in goat of contrasting genotypes. *Australian Journal of Agricultural Research* 46, 1273–1283.

Rubino, R., Di Trana, A. and Sepe, L. (2000) First appearance of Cashmere in a local goat breed in Italy. In: Gruner, L. and Chabert, Y. (eds) *Proceedings of the 7th International Conference on Goats*. Institut de l'Elevage and Institut National de la Recherche Agronomique, Paris, p. 656.

Russel, A.J.F. (1990) Nutrition of cashmere goats. In: Russel, A.J.F. (ed.) *Scottish Cashmere – The Viable Alternative*. D. & J. Croal Ltd, Haddington, UK, pp. 32–46.

Russel, A.J.F. (1995) Current knowledge on the effects of nutrition on fibre production. In: Laker, J.P. and Russel, A.J.F. (eds) *Proceedings of the European Fine Fibre Network, Occasional Publication 3*. Macaulay Land Use Research Institute, Aberdeen, UK, pp. 3–22.

Ryder, M. (1989) *Cashmere, Mohair and Other Luxury Fibres*, 1st edn. Itchen Printers Ltd, Southampton, UK.

Sahlu, T., Fernandez, M.J., Carneiro, H. and El Saher, H.M. (1992a) Partitioning of nutrient between milk and mohair in Angora goats. In: Lokeshwar, R.R. (ed.) *Proceedings of Recent Advances in Goats Production*. Nutan Printers, New Delhi, p. 1634.

Sahlu, T., Fernandez, M.J., Lu, C.D. and Manning, R. (1992b) Dietary protein level and rumen degradability for mohair production in Angora goats. *Journal of Animal Science* 70, 1526–1533.

Sauvant, D., Morand-Fehr, P. and Giger-Reverdin, S. (1991) Dry matter intake of adult goats. In: Morand-Fehr, P. (ed.) *Goat Nutrition*. EAAP Publication No. 46. Pudoc, Wageningen, The Netherlands, pp. 25–36.

Sepe, L., Fedele, V., Rubino, R. and Claps, S. (1996) Preliminary results on the adaptation of cashmere goats to the environment in Southern Italy. In: Holst, P.J., Rubino,

R., Galal, E., Soedjana, T., Haenlein, G.F., Gatenby, R., Morand-Fehr, P., Lebbie, S.H. and Restall, B. (eds) *Proceedings of the VI International Conference on Goats.* International Academic Publishers, Beijing, pp. 329–332.

Shahjalal, M., Galbraith, H., Topps, J.H. and Copper, J.M. (1991) Effect of level of protein supplementation on growth, body composition and fibre characteristics of British Angora goats. *Animal Production* 52, 608.

Shahjalal, M., Galbraith, H. and Topps, J.H. (1992) The effect of changes in dietary protein and energy on growth, body composition and mohair fibre characteristics of British Angora goats. *Animal Production* 54, 405–412.

Shelton, M. (1981) Fiber production. In: Gall, C. (ed.) *Goat Production.* Academic Press Inc. Ltd, London, pp. 379–409.

Souri, M., Galbraith, H. and Scaife, J.R. (1998) Comparison of the effect of genotype and protected methionine supplementation on growth, digestive characteristics and fibre yield in Cashmere and Angora goats. *Animal Science* 66, 217–223.

Trabalza Marinucci, M., Acuti, G., Antonini, M. and Olivieri, O. (2003) Fibre characteristics and follicle activity in Angora kids as affected by diet. *Italian Journal of Animal Science* 2(Suppl. 1), 545–547.

van der Westhuysen, J.M. (1982) Mohair as a textile fibre. In: Ayers, J.L. and Foote, W.C. (eds) *Proceedings of the III International Conference on Goat Production and Disease*, Tucson, Arizona. Dairy Goat Journal Publishing Co., Scottsdale, Arizona, pp. 264–267.

Willingham, T.D., Thompson, P.V. and Shelton, J.M. (1988) Chemical castration in Angora goats. *Research Reproduction Sheep and Goat, Wool and Mohair*. PR-4577. San Angelo Agricultural Research and Extension Center, San Angelo, Texas, pp. 28–30.

Wynn, P.C., Wallace, A.L.C., Kirby, A.C. and Annison, E.F. (1988) Effects of growth hormone administration on wool growth in Merino sheep. *Australian Journal of Biological Science* 41, 177–187.

Zhang, F. and Shi, B. (1996) Effects of age on the cashmere production of goats. In: Holst, P.J., Rubino, R., Galal, E., Soedjana, T., Haenlein, G.F., Gatenby, R., Morand-Fehr, P., Lebbie, S.H. and Restall, B. (eds) *Proceedings of the VI International Conference on Goats.* International Academic Publishers, Beijing, pp. 102–103.

Zhang, W., Zhu, X.P., Lu, D.X., Hou, W.J. and Jia, Z.H. (2004) Study on dietary copper proper level of Inner Mongolian White Cashmere Goats during the cashmere growing period. *Journal of China Agricultural University* 9, 36–40.

12 Metabolic and Nutritional Diseases

C. STELLETTA, M. GIANESELLA AND M. MORGANTE

Dipartimento di Scienze Cliniche Veterinarie, Università degli Studi di Padova, Viale dell'Università 16, 35020 Legnaro (PD), Italy

Disorders in Energy Metabolism

Introduction

Goats adapt well to different conditions, being particularly suited to tough environments, where they do better than sheep and cattle. The adaptive ability of goats is mainly due to: (i) their feeding behaviour, characterized by highly selective feeding and good digestibility of high-fibre forages; and (ii) their optimal ability to store and mobilize body reserves.

Appropriate storage and mobilization of body reserves by goats is crucial for their health and productive success, and varies with breed, feeding regime, length of lactation and environmental seasonal pattern.

For a good understanding of energy metabolism disorders in goats, knowledge of the physiology of this particular species is required. For example, the activation of seasonal physiological strategies, starting from the central nervous system and reaching peripheral tissue metabolism, is known to be involved in the adaptation of goats to different environmental conditions.

In this chapter, the disorders of energy metabolism in goats are dealt with by: (i) discussing the physiology of seasonal mammals; and (ii) describing the most important pathologies and their therapies.

Physiology of the energy metabolism in seasonal mammals

Seasonally breeding vertebrates show an extraordinary range of physiological and behavioural rhythms. The goat is a seasonal-breeding species and its adaptability depends on the synchronization of various animal and environmental factors.

High-yielding dairy ruminants experience dramatic changes in energy metabolism and feed intake during the periparturient period. Feed intake is reduced in the last period of pregnancy, and particularly during the last 2 or 3 weeks prepartum (Ingvartsen and Andersen, 2000), despite the increasing energy need for fetal and mammary gland development. With the onset of lactation, the added metabolic activities of the mammary gland increase the total energy requirements up to fourfold (Ingvartsen and Boisclair, 2001). A pronounced negative energy balance occurs because voluntary feed intake does not satisfy the increased energy requirements for milk production, which is then covered by mobilizing lipids from adipose tissue. Reduced appetite in ruminants has previously been ascribed primarily to physical limitations of the abdomen, but metabolic signals may play an equally important role. This hypothesis is based on the fact that intake reduction occurs simultaneously with changes in reproductive status, fat mass and metabolism. Specific signals responsible for intake reduction could include increased plasma levels of oestrogens and high rates of fat mobilization.

Among the hormones responsible for endocrine synchronization relative to environmental changes is leptin. In all studied species, including farm animals, leptin is primarily synthesized by adipose tissue (Chilliard *et al.*, 2005).

It is possible that the increases in plasma leptin occurring during early pregnancy do modulate maternal feed intake, distributing nutrients between maternal and fetal tissues as well as towards mammary and placental functions. After parturition, the lower concentration of plasma leptin could promote a faster increase in voluntary feed intake. Regulation of feed intake may not be the most important function of leptin. The main role of leptin could be to coordinate the adaptations required to survive periods of nutritional deprivation. In this context, the hypoleptinaemia associated with nutritional deficit triggers a series of centrally mediated adaptations promoting energy conservation. Another mechanism by which a reduction in plasma leptin could help early-lactation ruminants is by decreasing the response of peripheral tissues to insulin, a key adaptation promoting the partitioning of glucose to the mammary gland. Direct actions of leptin on liver, skeletal muscle, adipose tissue and pancreatic cells may also occur (Sauerwein *et al.*, 2006).

Central nervous system regulation of feed intake, energy expenditure and body weight (BW) gain is accomplished by activity in several brain sites. However, both genetic stimulation and lesion experiments suggest that the hypothalamus is critical to central integration of energy balance. Some trials on seasonal species have demonstrated interrelationships between environmental (light, temperature) and nutritional (feed availability) factors and metabolic signals (insulin, insulin-like growth factor-I, leptin, glucose, sex steroids) derived from peripheral organ activities (Lincoln *et al.*, 2001; Mercer *et al.*, 2001; Rousseau *et al.*, 2002; Yonezawa *et al.*, 2004). Such exchange of information leads to hypothalamic responses. Various hormones released by the hypophysis (growth hormone, adrenocorticotropic hormone, melanocyte-stimulating hormone, thyroid-stimulating hormone (TSH), follicle-stimulating hormone, luteinizing hormone) are subjected to corresponding hypothalamic releasing factors. The releasing hormones follow the metabolic activity and plasticity of hypothalamic neurones constituting different nuclei. Adaptation to productive (due to management decisions) and physiological variations seems to be related to central orchestrating functions. Therefore, more attention must be paid to regulatory mechanisms associated with adaptability and welfare.

Metabolic disorders

Ketosis

Ketosis is a disorder of carbohydrate and fat metabolism characterized by increased concentrations of ketone bodies in blood (ketonaemia), urine (ketonuria), milk (ketolactia) and other body fluids. Besides being converted into triglycerides, non-esterified fatty acids (NEFAs) are β-oxidized in mitochondria and peroxisomes in the liver. Under normal conditions, the peroxisomal β-oxidation is a minor pathway for fatty acid oxidation. However, this pathway is enhanced by fasting (Reddy *et al.*, 1999). Acetyl-CoA, the final product of the β-oxidation, is further oxidized in the tricarboxylic acid (TCA) cycle by binding with oxaloacetic acid (depleted from the mitochondria). The latter is an intermediate product of the TCA cycle and also an obligatory link between this cycle and the gluconeogenesis pathway. During early lactation, the demand for gluconeogenesis is conspicuously increased for the synthesis of milk lactose. Since gluconeogenesis (which occurs in the cytosol) increases the consumption of oxaloacetic acid, acetyl-CoA enters the pathway for ketogenesis rather than the TCA cycle (Krebs, 1966). The subsequent over-production of ketone bodies such as β-hydroxybutyric acid causes ketosis. The transport of NEFAs into mitochondria is controlled by carnitine palmitoyltransferase, which is regulated by malonyl-CoA, also playing a role in the development of ketosis (Brindle *et al.*, 1985). Ketosis is experimentally induced by non-feeding (Brumby *et al.*, 1975) and by administration of 1,3-butanediol, a ketone precursor eventually absorbed by the rumen (Veenhuizen *et al.*, 1991). In both cases, fatty liver precedes ketosis and seems to be the most common finding in naturally occurring ketosis (Oikawa *et al.*, 1997). Epidemiologically, overfeeding is suspected to be a major cause of ketosis, similar to fatty liver (Markusfeld, 1985). In addition to the increase in NEFA concentration, cholesteryl ester and phospholipid concentrations may decrease (Nakagawa and Katoh, 1998; Katoh, 2002).

In goats, two clinical forms of ketosis have been described: (i) pregnancy toxaemia during the last month of pregnancy; and (ii) primary ketosis during the first month of lactation (French, 1970). The following factors can favour this disease: (i) the presence of multiple fetuses; (ii) malnutrition during late gestation; (iii) hypernutrition or fat goat; and (iv) stress factors (e.g. concurrent diseases, dietary changes). Pregnancy toxaemia is a metabolic disease that causes significant economic losses in small ruminant herds. Animals are usually affected by pregnancy toxaemia during the last few weeks of gestation, usually when the uterus contains two or more well-developed fetuses. A low- quality diet, together with a relevant glucose demand from the rapidly growing fetuses, may lead to hypoglycaemia and subsequently to ketosis. Other factors such as stress, overfattening and pregnancy may also help to trigger the metabolic disorder.

Ketosis can be classified as: (i) subclinical or clinical; and (ii) a primary or secondary disease. Subclinical ketosis and clinical ketosis differ in prevalence, course of development and clinical findings. Ketosis causes economic losses to the dairy industry by: (i) impairing milk production; (ii) decreasing reproductive efficiency; (iii) increasing involuntary culling; and (iv) requiring treatment expenses. The losses caused by undiagnosed subclinical ketosis can exceed the losses caused by the clinical form. Initially, the affected animal tends to isolate itself and show: (i) mild depression; (ii) capricious appetite; (iii) blindness; (iv) low or no responses to threat; (v) ataxia; (vi) nystagmus; and (vii) rambling. Then, metabolic acidosis and, consequently, respiratory acidosis develop. A characteristic smell of acetone in urine, in milk and in exhaled air can be found. The animal hardly moves, sometimes remains in a decubitus position and shows no interest in its surroundings. Goats that survive ketosis show dystocia and high neonatal mortality. Some show nervous symptomatology characterized by short or intermittent periods of hypersensitivity. The animal may show: (i) tremors and contractions of the ears, snout and eyelids; (ii) grinding of the teeth; (iii) excessive licking; and (iv) respiratory noises due to the presence of excessive nasal drain. The most remarkable macroscopic lesion is the extensive fatty infiltration of the liver, and the adrenal cortex and the proximal tubular epithelium of the kidneys may be similarly affected. Focal degenerative changes in contractile and conductor myocardial cells may also occur (Tontis and Zwahlen, 1987).

The diagnosis of ketosis is based on the history, clinical signs and increase of haematic concentrations of one or more ketone bodies. The major ketone bodies are β-hydroxybutyrate (BHBA), acetocetate and acetone. In goat blood, BHBA concentrations lower than 1000 μmol/l are considered to be normal (Geishauser *et al.*, 2000). In dairy cows, BHBA in the range from 1000 to 3000 μmol/l in blood has been set as indicative levels of subclinical ketosis (Geishauser *et al.*, 2000). Since no specific reference values have been established for goats, the same values applied for cows could be used. The determination of ketone bodies in milk can make sampling easier and can provide an immediate result. In contrast to cows, a range of values for BHBA concentrations in milk has not been established for goats (Dirksen *et al.*, 1995; Geishauser *et al.*, 2000). A general recommendation would be to consider that the higher limit of the normal range for acetone concentration in milk is less than 100 μmol/l. Glucose values remain within the normal range (35–45 mg/dl), but hyperglycaemia may occur in the final stages of ketosis. The prevention of such pathology is based on the supply of appropriate diets. For example, a recommendation for pregnant goats would be to feed high-quality hay and at least 500 g of concentrate per head per day. The most appropriate therapy is the intravenous administration of glucose, eventually associated with oral administration of glucose, propylenic glycol or sodium propionate. The last two substances give origin to propionic acid, which is one of the main metabolic precursors of glycogen. At the first signs of primary ketosis, which can be observed in the first month of lactation, it is advisable to suspend milking for at least 48 h, in order to administer corticosteroids (for their hyperglycaemic effect) associated, if necessary, with drugs with antilipolitic action (e.g. insulin).

Forestomach fermentative disorders

Disorders in the activity of the composite stomach of ruminants are characterized by functional, biochemical and morphological changes. The term 'indigestion' regards digestive disorders in the forestomachs. Disorders originating directly in the forestomachs are usually primary dysfunctions, often caused by: (i) changes in activity and composition of the microflora and microfauna; (ii) inhibition of physiological and biochemical pathways; and (iii) decreased rumen motility. Secondary diseases do not originate in the reticulo-ruminal environment and frequently develop from pathologies of the autonomic nervous system and of the digestive tract (e.g. enteritis, peritonitis and hepatitis). Disorders may also arise from affliction of the autonomic nervous system of those organs that are reflexively linked to the forestomachs (Vrzgula, 1991).

Ruminal acidosis

Ruminal acidosis in domestic ruminants is a well-known disease attributed to wrong management. The onset of rumen acidosis is linked to the intake of low-fibre and high-energy diets, often combined with a ruminal environment not yet adapted to highly fermentable feeds. Rumen acidosis is characterized by: (i) a decrease in ruminal pH; (ii) an increase in total concentration of short-chain fatty acids (SCFAs); (iii) shifts in the ratio between acetic, propionic and butyric acids, with increases in the concentration of propionic and butyric acids; and (iv) an increase in the concentration of lactic acid in the rumen fluid. Development and role of the papillae of the ruminal wall have been well described by Dirksen *et al.* (1984). The ruminal papillae are of crucial importance in the adsorption of SCFAs and their proliferation is promoted by SCFAs released by fermentation. When the ruminal mucosa is not adapted, which frequently happens during the transition between late gestation and early lactation, the papillae are not fully developed. Consequently, often the reduced absorbing surface is unable to deal with a sudden increase of SCFA levels (Brand and Warner, 1996; Nocek, 1997). Other factors related to rumen acidosis are neuroendocrine, management and environment conditions. Among neuroendocrine factors, several neuropeptides which control voluntary feed intake and eating behaviour may result in the sorting out of feeds from the diet. In the literature, ruminal acidosis has been largely reported in sheep and cattle. There are fewer reports of ruminal acidosis in goats and such studies have dealt with acute (or lactic) rumen acidosis (Gnanaprakasam, 1970; Vihan *et al.*, 1982; Tanwar and Mathur, 1983; Cao *et al.*, 1987; Aslan *et al.* 1995; Nour *et al.*, 1998).

Rather than rumen acidosis, subacute rumen acidosis (SARA) represents one of the most important metabolic disorders in intensive dairy cow farms and affects rumen fermentations, animal welfare, productivity and, consequently, farm profitability. A lot is known about SARA in dairy cows, while little is known about the incidence, aetiology and pathogenesis of this disorder in goats. Based on the few available data, it could be hypothesized that goats are more resistant to SARA than cows, even if further studies are necessary. Precocious identification of SARA on goat farms might prevent the onset of associated pathologies.

Ruminal acidosis is often under-diagnosed because of: (i) lack of pathognomonic signs; (ii) diurnal fluctuations in rumen metabolism; and (iii) difficulties in obtaining representative rumen fluid samples (Jorgenson *et al.*, 1993; Nordlund and Garrett, 1994). Clinical signs of ruminal acidosis are non-specific and include anorexia, ruminal stasis, diarrhoea, weakness, lack of coordination and recumbency (Elam, 1976; Blood and Radostits, 1989). Moreover, polioencephalomalacia is often associated with ruminal acidosis in goats as described later in this chapter.

Acute acidosis causes ruminal and systemic acidosis, prostration, coma and frequent death among severely affected animals. The rumen fluid is filled with contents characterized by sour smell and low pH. The rumen papillae are usually oedematous with erosions in between, identified as necrotic foci that can be several millimetres or even centimetres wide. The intestines are filled with watery fluid and the mucosa of the anterior parts is reddened (Vrzgula, 1991). The clinical signs are linked to rumen pathology, because of the very high osmotic capacity of the rumen and the increase in the absorption rate of ruminal products, which may have toxic and vasomotor effects. However, since the various factors associated with ruminal acidosis lead to ambiguous clinical signs, the latter cannot be used alone for definitive diagnosis of the disease. In fact, the determination of ruminal pH is a key factor for the diagnosis of ruminal acidosis (Morgante and Stelletta, 2004). In goats, rumen acidosis may be defined as the condition when rumen pH is below 5.5. Several methods are available for the collection of rumen fluid for pH analysis. One technique uses a weighted oro-ruminal probe and suction pump but this technique is susceptible to saliva contamination. Instead, as in cows (Garret *et al.*, 1999; Duffield *et al.*, 2004), rumenocentesis may be useful for the collection of rumen fluid for pH determination in goats (Morgante and Stelletta, 2004). However, rumenocentesis is a more invasive technique, involving surgical preparation of the centesis site as well as chemical and physical restraint, and shows a risk of localized abscesses or peritonitis. According to Morgante and Stelletta (2004), rumenocentesis is an invasive procedure which potentially endangers animal health and welfare by sequelae (e.g. peritonitis, and abscesses in the abdominal and ruminal wall), but its side effects depend on the materials and procedure used. In fact, the same authors have highlighted the importance of proper preparation of the area (shearing and disinfection) and goat handling during the procedure. First, the centesis area should be manually compressed at the beginning, to avoid a sudden reaction of the goat as the needle is introduced. Second, the needle should not be inserted during the contraction of the ventral sac of the rumen, and third the duration of the aspiration procedure must not exceed 1 or 2 min. Stelletta *et al.* (2004) successfully performed rumenocentesis to evaluate the effects of an acidogenic diet on female kids. In that study, due to an increasing utilization of urea, probably through a recycling mechanism, pH drop was less pronounced in goats than normally observed in cows.

Disorders in Mineral Metabolism

Introduction

Ca, P and Mg play a very important role in the metabolism of animal body tissues. Animals are often affected by disorders of mineral metabolism when the balance between intake and excretion of macronutrients is disturbed. Requirements for mineral elements depend not only on animal breed, climatic conditions, stocking intensity and digestibility, but also on unknown factors that might favour or inhibit the incorporation of these elements (Ebbesvik, 1993; Klimienė *et al.*, 2005). Deficiency of Co, Cu, Fe, I, Mn, Se or Zn can reduce production during growth, reproduction or lactation stages. Evaluation of trace elements may be difficult because many diseases will alter the blood analytes used to assess nutrient adequacy. Proper evaluation of the diet, and of animal condition and responses to supplementation, is necessary before diagnosing a trace element deficiency. Trace element deficiencies have been observed and studied for many years. Early studies (Lamand and Perigaud, 1973) showed that different hays were deficient in Cu, Co, Zn and Se, and, frequently, in Mn and I. Very low concentrations of these trace elements have been found in maize silage as well. Initially, recommendations for the supplementation of trace elements had focused on avoiding reduced productivity and clinical signs of deficiencies (Lamand and Perigaud, 1973). Successive studies have emphasized the role of trace elements in immunity, as recently reviewed by Spears (2000). Dietary mineral supplementation is very common in lactating dairy cattle, but is much less used in goats. The trace mineral status of animals depends not only on dietary allowance, but also on digestion and storage efficiency, which can be affected by interactions between trace elements and other feed constituents. Veterinary surgeons often hypothesize that low performance of a herd is caused by trace element deficiencies. In fact, when a mineral deficiency status is effectively determined, it is considered to be the likely cause of disorders.

Disorders in calcium and phosphorus metabolism

Generality

The metabolisms of Ca and P are closely interrelated. Ca is the most abundant mineral in the animal body. More than 98% of the Ca is found in the bones and teeth, mainly in the form of hydroxyapatite (phosphate and carbonate). Ca makes up 37% of the bone ash. Approximately 1.5% of Ca is found in soft tissues. In the blood, Ca is found exclusively in the serum. The reference values for Ca in the blood serum of goats are 8.9–11.7 mg/dl (Kaneko *et al.*, 1997). In blood serum, Ca is found in three forms: (i) indiffusible Ca (non-ionized and physiologically inactive form); (ii) ionized Ca (diffusible and physiologically active as the cation Ca^{2+}); and (iii) Ca in phosphate, bicarbonate and citrate complexes (physiologically inactive form) (Vrzgula, 1991; Smith and Sherman, 1994).

Most P is found in the bones and teeth, in the form of hydroxyapatite, representing up to 90% of the total P in the body of the animal. P constitutes about 50% of the bone ash and only 1% of soft tissues. The whole blood contains about 5.0–7.0 mg P/dl, most of which is found in the blood cells. In the blood serum, P occurs especially as inorganic phosphate (HPO_4^{2-}), which can be readily used in biochemical reactions. Phosphate levels depend not only on the intake and excretion of P, but also on the activity of osteoblasts and osteoclasts in the bones. In goats, a characteristic symptom of metabolism disorders of bone-building elements is decreased or increased calcaemia and phosphataemia, usually due to inadequate supplementation of these minerals in the feed or to regulatory factors (Sherman and Robinson, 1983; Vrzgula, 1991).

Hypocalcaemia

Clinical signs of hypocalcaemia in goats generally resemble those in cows (Radostits *et al.*, 1994; Yamagishi *et al.*, 1999). Affected goats show signs of ataxia, shivering and muscle trembling, followed by recumbency. Experimental studies on goats in which hypocalcaemia was caused by intravenous infusion of a 4% Na_2EDTA solution have shown that: (i) mild hypocalcaemia (more than 6 mg of Ca per decilitre of blood) caused staggering and restlessness; (ii) moderate hypocalcaemia (4 to 5 mg Ca/dl) caused depressed anal and palpebral reflexes and repeated reeling and falling; and (iii) severe hypocalcaemia (less than 2 mg Ca/dl) led to complete recumbency (Kan, 1979; Kan and Motoyoshi, 1979). Researches on hypocalcaemia have focused on blood Ca, ionized Ca, or total Ca measurements during periparturient hypocalcaemia, due to the immediate connection between blood Ca and milk fever disease (Oetzel, 1988; Radostits *et al.*, 1994).

Blood P status has also been studied due to: (i) the connection between P and Ca deposition; and (ii) the changes in Ca and P balances which occur at the onset of lactation. Hypophosphataemia may be caused by: (i) P loss in colostrum; (ii) inadequate P intake around parturition; (iii) inadequate P mobilization from bones; and/or (iv) P diuresis caused by high blood concentrations of parathyroid hormone (PTH) due to hypocalcaemia (Oetzel, 1988). In experimentally induced hypocalcaemia, it is suggested that hypophosphataemia depends on PTH action. The secretion of PTH is stimulated by the decreasing levels of plasma Ca^{2+}. High blood concentrations of PTH then lower plasma P concentrations, by increasing renal phosphate excretion and salivary phosphate secretion (Daniel and Moodie, 1979; Yamagishi *et al.*, 1999).

It is extremely important to obtain a definitive diagnosis of hypocalcaemia because this disorder affects animal health and dairy goat profits. Several other parameters, such as the mobilization of Ca from body stores, have also been studied. A possible mechanism by which early-lactation animals may overcome hypocalcaemia is by massive mobilization of Ca from the bones to the blood. Establishing parameters connected with bone resorption in the periparturient period helped to understand and describe the mechanisms behind the complex and delicate system known as calcium homeostasis. In practice, most diagnoses of parturient paresis are based on clinical observations made by veterinary practitioners. Among the parameters monitored most often are rectal temperature, external temperature, muscle shivering and ataxia, rumen motility and paresis. The obvious advantage of clinical examination is that diagnosis can be made immediately, which is fundamental for a successful therapy for acute milk fever. Subclinical conditions of milk fever, known as hypocalcaemia, may be difficult to diagnose only by standard clinical procedures. However, hypocalcaemia may be important at farm level. Finally, a very important parameter related to the onset of milk fever in goats is the dietary anion–cation balance. In fact, excess of fixed anions (found in maize silage and grains) causes subclinical metabolic acidosis, which increases the rate of Ca absorption but does not affect the size of the exchangeable Ca pool. Excess of fixed cations (associated with diets containing lucerne and buffers) causes subclinical metabolic alkalosis, which diminishes Ca absorption and urinary Ca excretion. The mild metabolic acidosis necessary to induce hypercalciuria represents the metabolic cost to maintain high prepartum Ca absorption rates and high flux of Ca through the exchangeable Ca pool, which may help to adjust to sudden Ca losses at parturition (Fredeen *et al.*, 1988; Stratton-Phelps and House, 2004).

Urolithiasis

Urinary calculus is a term used to describe the pathological concretion of mineral salts in the excretory passage of the kidney, a quite common condition in both humans and animals. Infections, defective drainage, partial obstructions, congenital malformations and urine pH can be predisposing factors. The calculi are most often found in the bladder, although urolithiasis can also occur in the renal pelvis and ureter (Ewoldt *et al.*, 2006). Calculi occur in both sexes, although obstruction occurs most frequently in males. The majority of calculi in goats are either struvite or silicate calculi (Gutierrez *et al.*, 2000; George *et al.*, 2007). The aetiology of urolithiasis is complex and multifactorial, being related to geographical, seasonal and dietary factors, as well as low water intake (Poole, 1989; Stratton-Phelps and House, 2004). A diet with a high content of P and Mg can increase the renal excretion of these minerals, increasing the presence of crystals in urine, above all if the urine is considerably alkaline (Stewart *et al.*, 1990). The acidification of urine, instead, makes these crystals solvable and therefore avoids their formation. The effects of environment on the incidence of urinary tract diseases such as urolithiasis may be more evident in hot and dry climates. Moreover, other regional peculiarities may also influence the rate of stone formation in kidneys and the urinary tract (Sahinduran *et al.*, 2007). The role of water hardness in the development of urolithiasis is still controversial. A number of studies have suggested that water hardness and Mg levels in drinking water can induce a greater incidence of urolithiasis (Medina-Escobedo *et al.*, 2002; Sahinduran *et al.*, 2007).

Although the occurrence of silica uroliths in ruminants has seemed inevitable in some studies simulating field conditions (Bezeau *et al.*, 1961), attempts to produce silica uroliths experimentally in ruminants through the feeding of inorganic silicates under controlled conditions have generally been unsuccessful (Stewart *et al.*, 1990). Based on a rat model and *in vitro* studies, Emerick and Lu (1987) have identified dietary factors such as low P concentrations, high Ca/P ratio and alkali-forming effects as contributors to the formation of silica uroliths in animals fed diets containing high silica concentration. Since urinary silica concentrations did not differ between treatments, the researchers concluded that the greater urine alkalinity and lower urinary P concentrations resulted from the consumption of the silicic acid solutions, which are important contributors to silica uroliths formation (Stewart *et al.*, 1990). There is a paradox in urolithiasis in small ruminants. On the one hand, the commonly used surgical procedures for relief of urinary tract obstruction (e.g. urethral process amputation, perineal urethrostomy and tube cystostomy) are relatively straightforward. On the other hand, successful long-term resolution of urethral obstruction can be difficult. The metabolic consequences of urinary tract obstruction and related anaesthetic risks contribute to short-term morbidity and mortality. Stricture of the perineal urethrostomy stoma, urinary bladder catheter dislodgement, or obstruction in tube cystostomy, and, most frequently, recurrent post-operative urethral obstruction by other uroliths often limit long-term survival and quality of post-operative life. Veterinary surgeons and internists have long been frustrated by the apparent disparity that exists in small ruminants between the moderate level of difficulty associated with surgical relief and the far more substantial difficulty of achieving long-term resolution (Van Metre and Fubini, 2006).

Disorders of magnesium metabolism

Generality

Among the body electrolytes, Mg^{2+} is the fourth most abundant cation after Ca^{2+}, Na^{+} and K^{+}. It constitutes up to 0.05% of the total BW. Mg^{2+} is the second most important cation, after K^{+}, in the intracellular fluid. Approximately 62% of the Mg content is deposited in the bones, 37% in the cells and 1% in the extracellular fluid. Blood plasma concentrations of Mg for goats are normally between 2.8 and 3.6 mg/dl. Part of plasma Mg is bound to non-diffusible proteins, another portion is incorporated into the molecules and the remaining part is in the free ionized form (Vrzgula, 1991; Smith and Sherman, 1994). Mg is an essential element for all animals, above all ruminants. A physiological deficiency of Mg (concentration lower than 1.1 mg/dl) results in hypomagnesaemic tetany. Typically, only female ruminants are affected, usually in early lactation. Mg works at three biochemical levels: (i) as a co-factor for enzymatic reactions; (ii) as a structural factor assembling ribosomes; and (iii) as a stabilizing force on the cell membrane. PTH can affect Mg metabolism by decreasing urinary Mg excretion and stimulating bone resorption, thus releasing Mg into the extracellular fluid. Renal excretion of Mg and its absorption from the gastrointestinal tract are both increased by 1,25-dihydroxyvitamin D_3. A number of dietary factors may depress Mg absorption in ruminants. For example, high dietary K has the greatest and most consistent effect in reducing Mg absorption. On the other hand, feeding good quantities of readily digested carbohydrates increases Mg absorption, but the mode of action is not clear. High concentrations of Al in forage are sometimes associated with high incidence of grass tetany, but this effect does not appear to be related to Mg absorption. Interrelationships of Al with Mg, Ca, P and PTH are implicated (Fontenot *et al.*, 1989).

Hypomagnesaemia

Hypomagnesaemia, known as grass tetany, stall tetany or transport tetany, is a so-called 'production disease' and one of the most frequently occurring metabolic disturbances of ruminants in large-scale production systems. Grass tetany occurs only in ruminants and especially in countries with a temperate climate. It may afflict dairy and beef cattle, calves, goats and sheep. Susceptibility to the disease increases with age, and incidence of the disease is correlated with the intensification of production. Hypomagnesaemia is characterized by decreased Mg levels in blood serum (lower than 1.1 mg/dl), with consequent enhanced neuromuscular irritability or tetany (Vrzgula, 1991; Smith and Sherman, 1994). Hypomagnesaemic animals may exhibit clinical signs such as decreased productivity, teeth grinding, salivation, ataxia, recumbency, tetany, seizures and death (Howard and Smith, 1999; Martens and Schweigel, 2000). The development of any form of hypomagnesaemia is dependent on the balance between absorption and losses of Mg, which is influenced by host and environmental factors. Host factors, such as decreased feed intake, Mg-deficient diets, lactation and altered absorption, lead to the depletion of intracellular Mg levels and, eventually, serum Mg. Mg is primarily absorbed in the forestomachs of ruminants. Rumen epithelium takes up Mg by paracellular (passive) and transcellular (active) mechanisms. Unlike the metabolism of many other essential minerals, there is no specific hormonal system to regulate Mg homeostasis (Herdt *et al.*, 2000; Martens and Schweigel, 2000). The kidney is able to excrete excess Mg, but it is unable to conserve enough Mg systemically in situations of deficiency. As a result, endogenous and exogenous losses (i.e. lactation) may exceed the intake of Mg, causing a drop of the extracellular concentration of Mg, while renal losses continue. When changing from winter rations to spring grass, decreased Mg absorption may occur. An increased intake of oral K results in a decrease in Mg absorption from the forestomachs, due to a decrease in the passive uptake of Mg by the epithelium. The decreased level of Na and high level of K in young spring grass lead to Na deficiency, decreased Na/K ratio in the rumen, and consequently increased level of K in ruminal fluid. This further worsens the imbalance by decreasing the absorption of Mg from the rumen. Adequate amounts of fermentable carbohydrates are important in maintaining serum Mg levels, since both Mg solubility and the absorptive surface area of rumen papilla increase with availability of SCFAs and lowered rumen pH. Stress may lead to clinical hypomagnesaemia, since sympathetic nervous system activation causes adrenaline release, resulting in a decreased plasma Mg concentration (Herdt *et al.*, 2000; Martens and Schweigel, 2000).

The Mg status of an animal is most conveniently determined by serum Mg concentration. Since it characterizes the state of Mg metabolism in the animal, it is a good basic indicator of both dietary adequacy and nutritional status (Sherman and Robinson, 1983; Smith and Sherman, 1994). It should be noted that mild low serum Mg levels may not predict clinical signs, since the neurological manifestations are reflective of cerebrospinal fluid Mg concentrations.

Prevention of hypomagnesaemia should aim at: (i) continuous provision of adequate levels of Mg in the diet; and (ii) maximizing absorption of this essential mineral. Possible choices of supplemental salts include $MgCO_3$, $MgSO_4$ and $MgCl_2$, but MgO is the most common choice, due to its economic and palatability qualities, and fewer laxative effects. Improving absorption of Mg by the ruminal epithelium can be attempted by lowering dietary K intake, even though this may be impractical. The following practices effectively prevent the risk of grass tetany: (i) using MgO supplementation to increase Mg intake; (ii) offering salt lick blocks to prevent Na deficiency; and (iii) providing adequate amounts of fermentable carbohydrates to increase total energy intake. These are particularly important when moving from normal winter rations to young spring grass pastures.

Selenium and vitamin E deficiency

Generality

Animal and human diseases due to insufficient or excessive amounts of Se in the diet have led to extensive studies on the metabolism, biochemistry and nutritional requirements of this mineral (Gutzwiller, 1991; Wichtel *et al.*, 1996). The metabolic pathways of Se in the body have been studied by administering labelled Se and determining the amount of tracer present in blood, tissues and excreta at different time intervals (Janghorbani *et al.*, 1990). Se is readily absorbed from the intestine and extracted rapidly by the liver and the erythrocytes. Although faeces and urine are the major excretory pathways of Se in all species, urinary excretion is responsible for homeostatic control in monogastric animals under conditions of deficiency. On the other hand, the relative importance of the faecal and urinary routes of excretion in the regulation of Se metabolism in ruminants is controversial (Krishnamurti *et al.*, 1997). Rumen microbes reduce Se compounds in the feed to insoluble selenides, which are then excreted in the faeces. Due to the presence of higher levels of Se in the faeces than in the urine of ruminants, it has been hypothesized that the faecal route may be more important in homeostatic regulation in ruminants, unlike monogastric animals (Krishnamurti *et al.*, 1997).

Among the various syndromes caused by insufficient dietary Se, together with a concurrent deficiency of vitamin E, the best known in ruminants is nutritional muscular dystrophy (NMD), or white muscle disease (Radostits *et al.*, 2000). Se deficiency in goats may be associated with seasonal variations (Khan *et al.*, 2005) as well as with: (i) increased occurrence of infectious disease; (ii) reproductive failure; (iii) reduced growth; and (iv) reduced productivity (Pherson, 1993). The importance of adequate vitamin E and Se levels in goat nutrition is well documented (Ramirez-Bribiesca *et al.*, 2001; Yang *et al.*, 2004). By contrast, the practical importance of adequate vitamin E and Se status for mastitis resistance and proper reproductive function is more controversial (Radostits *et al.*, 2000). Several authors have reported a beneficial effect of vitamin E and/or Se on reducing mastitis frequency (Sanchez *et al.*, 2007). Free ions of several metals, such as Cu, Se, Fe and Zn, are required for the body's antioxidant system, but induce oxidative stress if in excess. Soils and feeds are known to be deficient in Se, and NMD is still commonly diagnosed in goats. Vitamin E levels in feed vary depending on the grain and forage type, growth and storage conditions.

Nutritional muscular dystrophy

NMD or white muscle disease is an enzootic disease caused by a deficiency of vitamin E and/or Se (Bickhardt *et al.*, 1999; Kennedy, 2000). Factors associated with increased incidence of NMD are: (i) dietary polyunsaturated fatty acids; (ii) unusual exercise levels; and (iii) rapid growth. This disease may affect any farm animal species, especially rapidly growing calves, lambs, kids and foals

(Tontis, 1984; Radostits *et al.*, 2000). Both vitamin E and Se are important in the protection of cellular membranes from free radicals, which cause peroxidation of membrane lipids. Vitamin E is an antioxidant that decreases hydroperoxide formation and scavenges free radicals at extracellular or intracellular level. Se is an important biochemical component of the enzyme glutathione peroxidase, an intracellular enzyme that protects cellular membranes and organelles from peroxidative damage (Anke *et al.*, 1989). It inhibits and destroys the endogenous peroxides and, in conjunction with vitamin E, protects cellular membranes. When these mechanisms are inadequate, cell membranes become physiologically defective, resulting in the accumulation of Ca inside mitochondria and injury to these organelles. Injured mitochondria are then unable to maintain homeostasis, resulting in cell death or segmental necrosis.

Clinical signs of NMD include stiffness, weakness, and recumbency. Diagnosis is usually based on: (i) clinical findings; (ii) elevated levels of muscle enzymes (creatinine phosphokinase and aspartate aminotransferase); (iii) low levels of vitamin E and Se in the diet, tissue and serum; and (iv) muscle degeneration. At necropsy, affected muscles are usually pale, and histological examination reveals hyaline degeneration and segmental necrosis (Ogryzkov *et al.*, 1966). Congenital muscular dystrophy is rare and not well documented in kids, possibly because the developing fetal kid is able to sequester sufficient Se. Therefore, except under extremely deficient conditions, proper quantities of Se are normally available at birth to prevent the expression of congenital white muscle disease. In a rare case (Koller *et al.*, 1984), the clinical findings suggestive of the disease were recumbency and weakness, which may have predisposed to hypothermia, since the kid was unable to stand and suck. The slightly elevated heart and respiratory rates and the weak sucking reflex may have been indicators of some myocardial, diaphragmatic and tongue involvement. Traces of blood on urine dipstick could have been due to myoglobin (e.g. in case of myopathy), haemoglobin, or even intact red blood cells. The low levels of vitamin E and Se found in the blood strongly support a diagnosis of congenital muscular dystrophy. A muscle biopsy could have helped to confirm the diagnosis, but it was not done.

Early stages of NMD can be treated with parenteral injection of vitamin E and Se. This disease can be prevented by supplementation of vitamin E and Se in the diet, or by strategic oral and/or parenteral vitamin E and Se to pregnant dams or young animals on pasture (Kennedy, 2000; Van Metre and Callan, 2001; Menzies *et al.*, 2003).

Copper deficiency

There can be two types of Cu deficiency in goats, as in other ruminants. The primary Cu deficiency results from dietary inadequacy. The secondary one occurs when Cu absorption or metabolism is negatively influenced by factors other than Cu-deficient diets.

Various conditioning factors may decrease gastrointestinal absorption of Cu. Recommended minimum dietary Cu concentration is 10 ppm (Lamand, 1978), keeping the Cu/Mo ratio below 2:1 (Buck, 1986) as well. Cu supplementation might be increased to 14 ppm, when there are elements in the diet able to interfere with Cu absorption. Secondary Cu deficiency mainly occurs at high dietary levels of Mo, sulphate, Zn, Fe and other elements. Dietary excess of Mo may form copper molybdate in the rumen (Frank *et al.*, 2000a,b). This compound is highly insoluble and not absorbable by the gut. Besides, molybdate reacts with excess S and sulphate in feed and water, forming an insoluble complex (copper thiomolybdate) in the rumen. Since the clinical signs of Cu deficiency often have a multifactorial origin, it might be hard to diagnose the deficiency.

Cu is an essential component of many enzymes, such as superoxide dismutase (cytoplasmatic form, Zn- and Cu-dependent), cytochrome oxidase, lysyl oxidase, ascorbic acid oxidase and caeruloplasmin. Adequate Cu intake seems to be essential for Fe absorption and transport from liver to reticulo-endothelial system and for the manufacture of haemoglobin. Most of the syndromes due to Cu deficiency concern a physiopathology which is still not exactly known. Cu seems to have a very important role in avoiding damage linked to cellular oxidation and in the metabolism of Fe and S.

Signs of Cu deficiency are: (i) copious and watery diarrhoea; (ii) a drop in body weight; (iii) unhealthy appearance; (iv) anaemia; (v) change in quality or colour of the hair coat (acromotrichia); (vi) spontaneous bone fractures; (vii) progressive demyelination of peripheral nerves (caprine and ovine enzootic ataxia); and (viii) lameness (epiphysitis). Generally, only one of these syndromes prevails on a single farm. Kids and fetuses are more susceptible to Cu deficiency than adults. Compared with calves, kids born from mothers with Cu deficiency are more likely to show clinical signs. Milk is a poor Cu source; in fact it contains only 0.2–0.6 ppm Cu in normal sheep and goat, and 0.01–0.02 ppm in those with severe deficiency. Most of the body Cu is stored in the liver, where its normal concentration is about 80–200 μg/g (ppm) of dry matter (DM). Cu concentrations in blood may be normal even when hepatic levels of Cu drop below 35 μg/g. However, if these levels decrease further, a decrease in blood Cu concentrations may also occur. Cu concentration is usually 5% higher in plasma than in serum. Normal Cu concentrations in serum are about 0.7–1.2 μg/ml. Values below 0.4 μg/ml are considered extremely low, while those between 0.4 and 0.7 μg/ml are considered marginal, and their aetiology is hard to find.

Cu deficiency is usually treated by using injectable copper glycinate (30% Cu). Prognosis is often good but it depends on the seriousness of the lesions. In the endemic areas, clinical signs of Cu deficiency may be prevented by administering 35 mg $CuSO_4$ per head to kids, twice a week. In goats, Cu requirements vary from 5 to 10 mg/kg DM. Adult goats are less susceptible to Cu toxicity than sheep. Despite this, chronic Cu poisoning incidents occurred in Angora kids fed reconstituted milk for calves (Humphries *et al.*, 1987).

Zinc deficiency

Zn plays an essential role within many enzyme systems in several body processes. It is found in more than 70 enzymes, including several dehydrogenases, aldolases, phosphatases and peptidases. It is found as a co-factor in other enzymatic systems, such as endolase, decarboxylase, arginase and RNA-polymerase. Other enzymes are enabled by Zn, but they do not form permanent complexes with it. Normal Zn concentrations in serum are 0.6–2.7 μg/ml. In ruminants, Zn is absorbed by the whole gastrointestinal tract. Zn absorption depends on many factors such as age, feed composition and relationship between Zn and other elements. For example, Ca seems to affect the absorption of Zn in piglets, chickens, goats, laboratory animals and dogs. An excess of Ca interferes with Zn absorption by increasing phosphate and phytic acid availability in feed (Singer *et al.*, 2000). Vitamins also interfere with Zn absorption. It decreases rapidly with vitamin A deficiency, even if Zn content in the diet is adequate. Vitamin D deficiency also causes a decrease in Zn absorption and consequent Zn deficiency. The same is caused by vitamins B_1 and B_6. The organism uses about 10–30% of dietary Zn, but when the ration is poor in Zn, its use may increase up to 75%. After being absorbed, Zn is carried by the blood to the liver, spleen and pancreas, where it is stored in a metabolizable form. The highest Zn concentration is found in chorioides, the iris, skeleton and in epidermal structures such as hooves. Zn is mainly eliminated in faeces and in small amounts in urine.

Since Zn is important for the function of many enzymes, its deficiency concerns many apparatuses and organs. Zn deficiency is reported in swine, laboratory animals, chickens, and small and large ruminants. Common clinical signs are reduced appetite and growth, BW loss, lesions of epidermal tissues and skeletal alterations (Ray *et al.*, 1997). Zn deficiency also causes hypofertility, with males and young animals being more susceptible. In males, spermatogenesis is inhibited since sperm has high Zn concentration. In females, Zn deficiency causes slow development of secondary sexual characteristics and inhibition of the reproductive cycle at every stage, from heat to lactation. The BW loss is due to insufficient functioning of proteolytic enzymes. Skeletal diseases were found in mice and chickens. Embryos born to mothers with Zn deficiency are also affected, showing: (i) deformed skeletons; (ii) shorter long bones; (iii) fused vertebrae; and (iv) abnormal cranial development. Zn metabolism is linked to that of endocrine glands. In the testes and prostate gland, Zn levels are directly governed by pituitary gonadotrophic hormones. Zn deficiency may be induced in piglets by suppressing parathyroid gland hormonal activity, and by supplying dietary Ca excess or thyrostatic substances. Zn deficiency can be prevented by supplementation of $ZnCO_3$ or $ZnSO_4$ in the diet.

Cobalt deficiency

Ruminal microorganisms are able to synthesize vitamin B_{12}, which contains 4% Co. Therefore, in ruminants, Co deficiency appears as vitamin B_{12} deficiency, whose subsequent determination is a good index of deficiency of this element (McDowell, 1992). The small amount of Co usually stored in the body does not easily return to the rumen or gut for vitamin B_{12} synthesis. Consequently, goats, as other ruminants, need a continuous intake of dietary Co for adequate ruminal vitamin B_{12} synthesis. Therefore, Co parenteral administration is useless in cases of secondary vitamin B_{12} deficiency.

Goats affected by Co deficiency and the secondary deficiency of vitamin B_{12} show the following signs: (i) loss of appetite; (ii) ill-thrift; (iii) weakness; (iv) anaemia; (v) diarrhoea; and (vi) decreased production. The clinical disease is more frequent in young, growing animals.

In ruminants affected by Co deficiency, a primary defect is propionate metabolism inadequacy at the stage in which methylmalonyl-CoA mutase requires vitamin B_{12} and then catalyses the conversion of methylmalonyl-CoA to succinyl-CoA. As Co deficiency becomes severe, the rate of blood propionate utilization decreases and methylmalonyl-CoA overstocks. The quantity of methylmalonic acid in urine increases with severe Co deficiency. In animals with Co deficiency, the quantity of voluntary feed intake decreases as long as the propionate clearance half-life increases. Decreased growth rate, weight loss, impairment and anorexia are strictly correlated with anomalies in carbohydrate metabolism. Diarrhoea as a result of Co deficiency has not yet been explained, but it may be due to an increase of parasitosis susceptibility.

Anaemia develops late in the Co deficiency syndrome and is classified as normocytic and normochromic. Co deficiency causes deficiency of vitamin B_{12} enzymes containing 5-methyltetrahydrofolate-homocysteine methyltransferase. This interferes with methionine recycling and has a strong effect on folate metabolism. Furthermore, increasing anaemia, due to inefficient folate metabolism, and low activity of methyltransferase could result in methionine deficiency. This is a possible explanation for N retention, reduced BW and wool growth in affected animals.

Since the role of Co in ruminants is linked to the synthesis, absorption and use of vitamin B_{12}, the most significant clinical test is the analysis of vitamin B_{12} levels in tissues. However, the effect of abstinence from feed tends to enhance vitamin B_{12} concentrations in liver and kidneys. If Co deficiency is linked to other situations which cause anorexia, vitamin B_{12} levels in tissues may be normal. Fresh liver vitamin B_{12} concentrations lower than 0.07 μg/g are found in severe deficiency, between 0.07 and 0.1 μg/g in moderate deficiency, between 0.11 and 0.19 μg/g in mild deficiency and higher than 0.19 μg/g in normal intake conditions (Johnson *et al.*, 1999).

Serum vitamin B_{12} determination is not adequate in many clinical situations. Serum or plasma vitamin B_{12} levels have relevant daily variations. Serum vitamin B_{12} concentration can explain more clearly short-term Co intake and might be decreased even with adequate liver storage. Serum vitamin B_{12} levels usually range between 1 and 3 ng/ml with normal Co intake. Decreased levels at 0.3 ng/ml represent a threshold with marked clinical signs of Co deficiency. Severe Co deficiency causes urine excretion of methylmalonic acid and formiminoglutamic acid (FIGLU) (accumulated due to defect of cyclodeaminase and formiminotransferase enzymes). Levels of FIGLU ranging from 0.08 to 0.2 μmol/l might be representative of Co deficiency and might go back to zero after vitamin B_{12} administration. Since determinations of methylmalonic acid and FIGLU excretion in urine may be useful only during the last period of Co deficiency, the applicability of this method is limited.

Co deficiency occurs more often in some regions of the world and is associated with various soil types. The clinical form has been reported in New Zealand,

Australia, Brazil, the UK, Ireland, Scandinavia and North America. In practice, diets with 0.1 mg Co/kg DM are appropriate for ruminants. Co content in *Leguminosae* is relatively high. Fast-growing grass has a much lower Co concentration than *Leguminosae*, while cereal grains have low Co content. Cereal seed oil is a good source of Co.

Co in forage and soil is rarely measured. For this reason, the biggest evidence of Co deficiency is the increase of farming profits due to Co supplementation. The response to supplementation is usually rapid, with an increase in appetite in less than a week and subsequent improvement in nutritional status. However, recovery from anaemia is much slower (Blood and Radostits, 1989).

Iodine deficiency (hypothyroid goitre)

The most evident sign of I deficiency is thyroid enlargement or goitre, particularly in newborn animals. Sheep and goats seem to be more susceptible to this deficiency than cattle. However, adult animals may also be affected by subclinical and pernicious forms that lead to a decrease of productive performance, abortions and high neonatal mortality (Aguggini *et al.*, 1998). I deficiency is widespread on goat farms, since most of the forage consumed has lower I levels than those needed to satisfy animal requirements. Other factors such as high Ca intake, which limits I absorption, and intake of *Brassica* species, which contain cynogenic glucoside, may worsen I deficiency. Progressive impoverishment of some areas leads to an increased risk of I deficiency. Almost all I present in the body is stored in the thyroid in the form of thyroglobulin and thyroxine. Thyroid hormones such as triiodothyronine (T_3) and tetraiodothyronine (T_4) contain iodine. Among these hormones, only T_3 represents a free, physiologically active form. These hormones carry out important functions in the control of energetic exchanges, tissue metabolism and growth (Aguggini *et al.*, 1998).

I deficiency leads to goitre formation because low levels of T_4 induce thyroid stimulation by TSH, produced by the pituitary gland, causing thyroid hyperplasia (Fig. 12.1). A decrease of basal metabolism causes the onset of various non-specific clinical signs such as: (i) insufficient development in growing animals; (ii) low milk production; (iii) general weakness; and (iv) progressive decline in sexual functions. Specific diagnostic signs are spontaneous abortions and stillbirths (Fig. 12.2). These fetuses might have goitre and insufficient hair cover. An accurate diagnosis might be made by evaluating I levels in blood, urine and milk, but interpreting the results is not simple. This is due to the fact that, because this is a long-term deficiency, the thyroid gland is able to store a sufficient amount of I, which is available during long periods of inadequate intake. Milk I concentrations lower than 80 μg/l are indicative of I deficiency in goats. Once it has been established that there is I deficiency, a simple dietary correction may be enough to overcome this problem. Many complex feeds, such as iodized salt blocks, have an appropriate content of I. A daily I intake of 0.8 ppm/kg DM for lactating goats and 0.2 ppm/kg DM for kids and dry goats is sufficient to avoid deficiencies. Excessive I supplementation may induce clinical problems linked to I excess or iodism (Aguggini *et al.*, 1998).

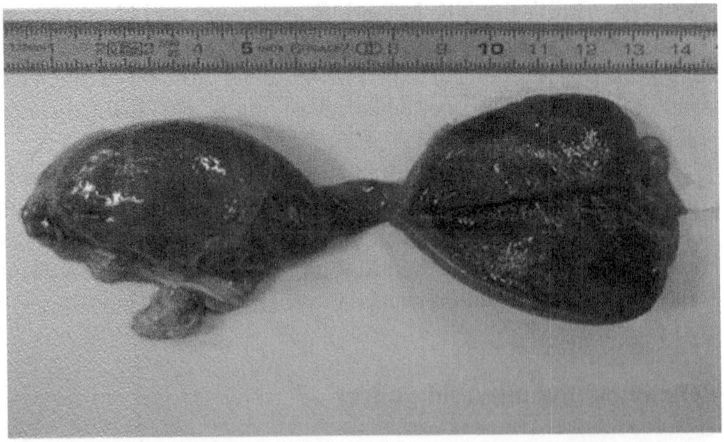

Fig. 12.1. Hyperplastic thyroid. The two lateral lobes are enlarged due to low levels of thyroxine, which induce thyroid stimulation by thyroid-stimulating hormone, produced by the pituitary gland.

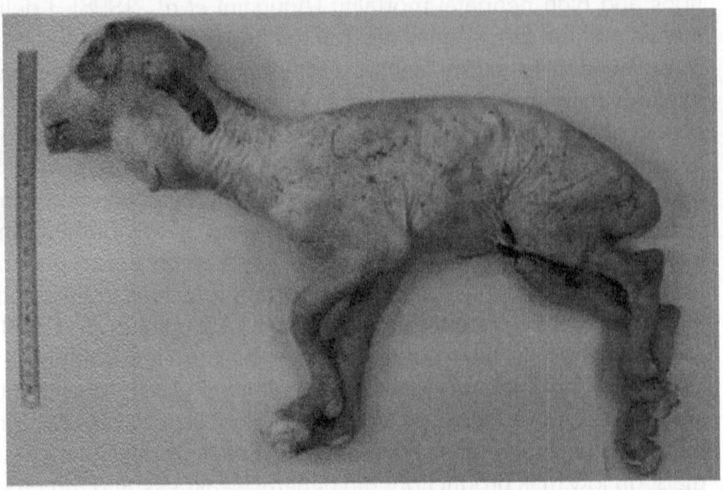

Fig. 12.2. Fetal abortion. Fetus with an evident goitre.

Vitamin B_1 deficiency

Polioencephalomalacia

Polioencephalomalacia is a central nervous system degeneration of ruminants resulting from a defect in thiamine metabolism. This disease might be primary or secondary to grain overload (ruminal acute acidosis) and is frequent in goats eating an excess of carbohydrates (Brent, 1976).

Clinical signs of polioencephalomalacia can develop quickly or slowly over several days. Early and specific signs of polioencephalomalacia are: (i) anorexia; (ii) diarrhoea; (iii) hyperesthesia; (iv) muscular trembling; (v) depression; (vi) sleepiness;

(vii) ataxia; (viii) proprioceptive deficit; (ix) blindness; (x) particular head postures (e.g. pushing against objects); (xi) dorsomedial cross-eye; (xii) excitement or aggressiveness; and (xiii) repetitive chewing movements. Usually, ocular palpebral reflexes are abnormal (Thomas *et al.*, 1987). Except for cross-eye, the function of the cranial nerves is normal during the first phases of the disease. In the more advanced stages, nervous signs, such as nystagmus, are due to central vestibular damage. Tonic–clonic convulsions can be easily induced by light stimulation. A high percentage of affected animals assume decubitus position and become hypertonic. The surviving ones remain practically decorticated and must be eliminated because of their poor productive performance. Rumen dilation and watery replenishment occur if the pathogenesis is related to a severe ruminal overload. The rectal temperature is normal unless in cases of excessive muscular cramps. Pulse and respiratory rates usually increase. Polioencephalomalacia has few post-mortem findings, with the exception of ruminal and erythrocyte transketolase. The microscopic analysis of ruminal content shows a decrease in Gram-positive cocci and Gram-negative coccobacilli, and an increase in Gram-positive bacilli. In practice, diagnosis is based mainly on clinical (especially blindness) and cortical signs in animals of a specific age and dietary condition. Measurements of ruminal or erythrocyte transketolase are affected by thiamine activity. Normally, mean normal transketolase activity varies between 0.3 and 2.9 μmol pentose/h per 10^9 erythrocytes. This activity is often reported as a percentage effect on phosphate thiamine. An increase of this activity is often linked to polioencephalomalacia. Also phosphate thiamine may be determined in erythrocytes and ruminal fluid. Other laboratory analysis might be blood concentrations of pyruvate, lactate and pyruvate kinase. Cephalorrhachidian liquid variations might occur, such as light pleocytosis (5–50 μm) and increased protein (higher than 50 mg/dl).

Thiamine has various functions in the body, with the diphosphatase thiamine being its active form. This molecule acts as a coenzyme for transketolase, pyruvate and α-ketoglutaratic dehydrogenase enzymes. Transketolase is present in erythrocytes and in nervous system cells, where depletion of this enzyme limits the rate of the pentose phosphate metabolic pathway. This pathway is essential for glucose metabolism in the nervous system. Other thiamine activities are decarboxylation of 2-oxoisovalerate, 2-oxoisocaproate, 2-oxoglutarate, etc. Loss of these functions leads to an increase in plasma concentration of α-ketoglutarate, phenylpyruvate and hydroxyphenylpyruvate in affected animals (Kinsella and Riley, 1999). Pyruvate and α-ketoglutarate promote normal functioning of the Krebs cycle, an important metabolic pathway which produces energy in the form of ATP in the nervous system. Therefore, the above-mentioned changes lead to a decrease of the activity of ATP-dependent water and Na transport in neurones, causing an intra-neuronal oedema. This leads to an increase in intracranial pressure and neuronal necrosis (Newsholme and O'Neill, 1985). As necrosis increases, oedema, endothelial capillary proliferation and macrophage cortical infiltration occur. The combination of increased intracranial pressure and neuronal oedema causes laminar necrosis and polioencephalomalacia.

Thiamine is also important in neurotransmitter metabolism. Studies on cattle have reported that healthy animals produce normal amounts of ruminal thiamine, while those in particular dietary conditions show lower ruminal thiamine concentrations. Possible mechanisms causing low ruminal thiamine concentrations are: (i) production of bacterial thiaminase; (ii) production or intake of thiamine-inactive analogues; (iii) intake of thiamine; (iv) insufficient absorption or phosphorylation of thiamine; (v) increase of thiamine excretion in faeces; or (vi) decrease of diphosphate thiamine production (Thomas *et al.*, 1987). In animals with polioencephalomalacia, faecal thiaminase and thiamine activity are negatively correlated, suggesting that the destruction of thiamine by ruminal flora is a very important factor determining the disease. Anaerobic germs may be the greatest thiamine producers in sheep, while the aerobic ones may be effective in goats and cattle. Two thiaminase types (I and II) have been reported. Thiaminase I is produced by thiaminelitic bacillus or *Clostridium sporogenes*; the enzyme catalyses thiamine hydrolysis at the methyl bridge between pyrimidinyl and the thiazolic ring. Thiamine II is produced by *Bacillus aneuricoliticus*, which proliferates in the case of grain overload; the enzyme catalyses hydrolysis of the methyl bridge between the two structural rings of thiamine. An animal affected by polioencephalomalacia might excrete 50% more thiamine in faeces than asymptomatic animals. In addition, these animals might also show a decrease in BW gain. Thus, single cases of polioencephalomalacia may be indicative of feeding problems of the herd. The ratio between ruminal and faecal thiaminase concentration, plasma and tissue thiamine concentrations, and the development of clinical signs of this disease have not been fully understood yet. For example, sometimes affected animals have normal thiamine levels in plasma but a strong decrease of thiamine levels in erythrocytes and tissues. This indicates that pathogenesis of the nervous system disease is caused by damage to intracellular transport mechanisms of thiamine.

Development of polioencephalomalacia is also affected by dietary factors, such as excess of sulphate, deficiency of Co, dietary excess of urea and molasses in ruminants, and uninterrupted feeding with reconstituted milk lacking in thiamine in preruminant animals. Dietary sulphate is able to interfere with ruminal flora thiamine synthesis. Polioencephalomalacia may occur at both individual and farming level (Rammell and Hill, 1988). Usually, it affects animals supplemented with grain seeds, but it might also affect grazing animals. In goats, this disease has been reported in animals from 3 weeks to 5 years of age, with a peak in animals older than 18 months. In some flocks, its incidence can be higher than 10%, being even greater in weaned lambs fed diets rich in cereals (Vrzgula, 1991).

Macroscopic lesions are characterized by oedema, cortical fragility and yellow coloration of the cerebral cortex. In the most severe cases, the cerebellum might be herniated in the forame magnum. Accumulation of lipofuscin pigments in necrotic areas might be identified by an ultraviolet lamp. Microscopic lesions include laminar necrosis of cortex grey matter, intra- and extracellular oedema, neuronal necrosis and gliosis (Rammell and Hill, 1988).

When the disease is diagnosed and treated before neuronal necrosis development, the animals respond well to thiamine administration. Animals with symptomatology may respond less and remain blind and depressed. Thiamine should be administered at the dose of 10–20 mg/kg BW by intramuscular or subcutaneous injection, three times per day. Since the destruction of all vitamins of the B-complex of the rumen may occur in affected animals, some authors suggest a treatment with B-vitamin complex (Vrzgula, 1991). Thiamine might be supplemented (from 3 to 10 mg/kg) in diets with low hay/concentrate ratio. Another recommendation to prevent polioencephalomalacia is the addition of beer yeast to the diet, and gradual introduction of cereal-rich diets to animals. A daily supplementation with 50–60 mg of thiamine per head is suggested, even if controlling goat diet might be enough.

References

Aguggini, G., Beghelli, V. and Giulio, L.F. (1998) *Fisiologia degli animali domestici con elementi di etologia*. UTET, Turin, Italy.

Anke, M., Angelow, L., Groppel, B., Arnhold, W. and Gruhn, K. (1989) The effect of selenium deficiency on reproduction and milk performance of goats. *Archiv für Tierernährung* 39, 483–490.

Aslan, V., Thamsborg, S.M., Jorgensen, R.J. and Basse, A. (1995) Induced acute ruminal acidosis in goats treated with yeast (*Saccharomyces cerevisiae*) and bicarbonate. *Acta Veterinaria Scandinavica* 36, 65–77.

Bezeau, L.M., Bailey, C.B. and Slen, S.B. (1961) Urine pH and the formation of uroliths in cattle. *Journal of Animal Science* 20, 924.

Bickhardt, K., Ganter, M., Sallmann, P. and Fuhrmann, H. (1999) Investigations on manifestations of vitamin E and selenium deficiency in sheep and goats. *Deutsche Tierärztliche Wochenschrift* 106, 242–247.

Blood, D.C. and Radostits, O.M. (1989) *Veterinary Medicine. A Text Book of the Diseases of Cattle, Sheep, Pigs, Goats and Horses*, 7th edn. Baillière Tindall, Philadelphia, Pennsylvania.

Brand, A. and Warner, M. (1996) Monitoring reproductive performance execution. In: Brand, A., Noordhuizen, J.P.T.M. and Schukken, Y.H. (eds) *Herd Health and Production Management in Dairy Practice*. Wageningen Press, Wageningen, The Netherlands, pp. 293–312.

Brent, B.E. (1976) Relationship of acidosis to other feedlot aliments. *Journal of Animal Science* 43, 930–935.

Brindle, N.P.J., Zammit, V.A. and Pogson, C.I. (1985) Regulation of carnitine palmitoyltransferase activity by malonyl-CoA in mitochondria from sheep liver, a tissue with a low capacity for fatty acid synthesis. *Biochemical Journal* 232, 177–182.

Brumby, P.E., Anderson, M., Tuckley, B., Storry, J.E. and Hibbitt, K.G. (1975) Lipid metabolism in the cow during starvation-induced ketosis. *Biochemical Journal* 146, 609–615.

Buck, W.B. (1986) Copper–molybdenum. In: Howard, J.L. (ed.) *Current Veterinary Therapy: Food Animal Practice*, 2nd edn. W.B. Saunders, Philadelphia, Pennsylvania, pp. 437–439.

Cao, G.R., English, P.B., Fillippich, L.J. and Inglis, S. (1987) Experimentally induced lactic acidosis in the goats. *Australian Veterinary Journal* 64, 367–370.

Chilliard, Y., Delavaud, C. and Bonnet, M. (2005) Leptin expression in ruminants: nutritional and physiological regulations in relation with energy metabolism. *Domestic Animal Endocrinology* 29, 3–22.

Daniel, R.C.W. and Moodie, E.W. (1979) Induced hypocalcaemia in cows and sheep. I. Changes in plasma inorganic phosphorus levels. *British Veterinary Journal* 135, 440–451.

Dirksen, G., Liebich, H.G., Brosi, H., Hagemeister, H. and Mayer, E. (1984) Morphologie der pansenschleimhaut und fettsäureresorption beim rind-bedeutende faktorem für gesundheit und leistung. *Zentralblatt für Phlebologie* 31, 414–430.

Dirksen, G., Breitner, W. and Berger, G. (1995) Ketosediagnostik: Semiquantitative Bestimmung von Beta-Hydroxybuttersäure in Kuhmilch mit einem neuen Trockentest. *Tierärztliche Umschau* 50, 239–244.

Duffield, T., Plaizier, J.C., Fairfield, A., Bagg, R., Vessie, G., Dick, P., Wilson, J., Aramini, J. and McBride, B. (2004) Comparison of techniques for measurement of rumen pH in lactating dairy cows. *Journal of Dairy Science* 87, 59–66.

Ebbesvik, M. (1993) Milk production in organic farming. Diet, feeding, health and yield. *Meieriposten* 11, 316–317.

Elam, C.J. (1976) Acidosis in feedlot cattle: practical observations. *Journal of Animal Science* 43, 898–901.

Emerick, R.J. and Lu, D. (1987) A possible synergism of dietary phosphate and urine-acidifying salts in preventing silica urolithiasis in a rat model. *Journal of Nutrition* 117, 1603–1608.

Ewoldt, J.M., Anderson, D.E., Miesner, M.D. and Saville, W.J. (2006) Short- and long-term outcome and factors predicting survival after surgical tube cystostomy for treatment of obstructive urolithiasis in small ruminants. *Veterinary Surgery* 35, 417–422.

Fontenot, J.P., Allen, V.G., Bunce, G.E. and Goff, J.P. (1989) Factors influencing magnesium absorption and metabolism in ruminants. *Journal of Animal Science* 67, 3445–3455.

Frank, A., Anke, M. and Danielsson, R. (2000a) Experimental copper and chromium deficiency and additional molybdenum supplementation in goats I. Feed consumption and weight development. *Science of the Total Environment* 249, 133–142.

Frank, A., Danielsson, R. and Jones, B. (2000b) Experimental copper and chromium deficiency and additional molybdenum supplementation in goats. II. Concentrations of trace and minor elements in liver, kidneys and ribs: haematology and clinical chemistry. *Science of the Total Environment* 249, 143–170.

Fredeen, A.H., DePeters, E.J. and Baldwin, R.L. (1988) Effects of acid–base disturbances caused by differences in dietary fixed ion balance on kinetics of calcium metabolism in ruminants with high calcium demand. *Journal of Animal Science* 66, 174–184.

French, M.H. (1970) *Observations on the Goat.* FAO Agricultural Studies No. 80. Food and Agriculture Organization of the United Nations, Rome, pp. 82–88.

Garrett, E.F., Pereira, M.N., Armentano, L.E., Goodger, W.J. and Oetzel, G.R. (1999) Diagnostic methods for detecting subacute ruminal acidosis in dairy cattle. *Journal of Dairy Science* 82, 1170–1178.

Geishauser, T., Leslie, K., Tenhag, J. and Bashiri, A. (2000) Evaluation of eight cowside ketone tests in milk for detection of subclinical ketosis in dairy cows. *Journal of Dairy Science* 83, 296–299.

George, J.W., Hird, D.W. and George, L.W. (2007) Serum biochemical abnormalities in goats with uroliths: 107 cases (1992–2003). *Journal of the American Veterinary Medical Association* 230, 101–106.

Gnanaprakasam, V. (1970) Rumen acidosis in goats. *Indian Veterinary Journal* 47, 904–910.

Gutierrez, C., Escolar, E., Juste, M.C., Palacios, M.P. and Corbera, J.A. (2000) Severe urolithiasis due to trimagnesium orthophosphate calculi in a goat. *The Veterinary Record* 146, 534.

Gutzwiller, A. (1991) The effect of the selenium status of goats on the resistance of erythrocytes to oxidative damage. *Schweizer Archiv für Tierheilkunde* 133, 157–161.

Herdt, T.H., Rumbeiha, W. and Emmett Braselton, W. (2000) The use of blood analyses to evaluate mineral status in livestock. *Veterinary Clinics of North America. Food Animal Practice* 16, 423–444.

Howard, J.L. and Smith, R.A. (1999) *Current Veterinary Therapy Food Animal Practice*, 4th edn. W.B. Saunders, Philadelphia, Pennsylvania, pp. 220–223.

Humphries, W.R., Morrice, P.C. and Mitchell, A.N. (1987) Copper poisoning in Angora goats. *The Veterinary Record* 121, 231.

Ingvartsen, K.L. and Andersen, J.B. (2000) Integration of metabolism and intake regulation: a review focusing on periparturient animals. *Journal of Dairy Science* 83, 1573–1597.

Ingvartsen, K.L. and Boisclair, Y.R. (2001) Leptin and regulation of food intake, energy homeostasis and immunity with special focus on periparturient ruminants. *Domestic Animal Endocrinology* 21, 215–250.

Janghorbani, M., Lynch, N.E., Mooers, C.S. and Ting, B.T.G. (1990) Comparison of the magnitude of the selenite exchangeable metabolic pool and whole body endogenous selenium in adult rats. *Journal of Nutrition* 120, 190–199.

Johnson, E.H., Muirhead, D., Annamalai, K., King, G.J., Al-Busaidy, R. and Shahul Hameed, M. (1999) Hepatic lipidosis associated with cobalt deficiency in Omani goats. *Veterinary Research Communications* 23, 215–221.

Jorgenson, R.J., Erdman, R., Murphy, M., Foldager, J., Norgaard, P., Moller, P., Andersen, P.H., Nielsen, E., Osteras, O., Harmoinen, P., Kodarik, P., Emanuelson, K. and Reinam, E. (1993) Rumen acidosis: areas of research. Summary of group discussion. *Acta Veterinaria Scandinavica Supplement* 89, 153–154.

Kan, J.B. (1979) Studies on calcium movement, electromyogram and electrocardiogram of experimental hypocalcemia in goats. PhD thesis, University of Tokyo, Tokyo.

Kan, J.B. and Motoyoshi, S. (1979) Studies on experimental hypocalcemia in goats. *Journal of the Japanese Veterinary Medicine Association* 32, 685–689.

Kaneko, J., Harvey, J.W. and Bruss, M.L. (1997) *Clinical Biochemistry of Domestic Animals*, 5th edn. Academic Press, New York.

Katoh, N. (2002) Relevance of apolipoproteins in the development of fatty liver and fatty liver-related peripartum diseases in dairy cows. *Journal of Veterinary Medical Science* 64, 293–307.

Kennedy, G.F. (2000) White muscle disease or stiff lamb disease. *International Sheep Letter* 20, 123–125.

Khan, Z.I., Ashraf, M., Hussain, A., Koyro, H.W. and Huchzermeyer, B. (2005) Seasonal variation in selenium status of different classes of grazing goats in a semi-arid region of Pakistan. *Deutsche Tierärztliche Wochenschrift* 112, 460–465.

Kinsella, L.J. and Riley, D.E. (1999) Nutritional deficiencies and syndromes associated with alcoholism. In: Goetz, C.G. and Papper, E.J. (eds) *Textbook of Clinical Neurology*. W.B. Saunders, Philadelphia, Pennsylvania, pp. 798–818.

Klimienė, I., Spakauskas, V. and Matusevicius, A. (2005) Correlation of different biochemical parameters in blood sera of healthy and sick cows. *Veterinary Research Communications* 29, 95–102.

Koller, L.D., Whitbeck, G.A. and South, P.J. (1984) Transplacental transfer and colostral concentrations of selenium in beef cattle. *American Journal of Veterinary Research* 45, 2507–2510.

Krebs, H.A. (1966) Bovine ketosis. *The Veterinary Record* 78, 187–192.

Krishnamurti, C.R., Ramberg, C.F. Jr, Shariff, M.A. and Boston, R.C. (1997) A compartmental model depicting short-term kinetic changes in selenium metabolism in ewes fed hay containing normal or inadequate levels of selenium. *Journal of Nutrition* 127, 95–102.

Lamand, M. (1978) Oligoelements. In: Jarrige, R. (ed.) *Alimentation des Ruminants*. INRA Publications, Versailles, France, pp. 143–159.

Lamand, M. and Perigaud, S. (1973) Carences en oligo-éléments en France. 1. Eléments d'enquête obtenus dans la pratique vétérinaire. *Annales de Recherches Veterinaires* 4, 513–534.

Lincoln, G.A., Rhind, S.M., Pompolo, S. and Clarke, I.J. (2001) Disruption of photoperiod-induced cycles in voluntary food intake, body weight, α-MSH, insulin and leptin secretion, and the induction of obesity in hypothalamo–pituitary disconnected rams. *American Journal of Physiology* 281, 76–90.

McDowell, L.R. (1992) Cobalt. In: McDowell, L.R. (ed.) *Minerals in Animals and Human Nutrition*. Academic Press, New York, pp. 205–223.

Markusfeld, O. (1985) Relationship between overfeeding, metritis and ketosis in high yielding dairy cows. *The Veterinary Record* 116, 489–491.

Martens, H. and Schweigel, M. (2000) Pathophysiology of grass tetany and other hypomagnesemias. Implications for clinical management. *Veterinary Clinics of North America. Food Animal Practice* 16, 339–368.

Medina-Escobedo, M., Zaidi, M., Real-de Leon, E. and Orozco-Rivadeneyra, S. (2002) Prevalence and risk factors of urinary lithiasis in Yucatan, Mexico. *Salud Pública de México* 44, 1–5.

Menzies, P.I., Boermans, H., Hoff, B., Durzi, T. and Langs, L. (2003) Survey of the status of copper, interacting minerals, and vitamin E levels in the livers of sheep in Ontario. *Canadian Veterinary Journal* 44, 898–906.

Mercer, J.G., Moar, K.M., Logie, T.J., Findlay, P.A., Adam, C.L. and Morgan, P.J. (2001) Seasonally inappropriate body weight induced by food restriction: effect on hypothalamic gene expression in male Siberian hamster. *Endocrinology* 142, 4173–4181.

Morgante, M. and Stelletta, C. (2004) Prelievo del liquido ruminale attraverso la ruminocentesi nella pecora e nella capra. In: *Proceedings of XVI SIPAOC Congress* (Società Italiana di Patologia e di Allevamento degli Ovini e dei Caprini), Siena, Italy, p. 310.

Nakagawa, H. and Katoh, N. (1998) Reduced activity of lecithin: cholesterol acyltransferase in the serum of cows with ketosis and left displacement of the abomasum. *Veterinary Research Communications* 22, 517–524.

Newsholme, S.J. and O'Neill, T.P. (1985). An outbreak of cerebrocortical necrosis (polioencephalomalacia) in goats. *Journal of the South African Veterinary Association* 56, 37–38.

Nocek, E.J. (1997) Bovine acidosis: implications on laminitis. *Journal of Dairy Science* 80, 1005–1028.

Nordlund, K.V. and Garrett, E.F. (1994) Rumenocentesis: a technique for collecting rumen fluid for the diagnosis of subacute rumen acidosis in diary herds. *The Bovine Practitioner* 28, 109–112.

Nour, M.M.S., Abusamra, M.T. and Hago, B.E.D. (1998) Experimentally induced lactic acidosis in Nubian goats: clinical, biochemical and pathological investigations. *Small Ruminant Research* 31, 7–17.

Oetzel, G.R. (1988) Parturient paresis and hypocalcemia in ruminant livestock. *Veterinary Clinics of North America. Food Animal Practice* 4, 351–364.

Ogryzkov, S.E., Amshokov, K.M. and Emel'ianova, M.V. (1966) Pathologico-histologic changes in kids in white muscle disease. *Veterinariia* 43, 64–65.

Oikawa, S., Katoh, N., Itoh, N., Miyamoto, T., Konno, M. and Kajita, T. (1997) Decreased serum apolipoprotein A-I concentrations in cows infected with *Salmonella typhimurium*. *Canadian Journal of Veterinary Research* 61, 182–186.

Pherson, B.O. (1993) Diseases and diffuse disorders related to selenium deficiencies in ruminants. *Norwegian Journal of Agricultural Science* 11, 79–93.

Poole, D.B.R. (1989) Observations on the role of magnesium and phosphorus in the aetiology of urolithiasis in male sheep. *Irish Veterinary Journal* 60–63.

Radostits, O.M., Blood, D.C. and Gay, C.C. (1994) Metabolic disease: parturient paresis (milk fever). In: *Veterinary Medicine*, 8th edn. Bailliere Tindall, London, pp. 1314–1328.

Radostits, O.M., Gay, C.C., Blood, D.C. and Hinchcliff, K.W. (2000) *Veterinary Medicine: A Textbook of the Diseases of Cattle, Sheep, Pigs, Goats and Horses*, 9th edn. W.B. Saunders, London, pp. 1515–1553.

Ramirez-Bribiesca, J.E., Tortora, J.L., Huerta, M., Aguirre, A. and Hernandez, L.M. (2001) Diagnosis of selenium status in grazing dairy goats on the Mexican plateau. *Small Ruminant Research* 41, 81–85.

Rammell, C.G. and Hill, J.H. (1988) Blood thiamine levels in clinically normal goats and goats with suspected polioencephalomalacia. *New Zealand Veterinary Journal* 36, 99–100.

Ray, S.K., Roychoudhury, R., Bandopadhyay, S.K. and Basu, S. (1997) Studies on 'zinc deficiency syndrome' in black Bengal goats (*Capra hircus*) fed with fodder (*Andropogon gayanus*) grown on soil treated with an excess of calcium and phosphorus fertilizer. *Veterinary Research Communications* 21, 541–546.

Reddy, S., Yang, W., Taylor, D.G., Shen, X.Q., Oxender, D., Kust, G. and Leff, T. (1999) Mitogen-activated protein kinase regulates transcription of the apoC-III gene. *Journal of Biological Chemistry* 274, 33050–33056.

Rousseau, K., Atcha, Z. and Loudon, A.S.I. (2002) Leptin and seasonal mammals. *Journal of Neuroendocrinology* 15, 409–414.

Sahinduran, S., Buyukoglu, T., Gulay, M.S. and Tasci, F. (2007) Increased water hardness and magnesium levels may increase occurrence of urolithiasis in cows from the Burdur region (Turkey). *Veterinary Research Communications* 31, 665–671.

Sanchez, J., Montes, P., Jimenez, A. and Andres, S. (2007) Prevention of clinical mastitis with barium selenate in dairy goats from a selenium-deficient area. *Journal of Dairy Science* 90, 2350–2354.

Sauerwein, H., Heintges, U., Bruhns, S.C., Hennies, M. and Gertler, A. (2006) Active immunization against leptin fails to affect reproduction and exerts only marginal effects on glucose metabolism in young female goats. *Journal of Animal Physiology and Animal Nutrition* 90, 278–288.

Sherman, D.M. and Robinson, R.A. (1983) Clinical examination of sheep and goats. *Veterinary Clinics of North America. Food Animal Practice* 5, 409–426.

Singer, L.J., Herron, A. and Altman, N. (2000) Zinc responsive dermatopathy in goats: two field cases. *Contemporary Topics in Laboratory Animal Science* 39, 32–35.

Smith, C.M. and Sherman, M.D. (1994) Fundamentals of goat practice. In: Smith, C.M. and Sherman, M.D. (eds) *Goat Medicine*. Lea and Febiger, Philadelphia, Pennsylvania, pp. 527–559.

Spears, J.W. (2000) Micronutrients and immune function in cattle. *Proceedings of the Nutrition Society* 59, 587–594.

Stelletta, C., Avondo, M., Lutri, L., Vazzana, I., Ferrantelli, V. and Morgante, M. (2004) Effetto della diversa concentrazione proteica del concentrato su alcuni parametri comportamentali, ematologici, e ruminali in caprette di razza girgentana. In: *Proceedings of XVI SIPAOC Congress* (Società Italiana di Patologia e di Allevamento degli Ovini e dei Caprini), Siena, Italy, pp. 327.

Stewart, S.R., Emerick, R.J. and Pritchard, R.H. (1990) High dietary calcium to phosphorus ratio and alkali-forming potential as factors promoting silica urolithiasis in sheep. *Journal of Animal Science* 68, 498–503.

Stratton-Phelps, M. and House, J.K. (2004) Effect of a commercial anion dietary supplement on acid–base balance, urine volume, and urinary ion excretion in male goats fed oat or grass hay diets. *American Journal of Veterinary Research* 65, 1391–1397. Erratum in: *American Journal of Veterinary Research* 65, 1700.

Tanwar, R.K. and Mathur, P.D. (1983) Studies on experimental rumen acidosis in goats. *Indian Veterinary Journal* 60, 499–500.

Thomas, K.W., Turner, D.L. and Spicer, E.M. (1987) Thiamine, thiaminase and transketolase levels in goats with and without polioencephalomalacia. *Australian Veterinary Journal* 64, 126–127.

Tontis, A. (1984) Occurrence of nutritional muscular dystrophy (NMD) in young goats in Switzerland. *Schweizer Archiv für Tierheilkunde* 126, 41–46.

Tontis, A. and Zwahlen, R. (1987) Pregnancy toxemia of small ruminants with special reference to pathomorphology. *Tierärztliche Praxis* 15, 25–29.

Van Metre, D.C. and Callan, R.J. (2001) Selenium and vitamin E. *Veterinary Clinics of North America. Food Animal Practice* 17, 373–402.

Van Metre, D.C. and Fubini, S.L. (2006) Ovine and caprine urolithiasis: another piece of the puzzle. *Veterinary Surgery* 35, 413–416.

Veenhuizen, J.J., Drackley, J.K., Richard, M.J., Sanderson, T.P., Miller, L.D. and Young, J.W. (1991) Metabolic changes in blood and liver during development and early treatment of experimental fatty liver and ketosis in cows. *Journal of Dairy Science* 74, 4238–4253.

Vihan, V.S., Wani, G.M. and Sahni, K.L. (1982) Observations on changes in blood serum in experimental rumen acidosis in goats. *Indian Veterinary Journal* 59, 998–1000.

Vrzgula, L. (1991) *Metabolic Disorders and their Prevention in Farm Animals. Developments in Animal and Veterinary Sciences*, Vol. 24. Elsevier, New York.

Wichtel, J.J., Thompson, K.G., Craigie, A.L. and Williamson, N.B. (1996) Short-term alteration in voluntary feed intake after selenium supplementation in Angora goat kids. *New Zealand Journal of Agricultural Research* 39, 107–110.

Yamagishi, N., Oishi, A., Sato, J., Sato, R. and Naito, Y. (1999) Experimental hyocalcemia induced by hemodialysis in goats. *Journal of Veterinary Medical Science* 61, 1271–1275.

Yang, D.Y., Chang, C.J., Peh, H.C. and Chen, M.T. (2004) Anti-peroxidation effects of vitamin E on low density lipoprotein and milk fat globule membrane of lactating goats: *in vivo* versus metal ion challenge *in vitro*. *Comparative Biochemistry and Physiology. A: Comparative Physiology* 139, 11–20.

Yonezawa, T., Mogy, K., Li, J.Y., Sako, R., Yamanouchi, K. and Nishihara, M. (2004) Modulation of growth hormone pulsatility by sex steroids in female goats. *Endocrinology* 146, 2736–2743.

Index

Printed and bound by CPI Group (UK) Ltd, Croydon, CR0 4YY

16/09/2026

14963056-0005